T0182182

Data Analytics

Series editors

Longbing Cao, Advanced Analytics Institute, University of Technology Sydney,
Broadway, NSW, Australia
Philip S. Yu, University of Illinois, Chicago, IL, USA

Aims and Goals:

Building and promoting the field of data science and analytics in terms of publishing work on theoretical foundations, algorithms and models, evaluation and experiments, applications and systems, case studies, and applied analytics in specific domains or on specific issues.

Specific Topics:

This series encourages proposals on cutting-edge science, technology and best practices in the following topics (but not limited to):

Data analytics, data science, knowledge discovery, machine learning, big data, statistical and mathematical methods for data and applied analytics,

New scientific findings and progress ranging from data capture, creation, storage, search, sharing, analysis, and visualization,

Integration methods, best practices and typical examples across heterogeneous, interdependent complex resources and modals for real-time decision-making, collaboration, and value creation.

More information about this series at http://www.springer.com/series/15063

Longbing Cao

Data Science Thinking

The Next Scientific, Technological and Economic Revolution

 Springer

Longbing Cao (ID)
Advanced Analytics Institute
University of Technology Sydney
Sydney, NSW, Australia

ISSN 2520-1859   ISSN 2520-1867   (electronic)
Data Analytics
ISBN 978-3-030-06975-9   ISBN 978-3-319-95092-1   (eBook)
https://doi.org/10.1007/978-3-319-95092-1

This Springer imprint is published by the registered company Springer Nature Switzerland AG
The registered company address is: Gewerbestrasse 11, 6330 Cham, Switzerland

To my family and beloved ones for their generous time and sincere love, encouragement, and support which essentially form part of the core driver for completing this book.

Preface

When you migrated to the twenty-first century, did you ever consider what today's world would look like? And what would inspire and drive the development and transformation of almost every aspect of our daily lives, study, work, and entertainment—in fact, every discipline and domain, including government, business, and society in general?

The most relevant answer may be data, and more specifically so-called "big data," the data economy, the science of data: *data science*, and data scientists. This is without doubt the age of big data, data science, data economy, and data profession.

The past several years have seen tremendous hype about the evolution of cloud computing, big data, data science, and now artificial intelligence. However, it is undoubtedly true that the volume, variety, velocity, and value of data continue to increase every millisecond. It is data and data intelligence that is transforming everything, integrating the past, present, and future. Data is regarded as the new Intel Inside, the new oil, and a strategic asset. Data drives or even determines the future of science, technology, economy, and possibly everything in our world today.

This desirable, fast-evolving, and boundless data world has triggered the debate about *data-intensive scientific discovery*—data science—as a new paradigm, i.e., the so-called "fourth science paradigm," which unifies experiment, theory, and computation (corresponding to "empirical" or "experimental," "theoretical," and "computational" science). At the same time, it raises several fundamental questions: What is data science? How does data science connect to other disciplines? How does data science translate into the profession, education, and economy? How does data science transform existing science, technologies, industry, economy, profession, and education? And how can data science compete in next-generation science, technologies, economy, profession, and education? More specific questions also arise, such as what forms the mindset and skillset of data scientists?

The research, innovation, and practices seeking to address the above and other relevant questions are driving *the fourth revolution* in scientific, technological, and economic development history, namely *data science, technology, and economy*.

These questions motivate the writing of this book from a high-level perspective.

There have been quite a few books on data science, or books that have been labeled in the book market as belonging under the data science umbrella. This book does not address the technical details of any aspect of mathematics and statistics, machine learning, data mining, cloud computing, programming languages, or other topics related to data science. These aspects of data science techniques and applications are covered in another book—*Data Science: Techniques and Applications*—by the same author.

Rather, this book is inspired by the desire to explore answers to the above fundamental questions in the era of data science and data economy. It is intended to paint a comprehensive picture of data science as a new scientific paradigm from the scientific evolution perspective, as data science thinking from the scientific thinking perspective, as a transdisciplinary science from the disciplinary perspective, and as a new profession and economy from the business perspective.

As a result, the book covers a very wide spectrum of essential and relevant aspects of data science, spanning the evolution, concepts, thinking and challenges, discipline and foundation of data science to its role in industrialization, profession, and education, and the vast array of opportunities it offers. The book is decomposed into three parts to cover these aspects.

In Part I, we introduce the evolution, concepts and misconceptions, and thinking of data science. This part consists of three chapters. In Chap. 1, the evolution, characteristics, features, trends, and agenda of the data era are reviewed. Chapter 2 discusses the question "What is data science?" from a high-level, multidisciplinary, and process perspective. The hype surrounding big data and data science is evidenced by the many myths and misconceptions that prevail, which are also discussed in this chapter. Data science thinking plays a significant role in the research, innovation, and applications of data science and is discussed in Chap. 3.

Part II introduces the challenges and foundations of doing data science. These important issues are discussed in three chapters. First, the various challenges are explored in Chap. 4. In Chap. 5, the methodologies, disciplinary framework, and research areas in data science are summarized from the disciplinary perspective. Chapter 6 explores the roles and relationships of relevant disciplines and their knowledge base in forming the foundations of data science. Lastly, Chap. 7 summarizes the main research issues, theories, methods, and applications of analytics and learning in the various domains and applications.

The last part, Part III, concerns data science-driven industrialization and opportunities, discussed in four chapters. Data science and its ubiquitous applications drive the data economy, data industry, and data services, which are explored in Chap. 8. Data science, data economy, and data applications propel the development of the data profession, fostering data science roles and maturity models, which are highlighted in Chap. 10. The era of data science has to be built by data scientists and engineers; thus the required qualifications, educational framework, and capability set are discussed in Chap. 11. Lastly, Chap. 12 explores the future of data science.

As illustrated above, this book on data science differs significantly from any book currently on the market by the breadth of its coverage of comprehensive data science, technology, and economic perspectives. This all-encompassing intention makes compiling a book like this an extremely challenging and risky venture. Basic theories and algorithms in machine learning and data mining are not discussed, nor are most of the related concepts and techniques, as readers can find these in the book *Data Science: Techniques and Applications*, and other more dedicated books, for which a rich set of references and materials is provided.

The book is intended for data managers (e.g., analytics portfolio managers, business analytics managers, chief data analytics officers, chief data scientists, and chief data officers), policy makers, management and decision strategists, research leaders, and educators who are responsible for pursuing new scientific, innovation, and industrial transformation agendas, enterprise strategic planning, or next-generation profession-oriented course development, and others who are involved in data science, technology, and economy from a higher perspective. Research students in data science-related disciplines and courses will find the book useful for conceiving their innovative scientific journey, planning their unique and promising career, and for preparing and competing in the next-generation science, technology, and economy.

Can you imagine how the data world and data era will continue to evolve and how our future science, technologies, economy, and society will be influenced by data in the second half of the twenty-first century? To claim that we are data scientists and "doing data," we need to grapple with these big, important questions to comprehend and capitalize on the current parameters of data science and to realize the opportunities that will arise in the future. We thus hope this book will contribute to the discussion.

Sydney, NSW, Australia   Longbing Cao
July 2018

Acknowledgments

Writing a book like this has been a long journey requiring the commitment of tremendous personal, family, and institutional time, energy, and resources. It has been built on a dozen years of the author's limited, evolving but enthusiastic observations, thinking, experience, research, development, and practice, in addition to a massive amount of knowledge, lessons, and experience acquired from and inspired by colleagues, research and business partners and collaborators. The author would therefore like to thank everyone who has worked, studied, supported, and discussed the relevant research tasks, publications, grants, projects, and enterprise analytics practices with him since he was a data manager of business intelligence solutions and then an academic in the field of data science and analytics.

This book was particularly written in alignment with the author's vision and decades of effort and dedication to the development of data science, culminating in the creation and directorship of the Advanced Analytics Institute (AAi) at the University of Technology Sydney in 2011. This was the first Australian group dedicated to big data analytics, and the author would thus like to thank the university for its strategic leadership in supporting his vision and success in creating and implementing the Institute's Research, Education and Development business model, the strong research culture fostered in his team, the weekly meetings with students and visitors which significantly motivated and helped to clarify important concepts, issues, and questions, and the support of his students, fellows, and visiting scholars. Many of the ideas, perspectives, and early thinking included in this book were initially brought to the author's weekly team meetings for discussion. It has been a very great pleasure to engage in such intensive and critical weekly discussions with young and smart talent. The author indeed appreciates and enjoys these discussions and explorations, and thanks those students, fellows, and visitors who have attended the meetings over the past 10+ years.

In addition, heartfelt thanks are given to my family for their endless support and generous understanding every day and night of the past 4 years spent compiling this book, in addition to their dozens of years of continuous support to the author's research and practice in the field.

The author is grateful to professional editor Ms. Sue Felix who has made significant effort in editing the book.

Last but not least, my sincere thanks to Springer, in particular Ms. Melissa Fearon at Springer US, for their kindness in supporting the publication of this monograph in its *Book Series on Data Analytics*, edited by Longbing Cao and Philip S Yu.

Writing this book has been a very brave decision, and a very challenging and risky journey due to many personal limitations. There are still many aspects that have not been addressed, or addressed adequately, in this edition, and the book may have incorporated debatable aspects, limitations, or errors in the thinking, conceptions, opinions, summarization, and proposed value and opportunities of the data-driven fourth revolution: data science, technology, and economy. The author welcomes comments, discussion, suggestions, or criticism on the content of the book, including being alerted to errors or misunderstandings. Discussion boards and materials from this book are available at www.datasciences.info, a data science portal created and managed by the author and his team for promoting data science research, innovation, profession, education, and commercialization. Direct feedback to the author at Longbing.Cao@gmail.com is also an option for commenting on possible improvements to the book and for the benefit of the data science discipline and communities.

Contents

Part I
Concepts and Thinking

Chapter 1
The Data Science Era

1.1 Introduction

We are living in the age of big data, advanced analytics, and data science. The trend of "big data growth" [29, 106, 266, 288, 413] (data deluge [210]) has not only triggered tremendous hype and buzz, but more importantly presents enormous challenges, which in turn have brought incredible innovation and economic opportunities.

Big data has attracted intense and growing attention from major governmental organizations, including the United Nations [399], USA [407], EU [101] and China [196], traditional data-oriented scientific and engineering fields, as well as non-traditional data engineering domains such as social science, business and management [91, 252, 265, 472].

From the disciplinary development perspective, recognition of the significant challenges, opportunities and values of big data is fundamentally reshaping traditional data-oriented scientific and engineering fields. It is also reshaping non-traditional data engineering domains such as social science, business and management [91, 252, 265, 472]. This paradigm shift is driven not just by data itself but by the many other aspects of the power of data (simply *data power*), from data-driven science to data-driven economy, that could be created, invented, transformed and/or adjusted by understanding, exploring and utilizing data.

This trend and its potential have triggered new debate about *data-intensive scientific discovery* as a new paradigm, the so-called "fourth science paradigm", which unifies experiment, theory and computation (corresponding to *empirical science* or *experimental science*, *theoretical science* and *computational science*) [198, 209], as shown in Fig. 1.1. Data is regarded as the new Intel Inside [319], or the new oil and strategic asset, and is driving—even determining—the future of science, technology, the economy, and possibly everything else in our world.

In 2005 in Sydney, we were asked a critical question at a brainstorming meeting about data science and data analytics by several local industry representatives from

© Springer International Publishing AG, part of Springer Nature 2018
L. Cao, *Data Science Thinking*, Data Analytics,
https://doi.org/10.1007/978-3-319-95092-1_1

Fig. 1.1 Four scientific
paradigms

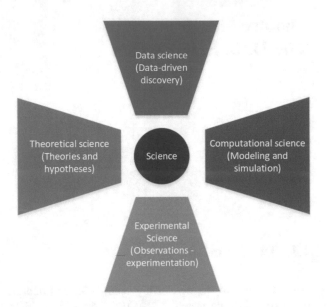

major analytics software vendors: "Information science has been there for so long,
why do we need data science?" Related fundamental questions often discussed in
the community include "What is data science?" [279], and "Is data science old
wine in new bottles?" [2]. Data science and associated topics have become the
key concern in panel discussions at conferences in statistics, data mining, and
machine learning, and more recently in big data, advanced analytics, and data
science. Typical topics such as "grand challenges in data science", "data-driven
discovery", and "data-driven science" have frequently been visited and continue to
attract wide and increasing attention and debate. These questions are mainly posited
from research and disciplinary development perspectives, but there are many other
important questions, such as those relating to data economy and competency, that
are less well considered in the conferences referred to above.

A fundamental trigger for these questions and many others not mentioned here
is the exploration of new or more complex challenges and opportunities [54,
64, 233, 252] in data science and engineering. Such challenges and opportunities
apply to existing fields, including statistics and mathematics, artificial intelligence,
and other relevant disciplines and domains. They are issues that have never been
adequately addressed, if at all, in classic methodologies, theories, systems, tools,
applications and economy. Such challenges and opportunities cannot be effectively
accommodated by the existing body of knowledge and capability set without the
development of a new discipline.

On the other hand, data science is at a very early stage and, apart from
engendering enormous hype, it also causes a level of bewilderment, since the issues
and possibilities that are unique to data science and big data analytics are not clear,
specific or certain. Different views, observations, and explanations—some of them
controversial—have thus emerged from a wide range of perspectives.

There is no doubt, nevertheless, that the potential of data science and analytics to enable data-driven theory, economy, and professional development is increasingly being recognized. This involves not only core disciplines such as computing, informatics, and statistics, but also the broad-based fields of business, social science, and health/medical science. Although very few people today would ask the question we were asked 10 years ago, a comprehensive and in-depth understanding of *what data science is*, and *what can be achieved with data science and analytics research, education, and economy*, has yet to be commonly agreed.

This chapter therefore presents an overview of the *data science era*, which incorporates the following aspects:

- Features of the data science era;
- The data science journey from data analysis to data science;
- The main driving forces of data-centric thinking, innovation and practice;
- The interest trends demonstrated in Internet search;
- Major initiatives launched by governments; and
- Major initiatives on the scientific agenda launched by scientific organizations.

The goal of this chapter is to present a comprehensive high level overview of what has been going on in communities that are representative of the data science era, before addressing more specific aspects of data science and associated perspectives in the remainder of the book.

1.2 Features of the Data Era

1.2.1 Some Key Terms in Data Science

Before proceeding to discuss the many aspects of data science, we list several key terms that have been widely accepted and discussed in relevant communities in relation to the data science era: data analysis, data analytics, advanced analytics, big data, data science, deep analytics, descriptive analytics, predictive analytics, and prescriptive analytics. These terms are highly connected and easily confused, and they are also the key terms widely used in the book. Table 1.1 thus lists and explains these terms.

A list of data science terminology is available at www.datasciences.info.

1.2.2 Observations of the Data Era Debate

With their emergence as significant new areas and disciplines, big data [25, 288] and data science [388] have been the subject of increased debate and controversy in recent years.

Table 1.1 Key terms in data science

Key terms	Description
Advanced analytics	Refers to theories, technologies, tools and processes that enable an in-depth understanding and discovery of actionable insights in big data, which cannot be achieved by traditional data analysis and processing theories, technologies, tools and processes
Big data	Refers to data that are too large and/or complex to be effectively and/or efficiently handled by traditional data-related theories, technologies and tools
Data analysis	Refers to the processing of data by traditional (e.g., classic statistical, mathematical or logical) theories, technologies and tools for obtaining useful information and for practical purposes
Data analytics	Refers to the theories, technologies, tools and processes that enable an in-depth understanding and discovery of actionable insight into data. Data analytics consists of descriptive analytics, predictive analytics, and prescriptive analytics
Data science	The science of data
Data scientist	A person whose role very much centers on data
Descriptive analytics	Refers to the type of data analytics that typically uses statistics to describe the data used to gain information, or for other useful purposes
Predictive analytics	Refers to the type of data analytics that makes predictions about unknown future events and discloses the reasons behind them, typically by advanced analytics
Prescriptive analytics	Refers to the type of data analytics that optimizes indications and recommends actions for smart decision-making
Explicit analytics	Focuses on descriptive analytics, by involving observable aspects, typically by reporting, descriptive analysis, alerting and forecasting
Implicit analytics	Focuses on deep analytics, by involving hidden aspects, typically by predictive modeling, optimization, prescriptive analytics, and actionable knowledge delivery
Deep analytics	Refers to data analytics that can acquire an in-depth understanding of why and how things have happened, are happening or will happen, which cannot be addressed by descriptive analytics

After reviewing [63] a large number of relevant works in the literature that directly incorporate data science in their titles, we make the following observations about the big data buzz and data science debate:

- Very comprehensive discussion has taken place, not only within data-related or data-focused disciplines and domains, such as statistics, computing and informatics, but also in non-traditional data-related fields and areas such as social science and management. Data science has clearly emerged as an inter-, cross- and trans-disciplinary new field.
- In addition to the thriving growth in academic interest, industry and government organizations have increasingly realized the value and opportunity of data-driven innovation and economy, and have thus devised policies and initiatives to promote data-driven intelligent systems and economy.

- Although many discussions and publications are available, most (probably more than 95%) essentially concern existing concepts and topics discussed in statistics, artificial intelligence, pattern recognition, data mining, machine learning, business analytics and broad data analytics. This demonstrates how data science has developed and been transformed from existing core disciplines, in particular, statistics, computing and informatics.
- While data science as a term has been increasingly used in the titles of publications, it seems that a great many authors have done this to make the work look 'sexier'. The abuse, misuse and over-use of the term "data science" is ubiquitous, and essentially contribute to the buzz and hype. Myths and pitfalls are everywhere at this early, and somehow impetuous, stage of data science.
- Very few thoughtful articles are available that address the low-level, fundamental and intrinsic complexities and problematic nature of data science, or contribute deep insights about the intrinsic challenges, directions and opportunities of data science as a new field.

It is clear that we are living in the era of big data and data science—an era that exhibits iconic features and trends that are unprecedented and epoch-making.

1.2.3   Iconic Features and Trends of the Data Era

In the era of data science, an essential question to ask is *what typifies this new era?* It is critical to identify the features and characteristics of the data science era. However, it is very challenging to provide a precise summary at this early stage.

To give a fair summary, the main characteristics of the data science era are discussed from the perspective of the transformation and paradigm shift caused by data science, the core driving forces, and the status of several typical issues confronting the data science field.

A data-centric perspective is taken to summarize the main characteristics of data science-related government initiatives, disciplinary development, economy, and profession, as well as the relevant activities in these fields, and the progress made to date.

We summarize eight data era features in Table 1.2 which represent this new age of science, profession, economy and education.

Data existence—Datafication is ubiquitous, and data quantification is ever-increasing: Data is physically, increasingly and ubiquitously generated at any time by any means. This goes beyond the traditional main sources of datafication [19]: sensors and management information systems. Today's datafication devices and systems are everywhere, involved in and related to our work, study, entertainment, socio-cultural environment, and quantified personal devices and services [96, 143, 160, 363, 377, 462]. In addition, *data quantification is ever-increasing*: The data deluge features an exponential increase in the volume and variety of data at a speed

Table 1.2 Key features and trends of the data science era

Landmark	Significance
Physical existence	Datafication is ubiquitous, and data quantification is ever-increasing
Complexities	Data complexities cannot be handled by classic theories and systems
Strategic values	Data becomes a strategic asset
	Openness becomes a new paradigm and fashion
Research and development	Data science research and innovation drive a new scientific agenda
Startup business	Data-driven strategic data initiatives and startups start to dominate new business
Job market	Data scientist becomes a new profession
Business and economy	Data drives both the new data economy and traditional industry transformation
Disciplinary maturity	Data science becomes a new discipline, and data science is interdisciplinary

and in forms that were not previously imaginable and cannot be precisely predicted. There is apparently no end to this ever-increasing data quantification trend.

Data complexities—Data complexities cannot be handled by classic theories and systems: Data that is substantially complex cannot be well addressed by existing statistical and mathematical theories and systems, information theories and tools, analytics and learning systems.

Data value—Data becomes a strategic asset, and openness becomes a new paradigm and fashion: The strategic value of data is increasingly recognized by data owners and data generators, who treat data as a strategic asset for commercialization and other purposes. At the same time, there is a strong push for data to be open. Open source software, services and applications, and free repositories and services are a highlight of this data era. To a certain extent, open data and the open source environment are key drivers of the big data and data science era, propelling the spread of open data, open access, and open source to open innovation and open science, creating a new paradigm for research and science.

Data research and development—Data science research and innovation drives a new scientific field: Due to the significant data complexities and data values that have not been addressed in existing scientific and innovation systems, data science research and innovation is high on the current scientific agenda. More and more national science foundations, science councils, research foundations, and research and innovation policy-makers are increasing their funding support for data science innovation and basic research in both general scientific disciplines and specific areas such as health informatics, bioinformatics, and brain informatics.

Data startup—Data-driven strategic data initiatives and startups start to dominate the new business: We are seeing rapidly increasing strategic initiatives established by increasing numbers of governments, vendors, professional bodies, and large and small businesses. Data industrialization is driving the new wave of economic transformation and startups.

Data science job positioning—Data scientist becomes a new profession: Data science jobs dominate the job market, demonstrating a rapidly increasing trend which is marked by a high average salary. New data professional communities are formed, as evidenced by the creation of new chief officer roles such as chief data officer, chief analytics officer, and chief data scientist, as well as multiple roles which are broadly termed 'data scientist'. This leads to a business-driven, fast-growing, open data science community, and the development of various analytics customized for specific domains, such as agricultural analytics and social analytics.

Data economy—Data is driving both the new data economy and traditional industry transformation: This is not only represented by the emergence of data-focused companies and startups such as Google, Facebook, and Cloudera, but also by the data-driven transformation of traditional industry and core business, in particular, banking, capital markets, telecommunication, manufacturing, the food industry, healthcare business, medicine and medical services, and the educational sector. In addition to the above typical data-driven businesses, data industrialization is changing the Internet landscape, driving new data products, data systems, and data services that are embedded in social media, mobile applications, online business, and the Internet of Things (IoT). In core business and traditional industry, the changes result from data-based competition, productivity elevation, service enhancement, and decision efficiency and effectiveness, which, while not as visible as the new data economy, are just incredible and hitherto unimaginable.

Data science discipline—Data science becomes a new discipline, and data science is interdisciplinary: Universities, research institutions, vendors and commercial companies have rapidly recognized data science as a new discipline and are establishing an enormous number of awarded degrees, training courses, and online courses which are combined with existing interdisciplinary subjects from undergraduate level to doctoral level, or from non-award training programs. The last 5 years have seen a rapid increase in the creation of institutes, centers, and departments focusing on data science research, teaching and engagement across a broad range of international communities and research, government and industry agendas.

1.3   The Data Science Journey

This section summarizes the findings of a comprehensive survey in [63] and other related work, such as in [129, 172, 330], of the data science journey from data analysis to data science and the evolution of the interest in data science.

When was "data science" as a term first introduced? It is likely that the first appearance of "data science" as a term in literature was in the preface to Naur's book "Concise Survey of Computer Methods" [301] in 1974. In that preface, *data science* was defined as "the science of dealing with data, once they have been established, while the relation of the data to what they represent is delegated to other fields and sciences." Another term, "datalogy", had previously been introduced in 1968 as "the science of data and of data processes" [300]. These definitions are clearly more specific than those we discuss today. However, they have inspired today's significant move toward the comprehensive exploration of scientific content and development.

The past 50+ years have seen the transformation from data analysis to data science, and the trend is becoming more evident, widespread and profound. This evolutionary journey from data analysis [216] to data science started in the statistics and mathematics community in 1962. At that time, it was stated that "data analysis is intrinsically an empirical science" [387]. (On this basis, David Donoho argued that data science had existed for 50 years and questioned how/whether data science is really different from statistics [129]).

Data processing quickly became a critical part of the research agenda and scientific tasks, especially in statistical and mathematical domains. Typical original work on promoting data processing included *information processing* [298] and *exploratory data analysis* [388]. These works suggested that more emphasis needed to be placed on using data to suggest suitable hypotheses for testing.

Our understanding of the role of data analysis in those early years extended beyond data exploration and processing to the aspiration to "convert data into information and knowledge" [217]. The development of data processing techniques and tools has significantly motivated the proposal of the later term of "data-driven discovery" used in the first Workshop on Knowledge Discovery in Databases in 1989 [245].

Several statisticians have pushed to transform statistics to data science. For example, in 2001, an action plan was suggested in [97], in which it was suggested that the technical areas of statistics should be expanded into data science.

Prior to data science being seriously adopted in multiple disciplines, as it is today, a major analytics topic in statistics was *descriptive analytics* (also called *descriptive statistics* in the statistics community) [373]. *Descriptive analytics* quantitatively summarizes and/or describes the characteristics and measurements of a data sample or data set. Today, descriptive analytics forms the foundation for the default analytical tasks and tools in typical data analysis projects and systems.

More than 20 years after this thriving period of descriptive analytics, the desire to convert data to information and knowledge fostered the origin of the currently popular community of the ACM SIGKDD conference, specifically the first workshop on Knowledge Discovery in Databases (KDD for short) with IJCAI'1989 [245], in which "data-driven discovery" was adopted as one of three themes of the workshop.

Since the establishment of KDD, key terms such as "data mining", "knowledge discovery" [161] and *data analytics* [339] have been increasingly recognized not only in IT but also in other areas and disciplines. *Data mining* (or *knowledge discovery*) denotes the technologies and processes of discovering hidden and interesting knowledge from data.

The concept of *machine learning* was probably firstly coined by Arthur Samuel at IBM who created a checkers-playing program and defined machine learning as "a field of study that gives computers the ability to learn without being explicitly programmed" [187, 447].

In the history of the development of the data science community, several other major data-driven discovery-focused conferences were established in addition to the establishment of the KDD workshop in 1989. Of particular importance were the International Conference on Machine Learning (ICML) in 1980, and the Neural Information Processing Symposium (NIPS) in 1987. Since then, many regional and international conferences and workshops on data analysis, data mining, and machine learning have been created, ostensibly making this the fastest growing and most popular computer science community.

Today, in addition to well-recognized events like KDD, ICML, NIPS and JSM, many regional and international conferences and workshops on data analysis and learning have been conceived. The latest development is the creation of global and regional conferences on data science, especially the IEEE International Conference on Data Science and Advanced Analytics [135]. Data Science and Advanced Analytics has received joint support from IEEE, ACM and the American Statistical Association, in addition to industry sponsorship. These efforts have contributed to making data science the fastest growing and most popular element in computing, statistics and interdisciplinary communities.

The development of data mining, knowledge discovery, and machine learning, together with the original data analysis and descriptive analytics from the statistical perspective, forms the general concept of "data analytics". Initially, data analysis focused on processing data. *Data analytics* is the multi-disciplinary science of quantitatively and qualitatively examining data for the purpose of drawing new conclusions or insights (exploratory or predictive), or for extracting and proving confirmatory or fact-based hypotheses about that information for decision making and action.

The value of data analysis and data analytics has been increasingly recognized by business and management. As a result, analytics has become more data characteristics-based, business-oriented [259], problem-specific, and domain-driven [77]. Data analysis and data analytics now extend to a variety of data and domain-specific analytical tasks, such as business analytics, risk analytics, behavior analytics [74], social analytics, and web analytics. These various types of analytics are generally termed "X-analytics". Today, data analytics has become the keystone of data science.

Figure 1.2 summarizes the data science journey by capturing the representative moments and major aspects of disciplinary development, government initiatives, scientific agendas, typical socio-economic events, and education in the evolution of data science.

In Sect. 6.5.3, we discuss the evolution from processing and analysis to the broad and deep analytics of data science. Figure 6.5 demonstrates the evolutionary path from analysis to analytics and data science.

1962 ● Data analysis

1968 ● Datalogy

1974 ● Data science

1980 ● ICML

1987 ● NIPS

1989 ● Data-driven Discovery & KDD

1996 ● IFCS Conference on Data Science, Classification, and Related Methods
1997 ● Jeff Wu "Statistics = Data Science?"

2001 ● 3Vs of Big Data
2002 ● Data Science J., J. Data Science

2004 ● Open Access/Open Data, Yahoo! Chief Data Officer/MapReduce

2007 ● Data Science & Knowledge Discovery Lab/Master of Science in Analytics

2011 ● Master of Analytics (Research)/PhD Thesis: Analytics at UTS
2012 ● US NSF Big Data Initiative
2013 ● IEEE Task Force on Data Science and Advanced Analytics
2014 ● IEEE/ACM Conf. on Data Science and Advanced Analytics
2015 ● U.S. Chief Data Scientist/J. Data Science and Analytics/IEEE Trans. Big Data
2016 ● Google AlphaGo beats Lee Se-dol In five game match

Fig. 1.2 Timeline of the data science journey

1.3.1   New-Generation Data Products and Economy

The disciplinary paradigm shift and technological transformation enables the innovation and industrialization of new-generation technical and practical data products and data economy.

These new-generation data products and new data economy emerge in many technical areas including data creation and quantification, acquisition, preparation and processing, sharing and storage, backup and recovery, retrieval, transport, messaging and communication, management, and governance. The dominant areas are probably the generation of new data services, such as cross-media recommender systems and cross-market financial products, as well as new data products and data systems for in-depth understanding of complex business problems that cannot be handled by existing data-driven reporting, analytics, visualization, and decision support, such as a trustful global online market supporting e-commerce of any product by anyone in any country, cross-organization data integration and analytical tools, and autonomous algorithms and automated discovery.

Another important innovation lies in the generation of domain-specific data products (including systems, applications, and services). This is typically highlighted by social media websites such as Twitter and Facebook, mobile health service and recommendation applications, online property pricing valuation and recommendation, tourism itinerary planning and booking recommendation, and personalized behavior insight understanding and treatment strategy planning.

Existing data-driven design, technologies and systems are significantly challenged by real-world human needs, which are typically intent-driven, mental, personalized, and subjective. This is reflected in online queries, preferences and demand in recommendation, online shopping and social networking. New technological innovation has to cater for these fundamental needs in the next generation of artificial intelligence and intelligent systems.

In the data and analytics areas, innovative data products, data services, and data systems may be generated in the following typical transformations:

- from a core business-driven economy to a digital and data economy;
- from closed organizations to open operations and governments;
- from traditional e-commerce to data-enabled online business;
- from landline telecommunication services to smart phone-based service mixtures that combine telecommunication and Web-based e-payment, messaging, and entertainment;
- from the Internet to mobile network and social media-mixed services; and
- from objective (object-based) businesses to subjective (intent, sentiment, personality, etc.) services.

Extended discussion on data products can be found in Sect. 2.8 and on data economy and industrialization in Chap. 8.

1.4   Data-Empowered Landscape

The disciplinary paradigm shift, technological transformation, and production of new-generation data product are driven by core data power. Core *data power* includes data-enabled opportunities, data-related ubiquitous factors, and various complexities and intelligences embedded in data-oriented transformation, production and products.

1.4.1   Data Power

Data power refers to the facilities, contributions, values, and opportunities that can be enabled by data. Data power may be reflected in different ways for different purposes. Typically, data power can be instantiated as scientific, technical, economic, cultural, social, military, political, security-oriented, and professional power.

Examples of *scientific data power* are the theoretical breakthroughs in data science research, such as new theories for learning non-IID (non-independent and identically distributed) data and new architectures for deeply representing rich hidden relations in data. Other opportunities include data-driven scientific discovery in areas that have never been explored, or that have never been possible, such as the identification of new planets and activities based on observable universal data.

Technical data power is currently widely represented by the invention of new data technologies for processing, analyzing, visualizing, and presenting complex data, such as Spark technology and Cloudera technology. Technical data power will be epitomized by novel and effective data products, data services, and data solutions that extend beyond the traditional landscape and thinking; for example, biomedical devices that can communicate with patients and understand a patient's personality and requirements.

Economic data power is reflected by the data economy and new data designs and products. The economic value of data is implemented by data-enabled industrialization, industry transformation, and productivity lift. This may lead to the development of new services, businesses, business models, and economic and commercial opportunities. It will result in smarter decision-making, more efficient markets, more personalized automated services, and best practice optimization.

Social data power is typically evidenced by social media business and social networking, which will extend into every part of our social life and society. This power creates a virtual social society in Internet and mobile network-based infrastructure that is parallel to our physical social society. The interactions and synergy between virtual society and physical society are changing our social and interpersonal relationships and lifestyle, as well as our modes of study and work. The fusion between these two worlds is significant, triggering their co-evolution and the emergence of new societal and social forms, including the way we live.

Cultural data power is progressively embodied in social data power, cultural change, and the promotion and integration of cross-cultural interactions and development. Cultural data power is also reflected in the quantification and comparison of various historical cultures, enabling global cross-culture fusion and evolution.

Military data power is deeply reflected in data-enabled and data-empowered military thinking, devices, systems, services, and methodologies. Modern military theories, systems and decisions have essentially been built on—and rely on—data. A typical example is the design of Worldwide Military Command and Control Systems [185] and the Global Command and Control System [329]. Military areas will lead data innovation, especially in fusing multiple military, professional and public systems and repositories, and developing integral detection, analysis, intervention, and weapons systems for globalized decision-making and action.

Political data power refers to the values and impact of data on politics. Political impact is reflected in data-driven evidence-based decision making, the optimization of existing policies, evidence-based informed policy-making, and optimal government services and service objectives. Significant political and governmental challenges, such as increasingly complex cross-agency decision-making and globalization-based national strategic planning, will have to rely on data fusion and deep analytics.

Security-oriented data power assures the compliance of data products by enabling the security of products and the development of data security products for more secure networks, systems, services, and devices. Secure data products, user environments, operation, data residency and sovereignty can significantly benefit from data-driven security research and innovation, complementing the traditional scope of security on infrastructure, cryptography, and protocols.

The various aspects of data power illustrated above are relative, meaning that they can be either positive or negative, depending on what drives the design, how such power is generated, and how it is utilized.

Positive data power refers to the positive value and impact that can be engendered by data. For example, algorithmic trading can identify high frequency trading strategies which can be applied to trading to increase profit.

Negative data power refers to the negative value and impact created by data. The algorithmic trading in the above example could also be used for negative purposes; for example, to manipulate high frequency trading that will result in increased personal benefit but will harm market integrity and efficiency. In this case, risk management strategies for market surveillance need to be developed to detect, predict and prevent harmful algorithmic trading.

More broadly, data power may be underused, overused or misused. Underused data power results in less competitive advantage for the data owner, e.g., the noncompetitive positioning of a company when data power is not effectively and fully utilized. Strategies, thought leadership, plans, approaches and personnel that can take full advantage of data power are necessary. In contrast, overused and misused data power may generate misleading or even unlawful outcomes and impact. Assessment, prediction, prevention and intervention strategies, systems and capabilities must be developed to detect and manage negative data power.

How the power of data is recognized, fulfilled and valuated may determine the strategic position, competitive advantage, tools, and development of a data-intensive organization. The level at which this is conducted is critical for countries, enterprises and individuals. With the emergence of many new companies built on recognizing and utilizing specific data power, as is evident in the increasingly growing big data landscape, entities that ignore the strategic value of data power may significantly lag behind and be disadvantaged. The imbalanced development of a country, an enterprise, or an individual may be the result of ineffective and/or inefficient recognition of data power, and the consequent vision, and actions of achieving data power. Competing in the fourth revolution in data-driven science, technology and economy, a fundamental and strategic matter is to study data power, and create corresponding early-bird vision, strategies, initiatives, and actions to take advantage of data power from political, scientific, technological, economic, educational, and societal perspectives.

1.4.2   Data-Oriented Forces

Ubiquitous data-oriented driving forces can be seen from the viewpoint of both high- and low-level vision and mission, given the prevailing data, behavior, complexity, intelligence, service and opportunity perspectives.

Vision and mission determine the big picture and strategic objectives, and the view of what data will satisfy organizational strategic needs and requirements, and how. Strategic, forward-looking, long-term and big picture thinking is required. This is often challenging, as few people have the training, capability or mindset for such purposes.

Technical and pragmatic data driving forces directly involve data-oriented elements: data, behavior, complexity, intelligence, service and future.

- *Data* is ubiquitous, and includes historical, real-time, and future data;
- *Behavior* is ubiquitous, and bridges the gaps between the physical world and the data world;
- *Complexity* is ubiquitous, and involves the type and extent of complexity that differentiates one data system from another;
- *Intelligence* is ubiquitous, and is embedded in a data system;
- *Service* is ubiquitous, and is present in multiple forms and domains; and
- *Future* is unlimited with ubiquitous opportunities, because data enables enormous opportunity.

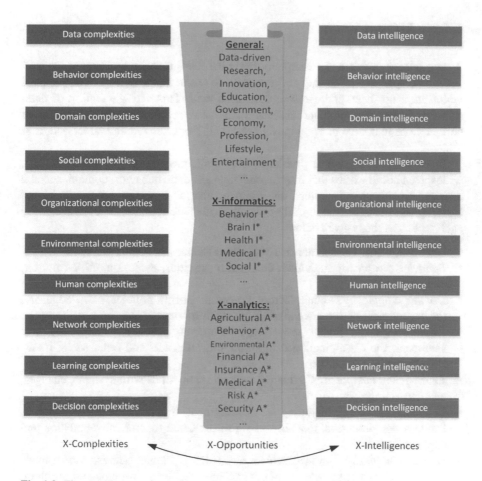

Fig. 1.3 The new X-generations: X-complexities, X-intelligence, and X-opportunities

1.5   New X-Generations

As a result of data business achieving a level of importance that is comparable to traditional, physical business, the world is experiencing a revolutionary migration of complexities, intelligences, and opportunities to their new X-generations. X-generations are embodied in both (1) fundamental driving areas: X-complexities (see Sect. 1.5.1) to be addressed and X-intelligence (see Sect. 1.5.2) to be involved or created, and (2) strategic potential: X-opportunities, such as X-analytics (see Sect. 7.6) and X-informatics (see Sect. 1.5.3) to be generated.

Figure 1.3 illustrates the aspects and perspectives related to X-complexities, X-intelligence, and X-opportunities, which are briefly explained below. Other X-generations are X-analytics and X-informatics, as discussed in Sect. 7.6.

1.5.1   X-Complexities

A data science problem is a complex system [62, 294] that has a variety of intrinsic system complexities. The study of data science has to tackle multiple complexities which have not been addressed or addressed well. This new generation of data-driven science, innovation and business relies on the exploration and utilization of complexities that have not previously been well characterized and addressed, if at all.

In complex data science problems, *X-complexities* [62, 64] refers to diverse, widespread complexities that may be embedded in data, behavior, domain, societal aspects, organizational matters, environment, human involvement, network, and learning and decision-making. These complexities are represented or reflected by such factors as those given below.

- *Data complexity* Comprehensive data circumstances and characteristics;
- *Behavior complexity* Individual and group activities, evolution, utility, impact, and change;
- *Domain complexity* Domain factors, processes, norms, policies, knowledge, and the engagement of domain experts in problem solving;
- *Social complexity* Social networking, community formation and divergence, sentiment, the dissemination of opinion and influence, and other social issues such as trust and security;
- *Environment complexity* Contextual factors, interactions with systems, changes, and uncertainty;
- *Learning complexity* Including the development of appropriate methodologies, frameworks, processes, models and algorithms, and theoretical foundation and explanation;
- *Human complexity* The involvement and roles of human beings, human intelligence and expert knowledge in data science problems, systems and problem-solving processes; and
- *Decision-making complexity* Methods and forms of deliverables, communications and decision-making actions.

More discussion about X-complexities from the data science challenge perspective will be conducted in Sect. 4.2.

1.5.2   X-Intelligence

A complex system is usually embedded with mixed intelligence, and in the data world, data systems are intelligence-based systems. In a complex data science problem, ubiquitous intelligence, called *X-intelligence* [62, 64], is often demonstrated.

X-intelligence is embedded with X-complexities and consists of data intelligence, behavior intelligence, domain intelligence, human intelligence, network intelligence, organizational intelligence, and environmental intelligence, which are briefly discussed below.

- *Data intelligence* highlights the interesting information, insights, and stories hidden in data about business problems and driving forces.
- *Behavior intelligence* demonstrates the insights of activities, processes, dynamics, impact, and the trust of individual and group behaviors by humans and action-oriented organisms.
- *Domain intelligence* includes domain values and insights that emerge from domain factors, knowledge, meta-knowledge, and other domain-specific resources.
- *Human intelligence* includes contributions made by the empirical knowledge, beliefs, intentions, expectations, critical thinking, and imaginary thinking of human individuals and group actors.
- *Network intelligence* results from the involvement of networks, the Web, and networking mechanisms in problem comprehension and problem solving.
- *Organizational intelligence* includes insights and contributions created by the involvement of organization-oriented factors, resources, competency and capabilities, maturity, evaluation, and dynamics.
- *Social intelligence* includes contributions and values generated by the inclusion of social, cultural, and economic factors, norms, and regulation.
- *Environmental intelligence* can be embodied in other intelligences specific to the underlying domain, organization, society, and actors.

X-intelligences in a data science system are mixed. They interact with each other and may not be easily decomposed. A good data product must effectively represent, incorporate and synergize core aspects of X-intelligence that play a fundamental role in system dynamics and problem-solving processes and systems.

More discussion about X-intelligences is available in Chap. 1 in book [68].

1.5.3 X-Opportunities

Our experience and literature review also confirm that data science enables unimagined general and specific opportunities, called *X-opportunities*, for

- *new research*: i.e., "what I can do now but could not do before";
- *better innovation*: i.e., "what I could not do better before but I can do well now."
- *new business*: i.e., "I can make money out of data."

X-opportunities from data may be general or specific. General X-opportunities are enormous and overwhelming. They extend from research, innovation and education to new professions, new ways of operating government, and new economy. In fact, as new models and systems of data-driven economy and research findings emerge, it is a matter of how our imagination can perceive these opportunities. New data products and services emerge as a result of identifying new data-driven business models and opportunities.

We highlight the directions for creative data-driven opportunities below.

*   *Research*, such as inventing data-focused breakthrough theories and technologies;
*   *Innovation*, such as developing cutting-edge data-based intelligent services, systems, and tools;
*   *Education*, such as innovating data-oriented courses and training;
*   *Government*, such as enabling data-driven evidence-based government decision-making and objective planning and execution;
*   *Economy*, such as fostering data economy, services, and industrialization;
*   *Lifestyle*, such as promoting data-enabled smarter living and smarter cities; and
*   *Entertainment*, such as creating data-driven entertainment activities, networks, and societies.

Data-driven opportunities are unlimited, especially in the scenario in which "I do not know what I do not know". Simply by recognizing some of the potential opportunities, the world could be incrementally or significantly changed. To have the capacity to recognize more data-driven opportunities, we need data science thinking.[1] Being creative and critical is important for detecting new opportunities.[2]

X-opportunities may be specified in terms of particular aspects, problems, and purposes. *X-informatics*, which refers to the creation and application of informatics for specific domain problems, is one instance. Another instance is *X-analytics*, which refers to the various opportunities discoverable by applying and conducting analytics on domain-particular data.

Examples of X-informatics are behavior informatics, brain informatics, health informatics, medical informatics, and social informatics. More discussion about informatics for data science can be found in Sect. 6.4.2.

Instances of X-analytics are agricultural analytics, behavior analytics, disaster analytics, environmental analytics, financial analytics, insurance analytics, risk analytics, transport analytics, and security analytics. More discussion about X-analytics is available in Chap. 3 in book [67].

1.6   The Interest Trends

Prior to the prevalence of big data, data analysis, data analytics, and data science were attracting growing attention from several communities, in particular statistics. In recent years, big data analytics, data science, and advanced analytics have become

---

[1]See more discussion about data science thinking in Chap. 3.
[2]Refer to Sect. 3.2.2 and in particular Sect. 3.2.2.3 for more discussion about creative and critical thinking in data science.

increasingly popular in not only the broad IT area but also in other disciplines and domains.

According to Google Trends [193], the online search interest over time in "data science" is similar to the interest in "data analytics", but is 50–100% less than the interest in "big data". However, the historical search interest in data science and analytics is roughly double the interest shown in big data about 10 years ago. Compared to the smooth growth of interest in data science and analytics, the interest in big data has experienced a more rapid increase since 2012. When we googled "data science", 83.8M records were returned, compared to 365M on "big data", and 81.8M on "data analytics".[3]

Although they do not reflect the full picture, the Google search results over the last 10 years, shown in Fig. 1.4, indicate that:

- Data science, data analysis, and data analytics have much richer histories and stronger disciplinary foundations than big data.
- The significant boom in big data has been fundamentally business-related, while data science has been highly linked with research and innovation.
- Data analysis has always been a top concern, although search interest has been flattened and diversified into other hot topics, including big data, data science and data analytics.
- Interestingly, the word "advanced analytics" has received much less attention than all other terms, reflecting the fact that knowledge of, and interest in, more general terms like data analytics is greater than it is for more specific terms such as advanced analytics.
- Compared to 10 years ago, scrutiny of the search trends in the past 4 years would find that big data saw significantly increasing interest from 2012 to 2015, followed by a period of less movement; however, the interest in data science and data analytics has consistently increased, although it has grown at a much lower rate (some one third of big data). Data analysis has maintained a relatively stable attraction to searchers during these 10 years.

1.7   Major Data Strategies by Governments

Governments play the driving role in promoting and operationalizing data science innovation, big data technology development, data industrialization and data economy formation. This section summarizes the representative data strategies and initiatives established by global governments and the United Nations [63].

---

[3]Note, these figures were collected on 15 November 2016.

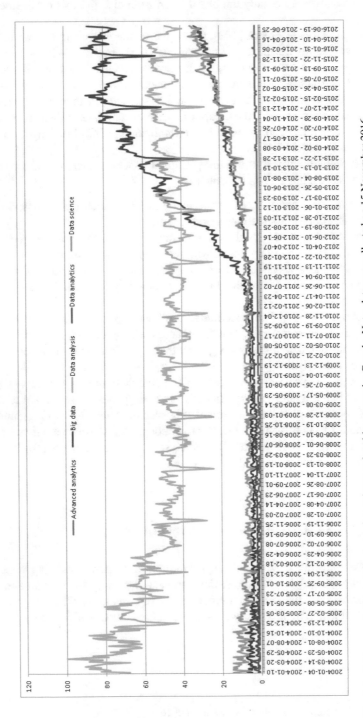

Fig. 1.4 Online search interest trends on data science-related key terms by Google. Note: data was collected on 15 November 2016

Table 1.3 Government initiatives in big data and data science

Government	Representative initiatives
Australia	Public Service Big Data Strategy [399], Whole-of-Government Centre of Excellence on Data Analytics [17]
Canada	Capitalizing on Big Data [50]
China	Big Data Guideline [196], China Computer Federation Task Force on Big Data [85], China National Science Foundation big data program [99]
EU	Data-driven Economy [102], European Commission Horizon 2020 Big Data Private Public Partnership [214]
UK	UK's Big Data and Energy Efficient Computing [395]
UN	UN Global Pulse Project [399]
US	US Big Data Research Initiative [407], Interagency Working Group on Digital Data [107], DARPA's XDATA Program [115], USA NSF Big Data Research Fund [407]

1.7.1   Governmental Data Initiatives

To effectively understand and utilize everywhere data, data DNA and its potential, increasing numbers of regional and global government data initiatives [356] are being introduced at different levels and on different scales in this age of big data and data science to promote data science research, innovation, funding support, policy making, industrialization, and economy.

Table 1.3 summarizes the major initiatives of several countries and regions.

1.7.2   Australian Initiatives

The Australian Government has published several papers on big data and analytics. In the Big Data Strategy—Issues Paper [4], major issues related to big data were discussed. The Australian Public Service Big Data Strategy paper set up its goal to address big data strategy issues and to "provide an opportunity to consider the range of opportunities presented to agencies in relation to the use of big data, and the emerging tools that allow us to better appreciate what it tells us, in the context of the potential concerns that this might raise" [4].

Australia's whole-of-government Centre of Excellence in Data Analytics [17] coordinates relevant government activities. It encourages respective government agencies and departments to think about, promote and accept data-driven approaches for government policy optimization, service improvement and cross-government operations.

As part of a critical government strategy, the Australian Research Council has granted approval to the Australian Research Council (ARC) Centre of Excellence for Mathematical and Statistical Frontiers [1] to conduct research on big data-based mathematical and statistical foundations.

Another recent effort made by the Australian Government was the establishment of Data61 [116], which consolidated data-related human resources in the former National Information and Communications Research Centre of Australia (NICTA) [308] and Commonwealth Scientific and Industrial Research Organisation (CSIRO) and aims to achieve a unified platform for data research and innovation, engagement with industry, government and academia, and software development.

1.7.3   Chinese Initiatives

The Chinese Government treats big data as an essential constituent in its government strategic plan, innovation strategies, and economic transformation. Big data has been an increasingly topical concept in central, state and city-based Chinese government policies, plans and activities. Increasing numbers of facilities, funding, and initiatives have been committed to this area. The Chinese Government is particularly interested in big data-driven innovation and economy.

Several guidance documents and plans have been issued by the central Chinese Government. China's Guidelines [196], for example, are aimed at boosting the development of big data research and applications, to "set up an overall coordination mechanism for big data development and application, speed up the establishment of relevant rules, and encourage cooperation between the government, enterprises and institutions."

China has also set up a national strategic plan for the IoT and big data [196]. Research and commercial-ready funding have been committed to national key research projects and key national labs for big data research, innovation and industrialization.

Many states and cities in China, such as Beijing [195], have launched national big data strategies and action plans for big data and cloud computing [3, 84]. A very early example in China was the Luoyang City-sponsored consulting project in 2011, for which we developed a strategic plan for the City's industrial transformation to a "data industry" [56].

Several data technology parks have been established in China. The Ministry of Education in China recently approved about 30 applications of opening undergraduate courses in data science and big data technology. China has experienced very fast conceptual shift in recent years, from the Internet of Things to cloud computing, and then to big data and now artificial intelligence.

1.7.4 European Initiatives

The European Union (EU) and its member countries consider data-driven economy as a new transformation opportunity and strategy. Several initiatives have been made, and some examples are given below.

The European Commission's (EC) EU communication Towards a Thriving Data-driven Economy [102] is "an action plan to bring about the data-driven economy of the future", which outlines "a new strategy on Big Data, supporting and accelerating the transition towards a data-driven economy in Europe. The data-driven economy will stimulate research and innovation on data while leading to more business opportunities and an increased availability of knowledge and capital, in particular for SMEs, across Europe."

In 2015, the European Data Science Academy, EDSA [154] was formed. EDSA will analyze the skillsets for data analysts, develop modular and adaptable curricula, and deliver multiplatform and multilingual training curricula.

Member countries in the EU establish their own big data strategies and initiatives. For example, the European Commission published a communication entitled "Towards a thriving data-driven economy" [102] aims to address the big data challenge by "sketching the features of the European data-driven economy of the future and setting out some operational conclusions to support and accelerate the transition towards it." The directive also sets out current and future activities in the field of cloud computing.

1.7.5 United States' Initiatives

The United States (US) has played the leadership role in the global promotion of big data and data economy. The US Government has initiated a government strategy and funding support for big data research and industrialization, and several of its initiatives are highlighted below.

Big Data Research Initiative [407] is directed toward "supporting the fundamental science and underlying infrastructure enabling the big data revolution." In 2005, the US National Science Board set the goal that it "should act to develop and mature the career path for data scientists" in its report "Long-lived Digital Data Collections: Enabling Research and Education in the 21st Century" [310].

In 2009, the Committee on Science of the National Science and Technology Council formed an Interagency Working Group on Digital Data which published a report [107] outlining the strategy to "create a comprehensive framework of transparent, evolvable, extensible policies and management and organizational structures that provide reliable, effective access to the full spectrum of public digital scientific data", which "will serve as a driving force for American leadership in science and in a competitive, global information society."

In addition, the Defence Advanced Research Projects Agency (DARPA) launched its XDATA Program [115], which aims to develop computational techniques and software tools for processing and analyzing large, imperfect and incomplete data. In 2012, the National Institute of Standards and Technology (NIST) introduced a new data science initiative [130], and in 2013, the US National Consortium for Data Science was established [405].

Many US vendors certainly drive the development of big data economy, as highlighted by the significant growth of new data-focused companies such as Google, Facebook, Spark, and Rapidminer, and traditional IT companies such as Microsoft and Oracle.

1.7.6   Other Governmental Initiatives

The United Nations (UN) Global Pulse Project is "a flagship innovation initiative of the United Nations Secretary-General on big data. Its vision is a future in which big data is harnessed safely and responsibly as a public good. Its mission is to accelerate the discovery, development and scaled adoption of big data innovation for sustainable development and humanitarian action." [399]

Most countries have made plans, or at least are making plans, to promote a data-driven economy. For example, Canada's policy framework Capitalizing on Big Data [50] aims at "establishing a culture of stewardship . . . coordination of stakeholder engagement . . . developing capacity and future funding parameters."

The United Kingdom's Big Data and Energy Efficient Computing initiative funded by Research Councils UK [395] aims to "create a foundation where researchers, users and industry can work together to create enhanced opportunities for scientific discovery and development."

Without any doubt, an increasing number of government efforts and investment will be made in big data, cloud computing, data science and artificial intelligence.

1.8   The Scientific Agenda for Data Science

An increasing number of scientific initiatives, activities and programs have been created by governments, research institutions, and educational institutions to promote data science as a new field of science.

1.8.1   The Scientific Agenda by Governments

The original scientific agenda of data science has been driven by both government initiatives and academic recommendations, building on the strong promotion of converting statistics to data science, and blending statistics with computing science in the statistics community [97, 128, 164, 197, 203, 205, 231, 467].

Today, many regional and global initiatives have been taken in data science research, disciplinary development and education, creating a data power-enabled strategic agenda for the data era. Several examples are given below.

- In Australia, a Group of Eight (Go8) report [45] suggested the incorporation of data as a keystone in K-12 education through statistics and science by such methods as creating data games for children.
- In China, the Ministry of Science and Technology very recently announced the establishment of national key labs in big data research as part of a strategic national agenda [98].
- In the EU, the High Level Steering Group (HLSG) report of the Digital Agenda for Europe "Riding the Wave" [211] and the Research Data Alliance (RDA) report "The Data Harvest" [212], urged the European Commission to implement the vision of creating "scientific e-infrastructure that supports seamless access, use, re-use, and trust of data" and to foster the development of data science university programs and discipline.
- In the US, a National Science Board report [310] recommended that the National Science Foundation (NSF) "should evaluate in an integrated way the impact of the full portfolio of programs of outreach to students and citizens of all ages that are 'or could be' implemented through digital data collections." Different roles and responsibilities were discussed for individuals and institutions, including data authors, users, managers and scientists as well as funding agencies. The report [107] from the US Committee on Science of the National Science and Technology Council suggested the development of necessary knowledge and skill sets by initiating new educational programs and curricula, such as "some new specializations in data tools, infrastructures, sciences, and management."

1.8.2   Data Science Research Initiatives

An increasing number of research streams, strengths and focused projects have been announced in major countries and regions. Examples include:

- The US NSF Big Data Research Fund [407],
- The European Commission Horizon 2020 Big Data Private Public Partnership [102, 214], and
- The China NSF big data special fund [99].

Each of these initiatives supports theoretical, basic and applied data science research and development in big data and analytics through the establishment of scientific foundations, high-tech programs and domain-specific funds such as health and medical funds. Significant investment has been made to create even faster high performance computers.

Many universities and institutions have either established or are creating research centers or institutes in data science, analytics, big data, cloud computing, and IoT. In Australia, for example, the author created the first data science lab: the Data Science and Knowledge Discovery Lab at UTS in 2007 [141], and the first Australian institute: the Advanced Analytics Institute [4, 409] in 2011 which implements the RED model of Research, Education and Development (RED) of big data analytics for many major government and business organizations. In the US, top universities have worked on building data science initiatives, such as the Institute for Advanced Analytics at North Carolina State University in 2007 [302], the Stanford Data Science Initiatives in 2014 [370], and the Data Science Initiatives at University of Michigan in 2015 [398].

1.9   Summary

As we stepped into the twenty-first century, significant opportunities and a fundamental revolution in economy and science were driven by the worldwide increase in big data, although few people recognized or anticipated the compelling changes that would result.

It is amazing to think that Yahoo created the important role of "Chief Data Officer" in early 2000, and that the term "data science" was created as long as 50 years ago. It is big vision that has driven this world revolution, yet although we are lucky to live at this exciting time, our limited imagination means that we have been slow to embrace and guide a different age: the data era.

Although this new era has been termed "the era of big data", "the age of analytics", and "the data science era", the data era effectively extends far beyond data science, big data, and advanced analytics. It has the capacity to fundamentally change our life, society, way of living, entertainment, and of course the way we work, study and do business. The data era opens the door to a new age of data-driven science and economy.

This chapter has somewhat limited the scope of the data era to one that is data science-based, but our thinking should exceed this limitation. The content of this book thus goes beyond the scientific component in several chapters to embrace concepts about the data economy, data industrialization, and data professions. It is hoped that this will complement the introduction in this chapter and compel the reader's thinking to another age: that of data-driven future science, economy, and society.

Chapter 2
What Is Data Science

2.1 Introduction

The art of data science [197] has increasingly attracted interest from a wide range of domains and disciplines. Communities or proposers from diverse backgrounds have often had contrasting aspirations, and have accordingly presented very different views or demonstrated contrasting foci. For example, statisticians may hold the view that data science is the new generation of statistics; people adopting a fresh approach may believe that data science is a consolidation of several interdisciplinary fields, or is a new body of knowledge; industry players may believe that it has implications for providing capabilities and practices for the data profession, or for generating business strategies.

In this chapter, these different views and definitions of data science are discussed. Key concepts including datafication, data quantification, data DNA, and data product are also discussed, as they build the foundation for data science and the world of data. Several definitions of data science are given, followed by discussions about myths and misconceptions about some key concepts and techniques related to data science.

2.2 Datafication and Data Quantification

Data is ubiquitous because *datafication* [19] and *data quantification* are ubiquitous. Datafication and data quantification contribute to the recognition of the big data era, and its corresponding challenges and prospects.

Datafication refers to how data is created, extracted, acquired or re-produced and rendered into specific data formats, from diverse areas, and through particular channels. Datafication takes the information in every area of business, working, daily life, and entertainment, renders it into data formats, and turns it into data.

© Springer International Publishing AG, part of Springer Nature 2018
L. Cao, *Data Science Thinking*, Data Analytics,
https://doi.org/10.1007/978-3-319-95092-1_2

Data quantification (or data quantitation) refers to the act of measuring and counting for the purpose of converting observations to quantities.

In addition to the commonly seen transactions acquired from business and operational information systems, increasingly popular and widespread datafication and data quantification systems and services are significantly contributing to the data deluge and big data realm.

As we have seen and can predict, datafication and data quantification may take place at any time in any place by anybody in any form in any way in a non-traditional manner, to a variable extent and depth, and at fluctuating speed.

- Quantification timing: *anytime quantification*, from working to studying, day-to-day living, relaxing, enjoying entertainment and socializing;
- Quantification places: *anyplace quantification*, from biological systems to physical, behavioral, emotional, cognitive, cyber, environmental, and cultural spaces, and in economic, sociological and political systems and environments;
- Quantification bodies: *anybody quantification*, from selves to others, connected selves, exo-selves [250] and the world, and from individuals to groups, organizations and societies;
- Quantification forms: *anyform quantification*, from observation to drivers, from objective to subjective, from physical to philosophical, from explicit to implicit, and from qualitative to quantitative forms and aspects;
- Quantification ways: *anysource quantification*, such as sources and tools that include information systems, digitalization, sensors, surveillance and tracking systems, the IoT, mobile devices and applications, social services and network platforms, and wearable devices [414] and Quantified Self (QS) devices and services; and
- Quantification speed: *anyspeed quantification*, from static to dynamic, from finite to infinite, and from incremental to exponential generation of data objects, sets, warehouses, lakes and clouds.

Examples of fast developing quantification areas are the health and medical domains. We are datafying both traditional medical and health care data and "omics" data (genomics, proteomics, microbiomics, metabolomics, etc.) and increasingly overwhelming QS-based tracking data [377] on personal, family, group, community, and/or cohort levels.

2.3   Data, Information, Knowledge, Intelligence and Wisdom

Before we discuss the definitions of data science, two fundamental questions to answer are:

- *What are data, information, knowledge, intelligence, and wisdom (DIKIW)?* and
- *What are the differences between data, information, knowledge, intelligence, and wisdom?*

This section discusses these relevant and important concepts.

It is challenging to quantify these concepts and their differences. An empirical understanding of the conceptualization has been provided in the so-called DIKW Pyramid [342, 425], its earlier variation "signal, message, information, and knowledge" [38], and various arguments and refinements of these conceptualizations [62].

These concepts, including intelligence, capture the different existing forms and progressive representations of objects (or entities) in a cognitive processing and production system.

Data, represents discrete or objective *facts*, *signals* (from sensors, which may be subjective or objective), or *symbols* (or signs) about an object (a physical or virtual entity or event). Data is at the lowest level of cognitive systems, can be subjective or objective, can be with or without meaning, and has a value. An example of data is "8 years old" or "young."

Information, represents a description of relevant data (objects) in an organized way, for a certain purpose, or having a certain meaning. Information can be structural (organized) or functional (purposeful), subjective (relevant to an intent) or objective (fact-based). For example, "Sabrina is 8 years old" is a piece of information which describes a structured relationship between two objects, "Sabrina" and "8 years old". Another example is "Sabrina is young."

Knowledge, represents the form of processed information in terms of an information mixture, procedural actions, or propositional rules. Knowledge can be subjective or objective, known or unknown, actionable or not, and reasonable or not. Examples of processed, procedural and propositional knowledge are "Year 3 students are mostly 8 years old", "Tea and medicine are not supposed to be taken at the same time," and "All 8 year old children should go to school."

Intelligence, representing the ability to inform, think, reason, infer, or process information and knowledge. Intelligence is either inbuilt or can be improved through learning, processing or enhancement. Intelligence can be high or low, hierarchical, general or specific. Examples of intelligence are: "Sabrina is probably in Year 3" (reasoning outcome based on the fact that she is 8 years old and a child of that age is usually at school) or "Sabrina is intelligent" (based on the information that she has always attained high marks at school.)

Wisdom, represents a high-level principle, which is the cognitive and thinking output of information processing, knowledge management, or simply inspiration gained through experiences or thinking. Wisdom indicates the superior ability, meta-knowledge, understanding, application, judgment, or decision-making inherent in knowing or determining the right thing to do at the right time for the right purpose. Wisdom can be non-material, unique, personal, intuitive, or mentally-inspired. Compared to knowledge, wisdom is timeless, comprehensive, general, and sentimental, being passed down in histories and cultures in the form of common sayings, quotations, or philosophical phrases. Examples of wisdom are "A young idler, an old beggar," and "The child is father of the man."

It is difficult to generate a simple framework to show the difference between data, information, knowledge, intelligence, and wisdom. Figure 2.1 illustrates the relationships between them and the path of progression from data to wisdom. In

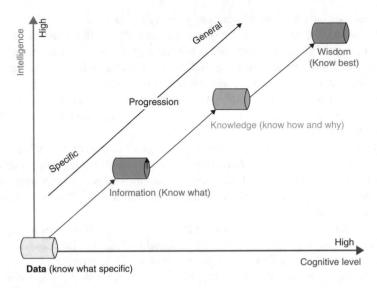

Fig. 2.1 Data-to-information-to-knowledge-to-intelligence-to-wisdom cognitive progression. Note: X-axis: the increase in cognitive level; Y-axis: the increase in intelligence

this framework, the data-to-wisdom path is a specific-to-general progressive journey according to the increase in cognitive level and intelligence.

- Data is about *the aspect of a subject*;
- Information describes *what is known about a subject*, i.e., *know what* (including *know who*, *know when*, *know where*, etc.) in relation to data;
- Knowledge concerns the *know how* and *know why* (or *why is*) about information;
- Wisdom is the intelligence to *know best* about *how to act* (or *why to act*) on the basis of a usually widely validated ability or understanding.

During the production and cognitive processing procedure, intelligence plays an enabling role for both progression and production.

In addition, as discussed in Sect. 4.2 about X-complexities and Sect. 4.3 about X-intelligence, the progression of data-to-information-to-knowledge-to-wisdom needs to involve and handle relevant complexities and intelligences.

2.4 Data DNA

2.4.1 What Is Data DNA

In biology, DNA is a molecule that carries genetic instructions that are uniquely valuable to the biological development, functioning and reproduction of humans and all living organisms.

As a result of data quantification, data is everywhere, and it is present in the public Internet; the Internet of Things (IoT); sensor networks; sociocultural, economic and geographical repositories; and quantified personalized sensors, including mobile, social, living, entertaining, and emotional sources. These form the "datalogical" constituent: *data DNA*, which plays a critical role in data organisms and performs a similar function to biological DNA in living organisms.

Definition 2.1 (Data DNA) *Data DNA* is the datalogical "molecule" of data, consisting of fundamental and generic constituents: entity (E), property (P), behavior (B), and relationship (R).

Here, "datalogical" means that data DNA plays a similar role in data organisms as biological DNA plays in living organisms. The four elements in data DNA, namely behavior, entity, relationship and property (BERP), represent diverse but fundamental aspects in data. *Entity* can be an object, instance, human, organization, system, or part of a subsystem, or environment. *Property* refers to the attributes that describe an entity. *Behavior* refers to the activities and dynamics of an entity or a collection of entities. *Relationship* corresponds to entity interactions and property interactions, including property value interactions.

2.4.2   Data DNA Functionalities

Entity, property, behavior and relationship have different characteristics in terms of quantity, type, hierarchy, structure, distribution, and organization. A data-intensive application or system often comprises many diverse entities, each of which has specific properties, and different relationships are embedded within and between properties and entities.

From the lowest to the highest levels, data DNA presents heterogeneity and hierarchical couplings across levels. On each level, it maintains *consistency* (the inheritance of properties and relationships) as well as *variations* (mutations) across entities, properties, and relationships, while supporting *personalized characteristics* for each individual entity, property, and relationship.

For a given data, its entities, properties, and relationships are instantiated into diverse and domain-specific forms which carry most of the data's ecological and genetic information in data generation, development, functioning, reproduction, and evolution.

In the data world, *data DNA* is embedded in the whole body of personal [417] and non-personal data organisms, and in the generation, development, functioning, management, analysis, and use of all data-based applications and systems.

Data DNA drives the evolution of a data-intensive organism. For example, university data DNA connects the data of students, lecturers, administrative systems, corporate services, and operations. The student data DNA further consists of academic, pathway, library access, online access, social media, mobile service, GPS, and Wifi usage data. Such student data DNA is both fixed and evolving.

In complex data, data DNA is embedded within various X-complexities (see detailed discussion in Sect. 1.5.1 and in [64] and [62]) and ubiquitous X-intelligence (more details in Sect. 1.5.2 and in [64] and [62]) in a data organism. This makes data rich in content, characteristics, semantics, and value, but challenging in acquisition, preparation, presentation, analysis, and interpretation.

2.5   Data Science Views

In this section, the different views of data science are discussed to create a picture of what makes data science a new science.

2.5.1   The Data Science View in Statistics

Statisticians have had much to say about data science, since it is they who actually created the term "data science" and promoted the upgrading of statistics to data science as a broader discipline.

Typical statistical views of data science can be reflected in the following arguments and recommendations.

In 1997, Jeff Wu questioned whether "Statistics = Data Science?". He suggested that statistics should be renamed "data science" and statisticians should be known as "data scientists" [467]. The intention was to shift the focus of statistics from "data collection, modeling, analysis, problem understanding/resolving, decision making" to future directions on "large/complex data, empirical-physical approach, representation and exploitation of knowledge".

In 2001, William S. Cleveland suggested that it would be appropriate to alter the statistics field to data science and "to enlarge the major areas of technical work of the field of statistics" by looking to computing and partnering with computer scientists [97].

Also in 2001, Leo Breiman suggested that it was necessary to "move away from exclusive dependence on data models (in statistics) and adopt a more diverse set of tools" such as algorithmic modeling, which treats the data mechanism as unknown [42].

In 2015, a statement about the role of statistics in data science was released by a number of ASA leaders [145], saying that "statistics and machine learning play a central role in data science." Many other relevant discussion is available in AMSTATNEWS [12] and IMS [473].

2.5.2 A Multidisciplinary Data Science View

In recent years, data science has been elaborated beyond statistics. This is driven by the fact that statistics cannot own data science, and the statistics community has realized the limitation of statistics-focused data science and the broader capability requirements that go beyond statistics.

A multidisciplinary view has thus been increasingly accepted not only by the statistics community, but also other disciplines, including informatics, computing and even social science. This reflects the progressive evolution of the concept and vision of data science, from statistics to informatics and computing, as well as other fields, and the interdisciplinary and cross-disciplinary nature of data science as a new science.

Intensive discussion has taken place in the research and academic communities about creating data science as an multidisciplinary academic field. As a new discipline [364], data science involves not only statistics, but also other disciplines. The concept of data science is correspondingly defined from the perspective of disciplinary and course development [470].

Although different communities may share contrasting views about what disciplines are involved in data science, statistics, informatics, and computing are three fields that are typically viewed as the keystones, making data science a new science.

In addition, some people believe a cross-disciplinary body of knowledge in data science includes informatics, computing, communication, management, and decision-making; while others treat data science as a mixture of statistics, mathematics, physics, computer science, graphic design, data mining, human-computer interaction, information visualization, and social science.

Today, there is increasing consensus that data science is inter-disciplinary, cross-disciplinary, and trans-disciplinary. We will further discuss the definition of data science from the disciplinary perspective in Sect. 2.6.2.

2.5.3 The Data-Centric View

Although there are different views or perspectives through which to define what makes data science a new science, a fundamental perspective is that *data science is data centric*. There are several aspects from which to elaborate on the data-centric view: hypothesis-free exploration, model-independent discovery, and evidence-based decision-making, to name three.

First, *hypothesis-free exploration* needs to be taken as the starting point of data understanding. There is no hypothesis before a data science task is undertaken. It is data that generates, indicates, and/or validates a new hypothesis. New hypothesis generation relies greatly on a deep understanding of the inbuilt data characteristics, complexities and intelligence of a problem and its underlying environment.

Second, *model-independent discovery*, also commonly called *data-driven discovery*, is the correct way to conduct data understanding; that is, to allow the data to tell a story. Models are not applied directly, since a model is built with certain embedded hypotheses and assumptions.

It is not easy to be fully data-driven, since the understanding of complex data characteristics and contexts is often challenging. A more feasible approach is therefore to combine data-driven discovery with model-based learning, which basically assumes that the data fits a certain assumption that can be captured by a model (usually a mathematical or statistical model) while data characteristics have to be deeply explored and then fed to the model. Section 3.5.3.2 discusses this approach.

Lastly, *evidence-based decision-making* is the outcome of data-centric scientific exploration. Here *evidence* constitutes the new hypotheses, insights, indicators, and findings hidden in data that were previously invisible to data modelers and decision makers.

Evidence is not simply the modeling of outputs or results; rather, it is the deep disclosure of the intrinsic nature, characteristics, complexities, principles, and stories inbuilt in the data and physical world. In essence, data-driven evidence captures the working mechanism, intrinsic nature, and solutions of the underlying problem.

The data-centric view is also largely derived from a large number of conceptual arguments; for example, that data-driven science is mainly interpreted in terms of the reuse of open data [299, 312], that data science comprises the numbers of our lives [290], or that data science enables the creation of data products [278, 279].

2.6 Definitions of Data Science

The various data science views discussed in Sect. 2.5 form the foundation for our discussion on what data science is. Below, we present several definitions of data science from different perspectives, building on the observations and insights we have gained from a comprehensive review of data science [63–66] and relevant experience in advanced analytics over the past decades, as well as inspirations from the relevant discussions about the concepts and scopes of data science (e.g., in [12, 19, 93, 97, 112, 128–130, 197, 205, 215, 231, 238, 279, 283, 285–287, 316, 327, 332, 364, 371, 461, 467]).

2.6.1 High-Level Data Science Definition

Taking the data-centric view outlined in Sect. 2.5.3, a high-level definition of data science can be adopted that accords with the usual type of umbrella definition found in any scientific field, as given below:

Definition 2.2 (Data Science[1]) Data science is the science of data, or data science is the study of data.

However, this high-level view does not provide a concrete explanation of what makes data science a new science, nor why or how. Therefore, in the following sections, more concrete and specific data science definitions are given.

2.6.2 Trans-Disciplinary Data Science Definition

As the multi- and trans-disciplinary views of data science indicate, *data science* is a new disciplinary field in which to study data and its domain with data science thinking (more discussion about data science thinking can be found in Chap. 3.).

Definition 2.3 (Data Science[2]) Data science is a new trans-disciplinary field that builds on and synthesizes a number of relevant disciplines and bodies of knowledge, such as statistics, informatics, computing, communication, management and sociology, to study data and its domain employing data science thinking.

Since a variety of multidisciplinary views exist, as discussed in Sect. 2.5.2, different people may have contrasting opinions of which disciplines constitute data science. In general, data science integrates *traditionally data-oriented disciplines* such as statistics, informatics, and computing with *traditionally data-independent fields* such as communication, management, and sociology.

As an example (illustrated in Fig. 2.2), a *discipline-based data science formula* is given below:

$$data\ science$$
$$\stackrel{def}{=} \{statistics \cap informatics \cap computing \cap communication$$
$$\cap sociology \cap management \mid data \cap domain \cap thinking\} \qquad (2.1)$$

where "|" means "conditional on."

Data science thinking is required to understand what data science is and to generate data products [64]. This thinking reflects data-driven methodologies and process, and the transformation from data analytics to knowledge generation and wisdom production.

In the data-to-information-to-knowledge-to-intelligence-to-wisdom progression (see Sect. 2.3), different disciplines play different roles.

In Chap. 6, we briefly discuss the roles played by relevant disciplines and the inter-disciplinary, cross-disciplinary and trans-disciplinary interactions between data science and the respective disciplines.

The definition of Data Science[2] also includes the new opportunities and development that exist through migrating or match-making existing disciplinary development to a data science-oriented stage. This highlights the potential for new data

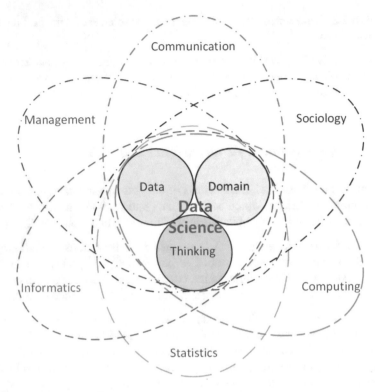

Fig. 2.2 Trans-disciplinary data science

science fields, such as behavioral data science, health data science, or even history-based data science.

2.6.3 Process-Based Data Science Definition

Generally speaking, *data science is the science (or study) of data* as defined in Data Science[1]. However, there are different ways of specifying what data science is; it may be object-focused, process-based, or discipline-oriented [64], as in Data Science[2]. From the data science process perspective, we offer the following definition, building on, involving and/or processing DIKIW.

Definition 2.4 (Data Science[3]) From the *DIKIW-processing* perspective, *data science* is a systematic approach to "thinking with wisdom", "understanding the domain", "managing data", "computing with data", "discovering knowledge", "communicating with stakeholders", "acting on insights", and "delivering products".

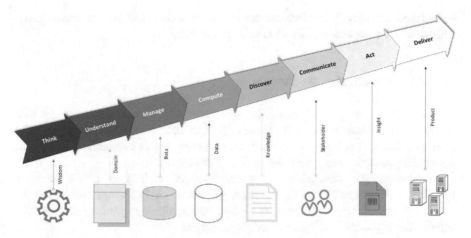

Fig. 2.3 Process-driven data science

As shown in Fig. 2.3, data science involves many important activities and corresponding resources and facilities related to DIKIW. Accordingly, a *process-based data science formula* is given below:

$$data\ science \overset{def}{=} \{think + understand + manage +$$

$$compute + discover + communicate + act + deliver | DIKIW\} \quad (2.2)$$

These components have specific meanings in the data science process, and these are discussed in the following subsections.

2.6.3.1 Thinking with Wisdom

Thinking with wisdom reflects the driving force, goal/objective and outcomes of data science that will transform data to wisdom. Data is the input, while wisdom reflects the data science outcomes.

Thinking with wisdom also reflects the value of data science, which is to produce wisdom for smarter, evidence-based, more informative decision-making. Thinking with wisdom reflects the challenge of data science, how much wisdom we can obtain from data-driven discovery, and what wisdom we can obtain through data science.

Thinking with wisdom is an output-driven thinking and purpose-based methodology. It differentiates data science from existing sciences such as statistics, computing and informatics in the sense that data science expects to produce deep insights that are not visible, and are evidence-based and valuable.

Initial wisdom comes from experience, expert knowledge, and preliminary hypothesis about the input data, problem, goal and path for problem-solving. These elements may be incorrect, and will also be fully domain-dependent. They will be verified, refined, updated or renewed with confirmed insights and intelligence following the completion of the data science process.

2.6.3.2   Understanding the Domain

Although there are general scientific and technological issues and systems, data science is often domain-dependent when the data science process is focused. For example, network science consists of data science that is customized to address networking complexities, characteristics, and objectives.

Domain-specific data science is necessary because data is usually domain-specific. This is similar to X-analytics, in which analytics are developed for specific domains; the same is true for X-informatics, although general analytics and informatics problems and theories are also essential.

In creating domain-specific data science, such as behavioral data science, domain complexities and characteristics need to be scrutinized and articulated. Specifying and generalizing domain-independent complexities and characteristics lead to general data science; specifying and generalizing domain-dependent complexities and characteristics lead to domain-specific data science.

As discussed in Chap. 6, domain-specific data science needs to consider many domain factors, domain complexities, and domain intelligence. In Sect. 2.6.2, the roles of and relations between data, domain and thinking in relation to what data science is are further discussed.

2.6.3.3   Managing Data

Managing data is a core component of data science. It involves the acquisition, management, storage, search, indexing and recovery of data.

The pre-processing of data is critical in managing data, and is related to data quality enhancement and the management of social issues.

Data quality enhancement involves the handling of data quality problems, such as missing values, errors, and fake data, and the processing of data complexities such as long-tail distributions and sparsity, bias and odd distributions, and non-IID characteristics.

Social and ethical issues are increasingly important in data science. Typical issues that need to be managed include privacy processing and protection, the security of data and systems, and the trust of data and data-based products. Detailed discussion about data quality, data quality issues, and data quality assurance is available in Sect. 4.7. Data social issues and data ethics, and the assurance of both, are discussed in Sect. 4.8.

In addition to data management, the role of management science in data science is also discussed in Sect. 6.8.

2.6.3.4   Computing with Data

Computing with data refers to how to manipulate data for what purposes and in what ways. Computing with data may consist of tasks such as feature engineering, data exploration, descriptive analysis, visualization, and data presentation.

Feature engineering consists of the understanding, selection, extraction, construction, fusion, and mining of features, which are fundamental for data use, knowledge discovery, and other data-driven learning.

Data exploration is to understand, quantify and visualize the main characteristics of data. This may be achieved by descriptive analysis, statistical methods, and visual analytics.

Descriptive analysis is typically statistical analysis. It may involve the quantitative examination, manipulation, summarization, and interpretation of data samples and data sets to discover underlying trends, changes, patterns, relationships, effects and causes.

Data visualization presents data in a visual way, i.e., in a pictorial or graphical format. Typically, charts, graphs, and interactive interfaces and dashboards are used to visualize patterns, trends, changes, and relationships hidden in data, business evolution, and problem development.

Data presentation involves the broad communication of data and data products. Typical data presentation tools and means include reports, dashboards, OLAP, and visualization.

In Sect. 6.6, more discussion about computing for data science is provided.

2.6.3.5   Discovering Knowledge

Knowledge discovery [161, 202] is a higher level of data manipulation that aims to identify hidden but interesting knowledge which discloses intrinsic insights about problems, problem evolution, causes, and effects.

Typical knowledge discovery tasks including prediction, forecasting, clustering, classification, pattern mining, and outlier detection. When knowledge discovery is applied to specific data issues, it generates respective knowledge discovery methods, tasks and outcomes. For example, climate change detection seeks to detect significant changes taking place in the climate system.

In data science, critical issues to consider in knowledge discovery are

- what knowledge is to be mined?
- is the knowledge actionable, trustful, and transformative? and
- how can the knowledge be converted to insight, intelligence and wisdom?

The *actionablity* [59, 77] of knowledge determines how useful the discovered knowledge will be for decision-making and game changing. The transformation from knowledge to intelligence and wisdom requires additional theories and tools, which could involve X-intelligence (see more in Sect. 4.3), including domain intelligence, social intelligence, organization intelligence, and human intelligence.

This is because intelligence and wisdom aims to achieve a high level and general abstraction of data values.

2.6.3.6 Communicating with Stakeholders

Communicating with stakeholders is particularly important for data science tasks. *Communicating with stakeholders* involves many important aspects, such as:

- Who are the stakeholders of a data science project?
- How can you communicate with each stakeholder?
- What is to be communicated to each stakeholder?
- What are the skills required for good communications in data science?

Stakeholders in a data science project may be of several types, at various levels: decision makers, project owners, data scientists, business analysts, business operators. They have different roles and responsibilities in an organization, and in the data science process and engineering function. Communications with each role and between roles may involve disparate objectives, channels, and content.

As data scientists are the core players in data science projects, it is critical for them to communicate with business owners and business operators about the business objectives, requirements, scope, funding models, priorities, milestones, expectations, evaluation methods, assessment criteria, implementation requirements, and deployment process of data science projects.

There are different means of initiating and undertaking communications with and between the various roles in a data science project. Executive reports are common for executives and decision makers. Business analysts may like more quantitative analytical reports with evidence and supporting justifications. Business operators may prefer dashboards and user-friendly "automated" systems.

Many skills may be required to achieve good communications. For example, summarization, abstraction, generalization, visualization, formalization, quantification, representation, and reporting may be used for different purposes on different occasions.

In Sect. 6.9, the roles and relations between communications (as a discipline) and data science are discussed.

2.6.3.7 Delivering Data Products

Data science needs to deliver outcomes, which we call *data products*. As defined in Sects. 2.8 and 1.3.1, data products refer to broad data-related deliverables.

Often data products are equivalent to the knowledge and findings discovered in a data analytical project, but this does not reflect the full picture of data science deliverables.

Data products are highly variable, depending on their purpose, the specific procedures of the data science process, the data science personnel who produce the deliverables, overall requirements and expectations, and so on.

Findings from data mining and machine learning are often presented in a technical manner and as half-way products, which need to be converted to final products, such as by incorporating business impact and being presented in business rules that are easily acceptable for business operations and decision support.

In business, data products extend far beyond data analytical models and algorithms. A data product may be presented as a mobile application, a service website, a social media system, a dashboard, an infrastructure-oriented system, a programming language, or even an analytical report.

See more discussion on the topic of data products in Sects. 2.8 and 1.3.1.

2.6.3.8 Acting on Insights

Insights in data science, i.e., *data insights*, are presented in two forms: analytical insight and deliverable insight.

The most valuable and difficult thing in data science is to extract insights from data and reflect such insights in data science deliverables, namely *deliverable insights*. *Deliverable insights* refer to the deep understanding and appropriate representation of intrinsic, fundamental, genuine and complete mechanisms, dynamics, principles, and driving forces in a business problem, and are reflected in the acquired data.

Deep analytics converts data into insights for decision-making. *Deep insights* discover original working mechanisms mapped from the physical world to the data world.

The second type of insight is the analytical insight, which is often overlooked in data science. Converting data to knowledge and wisdom requires deep insights, which drive and enable the deep understanding of data characteristics and complexities.

Analytical insights provide the appropriate perspectives to effectively capture the intrinsic and fundamental mechanisms, dynamics, principles and driving factors reflected in the data world. Naturally, analytical insights have to be extracted by X-analytics (more discussion in Sect. 7.6) by incorporating and utilizing X-complexities (see Sect. 4.2) and X-intelligence (see Sect. 4.3).

2.7 Open Model, Open Data and Open Science

The power of data continues to significantly transform the philosophy, methodology, mechanism and process of traditional data and science development, management, and evolution. The revolution in science, technology and economy is driven by an open model that drives many open movements and activities, in particular, open data and open science. These are built on the spirit of openness. This section discusses these critical enablers of the fourth revolution—data-driven science, technology, and economy.

2.7.1 Open Model

Openness is a critical system feature of open complex systems and their evolution, in which there are interactions and exchanges of information, energy, or materials between systems and outside environments [62, 333]. Systems with openness are open systems. Openness has shown that it plays a critical role in driving the faster development of an open system compared to a closed system without openness.

A key feature differentiating the data science era from the pre-data science era is the overwhelming adoption and acceptance of an open model rather than a closed one. The *open model* is enabled by adopting the openness principle, and advocates and enables public, free or shared, democratic, distributed, and collaborative modes in every aspect of science, technology, economy, society, and living. We outline the important features of each mode below.

- Public: is accessible to a wide community and audience;
- Free: is freely available without a financial paywall or for-profit publishing;
- Shared: is accessible with certain restrictions;
- Democratic: is contributed by and contributes to the public; and
- Collaborative: is jointly contributed by and jointly contributes to its audience.

Typical examples of open model-driven technological, economic, societal and lifestyle entities include the innovation of social media such as Facebook and LinkedIn, the migration of mobile to smart phone-embedded applications, and industrial transformations such as the migration of physical shop-based commerce to online businesses like Amazon and Taobao.

The openness principle and open model have been rapidly adopted and developed since Internet-based and smart phone applications have become widespread in everyday business. Many open model-based movements (open movements) and activities (open activities) have emerged, such as open source, open data, open access, open government, open education, and open science, which has fundamentally transformed their existing principles, mechanisms, operations, stakeholder relationships, and financial and accounting models. These open systems and mechanisms have been globalized and are represented by worldwide Internet-based movements, services, and practices, which interact with each other and their corresponding classic systems and mechanisms, driving the rapid evolution of our science, technology, economy, and society and living towards a more open and universal way of conducting business and communicating.

2.7.2 Open Data

Open data [452] is a fundamental principle and mechanism in the open model. It encourages and enables the free availability of data, particularly through the Internet, social media and mobile applications-based technologies, infrastructure, services, and operations. *Open data* offers the public access to, usage, dissemination and republishing of data without no copyright or financial paywall demands other than acknowledgement. Open data mechanisms and activities thus remove the control typically applied in classic closed systems, which include restrictions on access, intellectual property (including copyright, patents), and for-profit business models (e.g., licences).

There is an expanding variety of data that can be made open. Typical open data consist of scientific data (from every field of science), common and professional knowledge (from the public domain, literature and disciplinary research), policies (from institutions, government and professional bodies), and the public (about their activities and sentiment).

Open data and data sharing programs have been announced in many countries and domains. Major developments are the open government data initiatives supported by government organizations, and the corresponding policy support; for example, the US Government open data site [403], the UK open data project [396, 397], the Australian Government open government data site [15, 16] and Data Matching program [14], and the European Union Open Data Portal [155] and data sharing projects [212].

In addition, many Open Access schemes are increasingly being accepted by academic journals and funding bodies, and are included in institutional evaluations.

Efforts have also been made in diverse societies to create shareable data repositories, especially for science and research. Examples of open repositories are the global climate data [389], the global terrorism database [199], the Yahoo Finance data [469], the Gene Expression Omnibus [179], mobile data [194], the UCI repositories for machine learning [391], the Linguistic Data Consortium data for Natural Language Processing [270], the TREC data for text retrieval [309], Kaggle competition data [241], and the VAST challenge [411] for visual analytics, to name a few.

However, it is understandable that not all data is shareable, and exceptions need to be made for private data and other sensitive data. Open data also does not mean that the data can be used for unlawful purposes. The amount of non-open data is reducing, but issues related to the ethics, privacy and misuse of open data are emerging that could prove to be very serious, such as the 2018 Facebook scandal in which Cambridge Analytica improperly utilized Facebook user data in their commercial activities [24]. Relevant national and international laws, policies, agreements, and norms are not yet sufficiently well developed to regulate and support healthy open data activities and to protect the data creator/owner's information and rights from misuse. When powerful data is misused, negative outcomes and impact may result, so an awareness of the need to prevent such outcomes should be foremost in promoting open data and open activities. Urgent and systematic research, regulation and education on the risks and mitigation of open data and related open activities are critical from a legal, ethical, political, economic, and behavioral perspective.

2.7.3   Open Science

The scientific community has been one of the very first public domains to welcome, advocate and promote the openness principle and open model-driven activities in science, research, innovation, data management, dissemination of results, and intellectual property management. This has driven the formation of open science [455]. Open science takes the openness principle and implements the open model in all these areas, and in related activities (e.g., knowledge sharing, reviewing and evaluation, publishing and republishing, education, operations, and management).

Typically, open science consists of the following major open mechanisms, methods and activities: open research and innovation, open science data and open access, open source, open review and evaluation, and open education and training, as shown in Fig. 2.4.

Open research and innovation are the core activities in open science. They advocate and drive the undertaking of scientific research and innovation activities in an open fashion. Open research and innovation are more transparent, collaborative, and distributed than closed research which is undertaken within a more independent, confidential and private context. Open research highlights the spirit and value of conducting collaborative and shareable scientific activities, which drives the scientific activities of problem identification and statement, the formulation and

Fig. 2.4 Open science

resolution of scientific problems, and funding and projects to support research and innovation and the evaluation, dissemination and management of scientific results. Typical activities to support open science are crowdsourcing, international collaborative open projects, and research networks.

Open science data and open access are two necessary mechanisms for enabling open science and innovation. Science data are freely available to the public. Open access [451] is a principle and mechanism for enabling free access to scientific outputs (including peer-reviewed and non-peer-reviewed results) from scientific activities through scientific dissemination channels (including journals and conferences). Such freely accessible data are archived and managed by scientists, institutions or independent organizations. The authors of the scientific outputs control the copyright of their work when it is published or republished, the integrity of their work, and the right of their work to be lawfully used and acknowledged.

Open source [456] refers to the principle, methodology and mechanisms for creating, distributing, sharing, and managing software, also called *open source software*. The source codes are made freely available to the public, hence this is also known as *free software*. Open source software is typically associated with a licensing arrangement that allows the copyright holder to decide how the software is distributed, changed and used by others, and for what purpose. Open source software requires corresponding software development infrastructure; collaborative development models, methods, platforms and specifications; and copyright, laws, agreements, and norms for certification, distribution, commercialization and licensing, change, usage rights and risk management (e.g., security).

Open review and evaluation [454] are the review and evaluation, process and management mechanisms of scientific outputs by the public or peer reviewers whose names are disclosed to the authors. The review and evaluation process, commenting activities and reports, revision and responses between authors and reviewers may be open in a public (Internet-driven) or review management system (e.g., journal review system).

Open education and training [453] refers to the online provision of educational and training admission, course-offering, resource sharing, teaching-learning servicing, and accreditation. Open education and training exceeds the limitations of awarded courses and short courses that are traditionally offered through educational and training institutions. Open education and training changes the way that scientific knowledge and capabilities are transferred to learners by providing more ad hoc, flexible, and customizable study plans, channels, scheduling, course formation, and resources. It removes the restrictions on course availability, comparison, selection and change, lecturers, study modes, scheduling, and materials that are a feature of institution-based education and training, and enables learners to make choices and advance their learning through the global, fast and flexible approach enabled by Internet-based online courses. Open education thus encourages better teaching and learning quality and performance. The open science movement has motivated the emergence of many open movements and activities in a range of scientific disciplines, scientific research processes, enabling and support facilities, and in the assessment and refinement of scientific outputs and impact. Open science is significantly driving the development of data science as a new science.

2.8   Data Products

The outputs of data science are *data products* [278, 279]. Data products can be described as follows.

Definition 2.5 (Data Products) *Data products* are deliverables and outputs from data, or driven by data. Data products can be presented as discoveries, predictions, services, recommendations, decision-making insights, thinking, models, algorithms, modes, paradigms, tools, systems, or applications. The ultimate data products of value are knowledge, intelligence and wisdom.

The above definition of data products goes beyond technical product-based types and forms in the business and economic domain, such as social network platforms like Facebook, and recommender systems like Netflix. Producing data-based outputs in terms of products and business enables the generation of a new economy: *data economy*. Typically, the current stage of data economy is featured by social data business, mobile data business, online data business, and messaging-based data business.

Various data products enable us to explore new data-driven or data-enabled personalized, organizational, educational, ethical, societal, cultural, economic, political, cyber-physical forms, modes, paradigms, innovations, directions and ecosystems, or even thinking, strategies and policies. For example, there is a good possibility that large scale data will enable and enhance the transfer of subjective autonomy to objective autonomy, beneficence and justice in the social sciences [158], and will enable the discovery of indicators like Google Flu [269] which may not be readily predicted by domain-driven hypothesis and professionals.

These platforms deliver data products in various forms, ways, channels, and domains that are fundamentally transforming our academic, industrial, governmental, and socio-economic life and world. With the development of data science and engineering theories and technologies, new data products will be created. This creation is likely to take place at a speed and to an extent that greatly exceeds our current imagination and thinking, as demonstrated by the evolution to date of Internet-based products and artificial intelligence systems.

2.9   Myths and Misconceptions

Data science is still at an early stage, and it is therefore understandable to see different and sometimes contradictory views appearing in various quarters. However, it is essential to share and discuss the many myths, memes [129], and misconceptions about data science compared to the reality [234], and to ensure the healthy development of the field. This section draws on observations about the relevant communities, as well as the experiences and lessons learned in conducting data science and analytics research, education, and services, to list the relevant myths and misconceptions for discussion.

At the same time, it is important to debate the nature of data science, clarify the fundamental concepts and myths, and demonstrate the intrinsic characteristics and opportunities that data science has to offer.

Hence, this section lists various misunderstandings and myths about data science and also clarifies the reality. The common misconceptions are summarized in terms of the concepts of data science, data volume, infrastructure, analytics, capabilities and roles. Discussion and clarification are provided to present the actual status, intrinsic factors, characteristics, and features of data science, as well as the challenges and opportunities in data research, disciplinary development, and innovation.

2.9.1   Possible Negative Effects in Conducting Data Science

Big data and data science are fundamental drivers in the new generation of science, technologies, economy, and society. However, this does not mean that all applications and data science case studies will naturally have a positive effect, as might be expected.

While often overlooked, the improper use of big data and data science may result in negative effects. These may be caused such behaviors as

- the violation of assumptions made in applied theories, models, tools, and results;
- the improper alignment of theories and methods with their applicability (corresponding context and conditions);
- the violation of constraints (applicability) on data characteristics and complexities to fit the applied theories, models, and results;
- the quality issues associated with data (e.g., bias, see more in Sect. 4.7);
- the applicability and potential of data;
- the quality issues associated with features;
- the quality issues associated with theories and methods;
- the applicability and potential of theories and methods;
- the applicability and potential of evaluation systems;
- the availability of theoretical foundations and guarantee of the methods and results;
- the social issues and ethics associated with data science (see discussion in Sect. 4.8);
- the applicability of the resultant output and findings;
- the significance, transparency and learnability of underlying problems and their complexities;

- the effectiveness of computing infrastructure;
- the efficiency of computing infrastructure;

Any of the above scenarios could easily result in biased, misleading, inappropriate or incorrect applications and consequences. The use of biased findings, the overuse, underuse and misuse of modeling assumptions, treating unverifiable outcomes objectively, or enlarging the values of data-driven discovery, may cause defects and problems.

The possible negative effects of the improper application of data science may become particularly obvious and hard to address. One example is in the analysis of complex social problems. There are many sophisticated complexities to be considered in social data science, such as cultural factors, socio-economic and political aspects, in addition to data characteristics and complexities. Their working mechanisms, behaviors, dynamics, and evolution may be much more sophisticated and demonstrate characteristics different from those in business transactions collected from business operations and production.

2.9.2 Conceptual Misconceptions

Data science has typically been defined in terms of specific disciplinary foundations, principles, goals, inputs, algorithms and models, processes, tools, outputs, applications, and professions. Often, however, a fragmented statement may cause debate as a result of the failure to see the whole picture. In this section, we discuss some of the arguments and observations collected from the literature.

- Data science is statistics [44, 128]; "why do we need data science when we've had statistics for centuries" [461]? How does data science really differ from statistics [129]? (Comments: Data science provides systematic, holistic and multi-disciplinary solutions for learning explicit and implicit insights and intelligence from complex and large-scale data, and generates evidence or indicators from data by undertaking diagnostic, descriptive, predictive and/or prescriptive analytics, in addition to supporting other tasks on data such as computing and management.)
- Why do we need data science when information science and data engineering have been explored for many years? (Comments: Consider the issues faced in related areas by the enormity of the task and the parallel example of enabling a blind person to recognize an animal as large as an elephant (see more about blind knowledge and the parallel example in Sect. 4.4.2). Information science and data engineering alone cannot achieve this. Other aspects may be learned from the discussion about greater or fewer statistics; more in [87].)
- I have been doing data analysis for dozens of years; data science has nothing new to offer me. (Comments: Classic data analysis and technologies focus mostly on explicit observation analysis and hypothesis testing on small and simpler data.)

- Is data science old wine in a new bottle? What are the new grand challenges foregrounded by data science? (Comments: The analysis of the gaps between existing developments and the potential of data science (see Fig. 5.1) shows that many opportunities can be found to fill the theoretical gaps when data complexities extend significantly beyond the level that can be handled by the state-of-the-art theories and systems, e.g., classic statistical and analytical theories and systems were not designed to handle the non-IIDness [60] in complex real-life systems.)
- Data science mixes statistics, data engineering and computing, and does not contribute to breakthrough research. (Comments: Data science attracts attention because of the significant complexities in handling complex real-world data, applications and problems that cannot be addressed well by existing statistics, data engineering and computing theories and systems. This drives significant innovation and produces unique opportunities for generating breakthrough theories.)
- Data science is also referred to as data analytics and big data [6]. (Comments: This confuses the main objectives, features, and scope of the three concepts and areas. Data science needs to be clearly distinguished from both data analytics and big data.)
- Other definitions wrongly ascribed to data science are that it is big data discovery [121], prediction [126], or the combination of principle and process with technique [332].

It is also worth noting that the terms *big data*, *data science* and *advanced analytics* are often overused or improperly used by many communities and for various purposes, particularly because of the influence of media hype and buzz. Most Google searches on these keywords return results that are irrelevant to their intrinsic semantics and scope, or simply repeat familiar arguments about the needs of data science and existing phenomena. In many such findings [7, 45, 83, 89, 93, 112, 127, 159, 186, 200, 203, 208, 215, 244, 256, 279, 282, 290, 297, 316, 320, 331, 359, 371, 372], big data is described as being simple, data science is said to have nothing to do with the science of data, and advanced analytics is described as being the same as classic data analysis and information processing. There is a lack of deep thinking and exploration of why, what, and how these new terms should be defined, developed, and applied.

In [234], six myths were discussed:

- Size is all that matters;
- The central challenge of big data is that of devising new computing architectures and algorithms;
- Analytics is the central problem of big data;
- Data reuse is low hanging fruit;
- Data science is the same as big data; and
- Big data is all hype.

This illustrates the constituents of the ecosystem, but also shows the divided views within the communities.

These observations illustrate that data science is still young. They also justify the urgent need to develop sound terminology, standards, a code of conduct, statements and definitions, theoretical frameworks, and better practices that will exemplify typical data science professional practices and profiles.

2.9.3   Data Volume Misconceptions

There are various misconceptions surrounding data volume. For example,

- What makes data "big"? (Comments: It is usually not the volume but the complexities, as discussed in [62, 64], and large values that make data big.)
- Why is the bigness of data important? (Comments: The bigness of data âĂŞ which refers to data science complexities—heralds new opportunities for theoretical, technological, practical, economic and other development or revolution.)
- Big data refers to massive volumes of data. (Comments: Here, "big" refers mainly to significant data complexities. From the volume perspective, a data set is big when the size of the data itself becomes a quintessential part of the problem.)
- Data science is big data analytics. (Comments: Data science is a comprehensive field centered on manipulating data complexities and extracting intelligence, in which data can be big or small and analytics is a core component and task.)
- I do not have big data so I cannot do big data research. (Comments: Most researchers and practitioners do not have sizeable amounts of data and do not have access to big infrastructure either. However, significant research opportunities still exist to create fundamentally new theories and tools to address respective X-complexities and X-intelligence.)
- The data I can find is small and too simple to be explored. (Comments: While scale is a critical issue in data science, small data, which is widely available, may still incorporate interesting data complexities that have not been well addressed. Often, we see experimental data, which is usually small, neat and clean. Observational data from real business is live, complex, large and frequently messy.)
- I am collecting data from all sources in order to conduct big data analytics. (Comments: Only relevant data is required to achieve a specific analytical goal.)
- It is better to have too much data than too little. (Comments: While more data generally tends to present more opportunities, the amount needs to be relevant to the data required and the data manipulation goals. Whether bigger is better depends on many aspects.)

2.9.4 Data Infrastructure Misconceptions

Below, we list two misconceptions related to data infrastructure.

- I do not have big infrastructure, so I cannot do big data research. (Comments: While big infrastructure is useful or necessary for some big data tasks, theoretical research on significant challenges may not require big infrastructure.)
- My organization will purchase a high performance computer to support big data analytics (Comments: Many big data analytics tasks can be successfully undertaken without a high performance computer. It is also essential to differentiate between distributed/parallel computing and high performance computing.)

2.9.5 Analytics Misconceptions

There are many misconceptions relating to analytics. We list some here for discussion.

- Thinking data-analytically is crucial for data science. (Comments: Data-analytic thinking is not only important for a specific problem-solving, but is essential for obtaining a systematic solution and for a data-rich organization. Converting an organization to think data analytically gives a critical competitive advantage in the data era.)
- The task of an analyst is mainly to develop common task frameworks and conduct inference [42] from the particular to the general. (Comments: Analytics in the real world is often specific. Focusing on certain common task frameworks may trigger incomplete or even misleading outcomes. As discussed in Sect. 10.3.1, an analyst may take other roles, e.g., predictive modeling is typically problem-specific.)
- I only trust the quality of models built in commercial analytical tools. (Comments: Such tools may produce misleading or even incorrect outcomes if the assumption of their theoretical foundation does not fit the data, e.g., if they only suit imbalanced data, normal distribution-based data, or IID data.)
- Most published models and algorithms and their experimental outcomes are not repeatable. (Comments: Such works seem to be more hand-crafted rather than manufactured. Repeatability, reproducibility, open data and data sharing are critical to the healthy development of the field.)
- I want to do big data analytics, can you tell me which algorithms and program language I should learn? (Comments: Public survey outcomes give responses to such questions; see examples in [63]. Which algorithms, language and platform should be chosen also depends on organizational maturity and needs. For long-term purposes, big data analytics is about building competencies rather than specific functions).

- My organization's data is private, thus you cannot be involved in our analytics. (Comments: Private data can still be explored by external parties by implementing proper privacy protection and setting up appropriate policies for onsite exploration.)
- Let me (an analyst) show you (business people) some of my findings which are statistically significant. (Comments: As domain-driven data mining [77] shows, many outcomes are often statistically significant but are not actionable. An evaluation of those findings needs to be conducted to discover what business impact [79] might be generated if the findings they generate are operationalized.)
- Strange, why can I not understand and interpret the outcomes? (Comments: This may be because the problem has been misstated, the model may be invalid for the data, or the data used is not relevant or correct.)
- Your outcomes are too empirical without theoretical proof and foundation. (Comments: While it would be ideal if questions about the outcomes could be addressed from theoretical, optimization and evaluation perspectives, real-life complex data analytics are often more exploratory, and it may initially be difficult to optimize empirical performance.)
- My analysis shows that what you delivered is not the best for our organization. (Comments: It may be challenging to claim "the best" when a variety of models, workflows and data features are used in analytics. It is not unusual for analysts to obtain different or contradictory outcomes on the same data as a result of the application of different theories, settings and models. It may turn out to be very challenging to find a solid model that perfectly and stably fits the invisible aspect of data characteristics. It is important to appropriately check the relevance and validity of the data, models, frameworks and workflows available and used. Doing the right thing at the right time for the right purpose is a very difficult task when attempting to understand complex real-life data and problems.)
- Can your model address all of my business problems? (Comments: Different models are often are required to address diverse business problems, as a single model cannot handle a problem sufficiently well.)
- This model is very advanced with solid theoretical foundation, let us try it in your business. (Comments: While having solid scientific understanding of a model is important, it is data-driven discovery that may better capture the actual data characteristics in real-life problem solving. A model may be improperly used without a deep understanding of model and data suitability. Combining data driven approaches with model driven approaches may be more practical.)
- My analytical reports consist of lots of figures and tables that summarize the data mining outcomes, but my boss seems not so interested in them. (Comments: Analytics is not just about producing meaningful analytical outcomes and reports; rather, it concerns insights, recommendations and communication with upper management for decision-making and action.)
- It is better to have advanced models rather than simple ones. (Comments: Generally, simple is better. The key to deploying a model is to fit the model to the data while following the same assumption adopted by the model.)

- We just tuned the models last month, but again they do not work well. (Comments: Monitoring a model's performance by watching the dynamics and significant changes that may take place in the data and business is critical. Real-time analytics requires adaptive and automated re-learning and adjustment.)
- I designed the model, so I trust the outcomes. (Comments: The reproducibility of model outcomes relies on many factors. A model that is properly constructed may fall short in other aspects such as data leakage, overfitting, insufficient data cleaning, or poor understanding of data characteristics and business. Similarly, a lack of communication with the business may cause serious problems in the quality of the outcome.)
- Data science and analytics projects are just other kinds of IT projects. (Comments: While data projects share many similar aspects with mainstream IT projects, certain distinctive features in data, the manipulation process, delivery, and especially the exploratory nature of data science and analytics projects require different strategies, procedures and treatments. Data science projects are more exploratory, ad hoc, decision-oriented and intelligence-driven.)

2.9.6 Misconceptions About Capabilities and Roles

There are various misconceptions about data science capabilities and roles. For example:

- I am a data scientist. (Comments: Lately, it seems that everyone has suddenly become a data scientist. Most data scientists simply conduct normal data engineering and descriptive analytics. Do not expect omnipotence from data scientists.)
- "A human investigative journalist can look at the facts, identify what's wrong with the situation, uncover the truth, and write a story that places the facts in context. A computer can't." [256] (Comments: The success of AlphaGo and AlphaGo Zero [189] may illustrate the potential that a data science-enabled computer has to undertake a large proportion of the job a journalist does.)
- My organization wants to do big data analytics, can you recommend some of your PhD graduates to us? (Comments: While data science and advanced analytics tasks usually benefit from the input of PhDs, an organization requires different roles and competencies according to the maturity level of the analytics and the organization.)
- Our data science team consists of a group of data scientists. (Comments: An effective data science team may consist of statisticians, programmers, physicists, artists, social scientists, decision-makers, or even entrepreneurs.)
- A data scientist is a statistical programmer. (Comments: In addition to the core skills of coding and statistics, a data scientist needs to handle many other matters; see the discussion in [63].)

2.9.7   Other Matters

Some additional matters require careful consideration in relation to conducting data science and analytics. We list a few common remarks, with our comments in parentheses.

- Can big data address or even solve big questions, such as global issues, global warming, climate change, terrorism, financial crises? (Comments: Mixed sources of linked data and data matching enable the comprehensive analysis of big questions and global issues, for example, linking airfare booking, accommodation booking, transport information, and cultural background.)
- Can data science transform existing disciplines such as social science? (Comments: When data science meets social science, the synergy between the two has a good chance of promoting the upgrade and transformation of social science, in particular sociology, by providing new thinking, new methods, and new approaches, such as reductionism, data-driven discovery, the scaling-up of social experiments, deep analytics of social problems, and crowdsourcing-based problem-solving.)
- Garbage in, garbage out. (Comments: The quality of data determines the quality of output.)
- More complex data, a more advanced model, and better outcomes. (Comments: Good data does not necessarily lead to good outcomes; a good model also does not guarantee good outcomes.)
- More general models, better applicability. (Comments: General models may lead to weaker outcomes on a specific problem. It is not reasonable or practical to expect a single tool for all tasks.)
- More frequent patterns, more interesting. (Comments: It has been shown that frequent patterns mined by existing theories are generally not useful and actionable.)
- We're interested in outcomes, not theories. (Comments: Actionable outcomes may need to satisfy both technical and business significance [77].)
- The goal of analytics is to support decision-making actions, not just to present outcomes about data understanding and analytical results. (Comments: This addresses the need for actionable knowledge delivery [54] to recommend actions from data analytics for decision support.)
- Whatever you do, you can at least get some values. (Comments: This is true, but it may be risky or misleading. Informed data manipulation and analytics requires a foundation for interpreting why the outcomes look the way they do.)
- Many end users are investing in big data infrastructure without project management. (Comments: Do not rush into data infrastructure investment without a solid strategic plan of your data science initiatives, which requires the identification of business needs and requirements, the definition of reasonable objectives, the specification of timelines, and the allocation of resources.)
- Pushing data science forward without suitable talent. (Comments: On one hand, you should not simply wait for the right candidate to come along, but should

actively plan and specify the skills needed for your organization's initiatives and assemble a team according to the skill-sets required. On the other hand, getting the right people on board is critical, as data science is essentially about intelligence and talent.)

- No culture for converting data science insights into actionable outcomes. (Comments: This may be common in business intelligence and technically focused teams. Fostering a data science-friendly culture requires a top-down approach driven by business needs, making data-driven decisions that enable data science specialists and project managers to be part of the business process, and conducting change management.)
- Correct evaluation of outcomes. (Comments: This goes far beyond such technical metrics as Area Under the ROC Curve and Normalized Mutual Information. Business performance after the adoption of recommended outcomes needs to be evaluated [54]. For example, recent works on high utility analysis [471] and high impact behavior analysis [79] study how business performance can be taken into account in data modeling and evaluation. Solutions that lack business viability are not actionable.)
- Apply a model in a consistent way. (Comments: It is essential to understand the hypothesis behind a model and to apply a model consistent with its hypothesis.)
- Overthinking and overusing models. (Comments: All models and methods are specific to certain hypotheses and scenarios. No models are universal and sufficiently "advanced" to suit everything. Do not assume that if the data is tortured long enough, it will confess to anything.)
- Know nothing about the data before applying a model. (Comments: Data understanding is a must-do step before a model is applied.)
- Analyze data for the sake of analysis only. (Comments: This involves the common bad practice of overusing analytics.)
- What makes an insight (knowledge) actionable? (Comments: This is dependent on not only the statistical and practical values of the insight, but also its predictive power and business impact.)
- Do not assume the data you are given is perfect. (Comments: Data quality forms the basis for obtaining good models, outcomes and decisions. Poor quality data, the same as poor quality models, can lead to misleading or damaging decisions. Real-life data often contains imperfect features such as incompleteness, uncertainty, bias, rareness, imbalance and non-IIDness.)

The above is only a partial list of the misconceptions and misunderstandings that can be witnessed and heard in the current data science and analytics community.

It is not surprising to hear a range of diverse arguments, ideas, and beliefs about any new area. Data science is a highly non-traditional, interdisciplinary, cross-domain, transformative, fast-growing, and pragmatic scientific field. It is in fact a very positive sign to see the increased debate and emerging understanding of the intrinsic nature of data science as a new science, and data economy as the new generation of economy. Importantly, significant and deterministic efforts are required to substantially deepen and widen our understanding and repetitively reflect on our understanding of data science, data innovation, and data economy. In doing so, we will hopefully capture the true value of data science and guide its evolution to a substantial new stage.

2.10   Summary

The answers to the question "What is data science?" are varied, and sometime confusing and conflicting. Exploring data science understanding from a range of aspects and perspectives provides more opportunities to explore and observe the nature of data science.

This chapter first explores how data is quantified (namely by datafication and data quantification), the relationships between five key concepts: data, information, knowledge, intelligence and wisdom, introduces the concept of data DNA as the key datalogical "molecule" of data, and explains the outputs of data science: data products.

This knowledge lays the foundation for exploring the different meanings of data science, the multiple views on data science, and the various definitions of data science. Lastly, this chapter lists and comments on many myths and misconceptions appearing in literature.

The next important and challenging question is "What makes data science a new science?" Answering this question requires an in-depth understanding of the unique but fundamental abstraction in data science, otherwise known as "data science thinking". Accordingly, Chap. 3 explains the concept of data science thinking and discusses its implications.

Chapter 3
Data Science Thinking

3.1 Introduction

What makes data science essential and different from existing developments in data mining, machine learning, statistics, and information science?

This is a very important and challenging question to answer. To adequately answer this question, it is necessary to capture the unique and foundational features of data science.

Although important answers and observations could come from disciplinary, theoretical, technological, business, value, or even big data era perspectives, we feel that such answers reflect more on the tactical and specific aspects of data science as a new science.

A more strategic and foundational observation can be found in *data science thinking*. It is the *thinking in data science* that drives the emergence, development, revolution, and formation of data science as an independent discipline and science.

Definition 3.1 (Data Science Thinking) Data science thinking refers to the perspective on the methodologies, process, structure, and traits and habits of the mind in handling data problems and systems.

Data science thinking also drives, initiates, and ignites transformative, new and original research, innovation, development, and the industrialization of data and data-related businesses. This further upgrades and re-invents relevant disciplines and areas, as well as enhancing the lower level knowledge base, skill set, and capabilities.

Accordingly, this chapter explores and discusses notions of data science thinking from a number of aspects: general thinking in science, which forms the foundation for discussing data science thinking; the complexities in data science from a complex system perspective; extending the thinking in science to form a view and structure of data science; and critical thinking in engaging with data science.

© Springer International Publishing AG, part of Springer Nature 2018
L. Cao, *Data Science Thinking*, Data Analytics,
https://doi.org/10.1007/978-3-319-95092-1_3

3.2   Thinking in Science

In this section, the main concepts, types of thinking, mental functions and skills for thinking in science are discussed. We particularly focus on discussing comparative aspects in negative vs. positive thinking, and the habits and skills that are necessary for cultivating creative and critical thinking. We discuss these in the context of forming the foundation for developing data science thinking.

3.2.1   Scientific vs. Unscientific Thinking

Scientific thinking [324] is critical for training scientific thinkers, framing scientific activities, and ensuring scientific ways of doing science. Scientific thinking needs to be cultivated in scientist and engineers to enable them to follow scientific processes, undertake relevant scientific activities, and avoid unscientific thinking.

The key characterizations of thinking in science consist of

- curiosity: eagerness to explain observations;
- conceptualization: a conceptual system of science;
- experimental design and hypothesis testing;
- thinking with evidence; and
- evaluation: verification of the results both theoretically and experimentally.

Scientific thinking follows certain processes, as shown in Fig. 3.1, consisting of

- *phenomena identification*: observing and identifying phenomena that have not been well or fully explained,

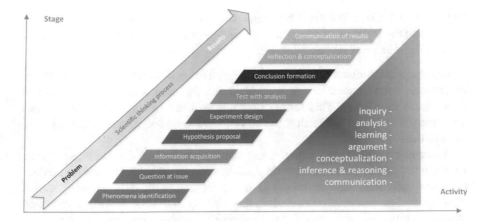

Fig. 3.1 Scientific thinking process with key activities

- *question at issue*: discovering and raising important scientific questions about identified phenomena,
- *information acquisition*: gathering relevant scientific data, facts, evidence, observations, experiences, reasons, and knowledge,
- *hypothesis proposal*: interpreting and refining these questions and problems, structuring a point of view, and proposing hypotheses and provisional explanations,
- *experimental design*: designing sufficient and necessary experiments (tests) to check and verify hypotheses and explanations,
- *test with analysis*: implementing, executing and reviewing experiments to test hypotheses on collected scientific data and information,
- *conclusion formation*: drawing scientific conclusions and solutions using scientific approaches that include reasoning, deduction/inference, evidence-based summarization and abstraction,
- *reflection and conceptualization*: assessing, validating or adjusting assumptions, implications and conclusions based on experimental results and theoretical support, forming definitions, theories, models, laws, or principles,
- *communication of results*: effectively communicating and disseminating the verified knowledge and outcomes.

In the above process, relevant scientific (cognitive and computational) activities are undertaken, including

- *learning* about existing scientific assumptions and knowledge,
- *inquiry* into problems and issues,
- *analysis* of scientific information,
- *inference and reasoning* of information analysis results and theoretical analysis,
- *conceptualization* of scientific point of view and findings,
- *argument* concerning implications and consequences,
- *communication* of outputs and outcomes.

In contrast, *unscientific thinking* consists of (1) tenacity (holding onto traditions, superstitions, myths or truisms), (2) common sense (perception of concepts shared generally among people), (3) personal experience, (4) intuition, (5) authority (expert knowledge and opinion), and (6) rationalism (logic or deduction).

To avoid unscientific thinking and assure scientific thinking and its respective activities, shown in Fig. 3.2, it is essential to create and follow certain codes of scientific conduct, scientific thinking, and scientific ethics. Examples of codes and guidelines for scientists can be found in [11, 152, 352].

Important ethical features of scientific codes of conduct are logic, rationality, creativity, accuracy, precision, clarity, sound evidence, good reasoning, relevance, transparency, fairness, reliability, testability, reproducibility, depth and breadth of analysis.

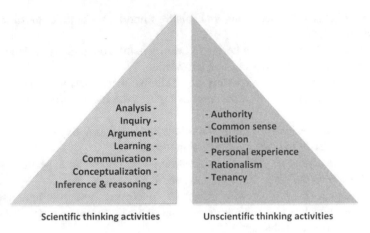

Scientific thinking activities   Unscientific thinking activities

Fig. 3.2 Scientific thinking activities vs. unscientific thinking activities

3.2.2   Creative Thinking vs. Logical Thinking

Creative thinking is a major mental function in scientific thinking. Historically, however, logical thinking has been predominately encouraged in science, education and engineering.

3.2.2.1   Logical Thinking

Logical thinking, otherwise known as *logical reasoning* [131], has conventionally been prioritized in scientific thinking on education and training, and in science and technology. *Logical thinking* is a process of reasoning from the truth of a premise to reach a conclusion. Logical thinking is a step-by-step process, logic-based, and often linear. It is the process of moving to the next step or drawing sound conclusions based on a series of rational and systematic procedures. Three types of logical thinking are inductive reasoning, deductive reasoning, and abductive reasoning or inference.

Inductive thinking attempts to derive a general conclusion or theory or rule from one to multiple supporting premises (observations, facts, concepts, examples, logics, or opinions). A bottom-up approach is often taken to transform specific observations into a general and merely likely (i.e., possibly) true conclusion. The logic process begins with specific observations, then identifies tentative hypotheses, and finally draws general and possibly true conclusions or theories. The bottom-up logic path is:

$$Inductive\ thinking : observation \rightarrow tentative\ hypotheses$$

$$\rightarrow conclusion\ (likely\ true) \qquad (3.1)$$

Deductive thinking narrows a general conclusion down to one or many true premises (hypotheses) and tests them to confirm the true conclusion. Deductive logic follows a general to specific logic process, often referred to as *top-down logic*. A conclusion or theory is first adopted and broken down into specific supporting hypotheses, further observations are collected to test these hypotheses, and lastly the conclusion (theory) is confirmed as true. Accordingly, the top-down logic path is:

$$Deductive\ thinking : conclusion \rightarrow hypotheses \rightarrow observation$$
$$\rightarrow conclusion\ confirmation\ (definitely\ true) \quad (3.2)$$

Abductive reasoning (or abductive inference) [240] is the process of inferring a general conclusion (or theory) which accounts for a specific observation or observations. It infers the best explanation, and the supporting premises do not guarantee the conclusion.

3.2.2.2 Creative Thinking

Creative thinking is the process of developing unique and innovative ideas, usually through the application of non-standard and non-conventional formats or processes. Creative thinkers may reject traditional patterns or outlooks, take divergent and multi-perspective views, and use trial and error and forward-thinking methods.

Creativity is

- a *mindset* that is active, open to, and curious about new and meaningful questions, perspectives, causes, intrinsic nature and characteristics;
- an *attitude* that is flexible, alternative, and appreciative of newness, changes, opportunities, ideas, and possibilities;
- an *ability* that can generate, imagine, and invent new and valuable ideas and things;
- the application of a *process* that iteratively questions, refines, and optimizes ideas and solutions; and
- the capacity to devise a *method* that drives the evolution of better ideas and solutions, synthesizes existing ideas and solutions, and changes the way problems are solved.

Uncreative thinking may include such scenarios and thinking habits as:

- I am not creative, only scientists can be creative. Not everyone can be creative.
- If I do this, what will people think? Will I appear to be odd?
- Everyone does it in that way, it is the default solution.
- I am not a creator, I cannot do it.
- The problem is too hard, it cannot be done.
- It is all fine, I cannot see any problems (though there are likely problems there).
- I tried it, but I failed. It would not work.
- It is too complex for people like me.
- That will never work, it is no good.

The above examples of negative, uncreative thinking may be associated with biased mental habits or restricted thinking traits. Examples of restricted thinking traits include closed mindset, prejudice, biased thinking, functional fixation, or other types of psychological obstruction.

Accordingly, it is very important to train our brain and mental systems, encourage a positive mindset and attitude, avoid or change uncreative thinking habits, and adopt, develop and encourage processes and methods that will enable creative thinking. Creative thinking may be trained and supported by cultivating the following thinking habits:

- mindset: keeping an active, relaxed and open mind,
- curiosity and questioning: sustaining curiosity about the causes and intrinsic nature of an issue, being open to new technologies and ideas, posing new questions and new perspectives, and tracking new ideas,
- diversity and alternativity: enjoying interdisciplinary and cross-disciplinary knowledge and ideas, encouraging divergent thinking, suspending judgment and assumptions, fostering alternative thinking and possibilities,
- depth: conducting deep thinking, exploiting sensitivity to distinguish exceptional opportunities, challenging assumptions, traditions and authorities, cultivating constructive discontent and self-confidence,
- criticism: stimulating comparison and criticism, appreciating debate and argument,
- perseverance: accepting mistakes and failures and persevering with trial and error, and
- perfection and imperfection: seeking higher quality and better outcomes, pursuing the bigger picture, greater opportunities, and higher standards, and seeing the good in the bad and vice versa.

3.2.2.3   Critical Thinking

Critical thinking refers to the ability to make reasoned judgments through and information and critique. *Critique* is enabled by the well-grounded, well-reasoned, well-judged, and detailed analysis, assessment, and evaluation of information.

There are a number of definitions of what constitutes critical thinking [239]. For example, the U.S. National Council for Excellence in Critical Thinking defines critical thinking as the "intellectually disciplined process of actively and skillfully conceptualizing, applying, analyzing, synthesizing, or evaluating information gathered from, or generated by, observation, experience, reflection, reasoning, or communication, as a guide to belief and action" [110].

Critical thinking activities include information acquisition, questioning standards, brainstorming, analysis, discrimination, logical reasoning, comparison, challenging assumptions, reflective skepticism, inter-disciplinary and cross-domain interest, adopting multiple perspectives, and forward thinking.

Many critical thinking traits and skills may be involved in the critical thinking process, and may include some or all of the following.

- *depth* of problem understanding and solutions,
- *breadth* of problem scope and solutions,
- *intuition* or insight into solution pathways,
- *imaginative thinking* about 'what if' scenario analysis and results,
- *creativity* in introducing new views and inventing better designs,
- *inquisitiveness* about possibilities and alternatives,
- *questioning* essential oversights, issues, limitations, reasons, foundations and opportunities,
- *rationality* in drawing conclusions,
- *relevance* of the nature, characteristics, and environment of a problem,
- *flexibility* in the approach to ideas, options and behaviors,
- *open-mindedness* to divergent ideas, new opportunities, and inspiration,
- *good reasons* for doing things in any given way,
- *confidence* in one's well-grounded capabilities and capacities,
- *perseverance* in overcoming obstacles to reach conclusions,
- *reasoning* about consequences and alternative options,
- *justification* for hypotheses and empirical summaries,
- *sound evidence* for arguments,
- *fairness* of assumptions and conclusions,
- *integrity* of the process and results in seeking truth, and
- *reflection* for better and deeper understanding, self-evaluation, and self-improvement.

Critical thinking is often driven by, and relevant to the underlying problem, objective and corresponding approaches. Compared to creative thinking, critical thinking is self-disciplined, self-directed, self-monitored, self-correcting, and self-improving. The key thinking activities and traits in creative and critical thinking are listed in Fig. 3.3.

3.2.2.4 Lateral Thinking

A more indirect or off-center way of thinking creatively is lateral thinking [35]. *Lateral thinking* seeks to solve problems in an indirect and creative way by considering and reasoning about new ideas that may not be obtainable through logical thinking.

Lateral thinking is concerned with using known ideas and information to create or restructure new ideas and insights. Four types of thinking tools were proposed in [35] to support lateral thinking: ideas-generating tools to break routine thinking patterns, focus tools to broaden the search for new ideas, harvesting tools to gain more value from the generation of ideas, and treatment tools to consider relevant constraints.

Fig. 3.3 Creative traits vs.
critical thinking traits

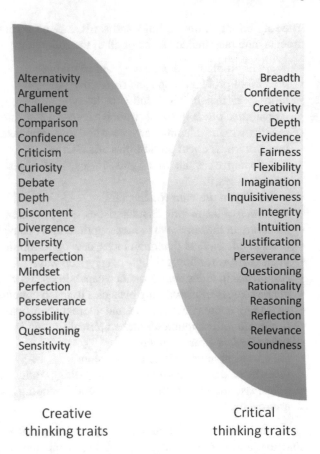

Creative	Critical
Alternativity	Breadth
Argument	Confidence
Challenge	Creativity
Comparison	Depth
Confidence	Evidence
Criticism	Fairness
Curiosity	Flexibility
Debate	Imagination
Depth	Inquisitiveness
Discontent	Integrity
Divergence	Intuition
Diversity	Justification
Imperfection	Perseverance
Mindset	Questioning
Perfection	Rationality
Perseverance	Reasoning
Possibility	Reflection
Questioning	Relevance
Sensitivity	Soundness

Creative
thinking traits

Critical
thinking traits

3.3   Data Science Structure

The review and discussion of thinking in science in Sect. 3.2 lays the foundation for us to discuss the thinking in data science.

Rather than highlighting the specificity and broad psychological and mental traits, skills and activities in discussion on thinking in science, this section focuses on the high-level disciplinary and functional aspects of forming and conducting data science. Hence, *thinking in data science* as it is discussed here refers to the way of viewing and structuring data science problems, and the process, activities, important aspects, and systems required to solve data problems. This will complement the discussion on critical thinking in data science in Sect. 3.2.2.3.

In Sect. 2.5, we have discussed several views of data science, including

- a statistical view of data science in Sect. 2.5.1,
- a multidisciplinary view of data science in Sect. 2.5.2, and
- a data-centric view of data science in Sect. 2.5.3.

The relevant discussion in Chap. 2 on what is data science further presents several aspects of data science as a new science.

With regard to the structure of data science, there are also different perspectives to explore it: a disciplinary perspective, a system perspective, a thinking perspective, a technical perspective, a process perspective, and a curriculum perspective.

- A disciplinary perspective of data science: data science is built on and involves many disciplines as its foundation and building blocks to form a new discipline, which is interdisciplinary and transdisciplinary. The relevant disciplinary modules forming the data science are discussed in Chap. 5.
- A system perspective of data science: if data science is viewed as a complex system, then we can create a system architecture of data science, which may consist of various data feeds, data connection, or data extraction modules; data cleaning, transformation, and preprocessing modules; data feature engineering modules; data modeling, analysis, and learning modules; output presentation, visualization, and decision-support; and data governance, management, and administration. There are many models and tools focusing on specific aspects of the above. Section 3.4 further presents a view of data science as a complex system.
- A thinking perspective of data science: this views data science as a cognitive and philosophical methodology to handle data and data-relate problems. This includes the methodology, principle, strategy, process, trait, habit, etc. for quantifying, understanding, making sense of, and evaluating data and data applications. These also involve both general thinking aspects and data-specific aspects of the above.
- A technical perspective of data science: this refers to the techniques required for handling data and problems along the life cycle of processing data and providing results of analyzing data. In Chap. 7, an overview of relevant data science techniques is given.
- A process perspective of data science: when a data and data application are given, we start a process of handling the data and problems. This process follows certain procedures, actions to be taken in each procedure, specifications, and connections between procedures. Section 2.6.3 discusses the relevant actions in the process-based data science.
- A curriculum perspective of data science: this refers to the structure of data science education. The education and training system of data science can be further decomposed in terms of different purposes and aspects, e.g., the course structures for each level of awarded degrees in a full hierarchical educational system from technician training to bachelors, masters, PhDs, and executive training. The curriculum-oriented data science systems are also highly connected with other perspectives, e.g., data science techniques, data science foundations, data science thinking, and data science applications. Therefore, we can further decompose and propose many versions of data science structures for each of these aspects. In Chap. 11, discussion is available on the framework and structures of awarded and non-awarded courses in data science.

- A professional perspective of data science: when data science is recognized as a profession, different roles, positions, and jobs are created. Specific responsibilities are assigned to respective roles and positions, which further form an ecosystem of the data science profession. In Chap. 10, data science profession is discussed.
- An economic perspective of data science: this refers to the value proposition, industrialization, and applications of data science in the real world. Accordingly, different landscapes of data technologies, data applications, and socioeconomic opportunities can be generated. Chapter 8 outlines opportunities for creating data economy, industry, and services. Chapter 9 further discusses the applications of data science in many traditional and contemporary domains.

With the deepening of our understanding of the challenges, solutions, and opportunities of data science, data science may be depicted in terms of many other perspectives. The respective data science structure will evolve when today's unknown world becomes known in the near future. Such knowledge growth and revolution will further disclose other issues and opportunities unknown to the then communities, and more deeply invisible areas and their intricate logics and structures.

3.4 Data Science as a Complex System

Although there are different views and arguments about what constitutes data science, as discussed in Chap. 2, taking a complex system perspective is appropriate when considering data science problems. Data science problems are, in general, complex; problem-solving could otherwise be achieved within the existing discipline of data analytics, including statistics, computing, and informatics.

A systematic view is therefore discussed in this section from the perspective that *data science is a complex system*. The key system characteristics of particular system complexities are discussed, following the introduction of a systematic view of data science.

3.4.1 A Systematic View of Data Science Problems

When a complex system perspective is adopted to understand data science, a systematic view of what data science is emerges. *A systematic view of data science* incorporates the belief that data science consists of a comprehensive range of aspects that include system complexity, structure, discipline, direction, profession, education, economy, and product.

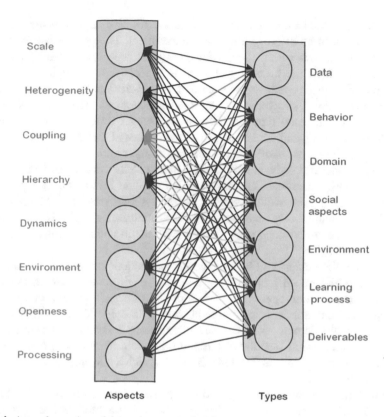

Fig. 3.4 A two-layer view of data science complexities

System complexities in data science consist of those complexities associated with data, behavior, domain, social aspects, environment, learning process, and deliverables. These form the so-called *X-complexities*, discussed in Sect. 4.2.

In addition to the various types of system complexities, data science problems may be complex in relation to the following significant aspects: scale, heterogeneity, coupling relationship, hierarchy, dynamics, environment, openness, and processing. We briefly discuss these significant aspects of system complexities in data science below.

Figure 3.4 presents the two-layered view of data science in terms of comprehensive aspects and system complexities.

Usually, a large or giant (peta or above) scale of entities, relationships, behaviors, processes, and contexts is involved in a complex data problem. For example, in environmental data science systems, the number of relevant entities (such as geographical locations, related animals and vegetation, etc.) are numerous on a giant scale. These relevant entities interact with each other and with their contexts in countless relations and couplings. The activities of diverse entities and their contexts are innumerable. The environmental system evolves by following semi-clear and often uncertain processes.

These entities, relationships, behaviors, processes, and environments on a giant scale are likely to be heterogeneous and hierarchical. They have diverse relationships and interactions within each of the significant aspects noted above, between aspects, and within the hierarchy of various aspects. For example, in a natural environmental system, entities may consist of different animals, various plants, diverse geographical structures and seasonal appearances. The relationships and interactions within each type of entity and between different types of entities are clearly colorful and multi-layered. Environmental behaviors and activities associated with relevant entities are sophisticated and multi-typed.

Complex data problems are typically dynamic. They evolve in terms of such aspects as entity quantities and types, relationships, behaviors, and processes. For example, environmental systems are always evolving, and behaviors and relationships continuously change. Often, the number of entities is also unfixed. This also applies to artificial social media systems, in which infinitely expanding networks evolve.

There are hierarchical and heterogeneous interactions between data entities (of micro, meso and macro granularities) and their environments, which make data problems open or semi-open. Closed data problems are simple and rarely seen in the real world. Granularity is an important issue in complex data problems. As most complex systems demonstrate, there are often multiple layers of sub-systems and systems within systems (see the discussion on system hierarchy in open complex intelligent systems in [62, 69, 70, 114, 333–335]), which may appear at micro, meso and/or macro levels.

These large scale and complex relationships contribute to complex structures in a complex data system which are often hidden, evolving, hierarchical, and heterogeneous. While some structures are observable and describable in the quantitative sense, such as by tree, graph or hash structures, it is often unclear what hidden structures exist in complex data systems. Discovering hidden structures is further affiliated with other complexities discussed above and below. From a mathematical or quantitative perspective, we need to align our existing understanding of data structures and distribution with the underlying data, probably with a mixture of structures and distributions. However, it is not clear whether a mixture of well-understood structures and distributions can genuinely and completely capture the structure complexities in a complex data problem.

Comprehensive disciplinary and research/development direction aspects are evident in inter-disciplinary, multi-level, multi-source, mixed-type, and cross-domain aspects, challenges, and opportunities. These are related to cross-disciplinary aspects and challenges, as discussed in Chap. 6, and data complexities such as non-IID learning, as discussed in Sect. 4.5.

Data science is driving the generation of a new profession, known as *data profession*, which broadly refers to any career, job role or responsibilities related to data. This data-driven professionalization is unprecedented and comprehensive, as discussed in Sect. 10.2. It pushes the creation of a variety of new data roles, data capability maturity, and organizational competency. More discussion on the data profession are available in Chap. 10.

The emergence of the data profession fundamentally challenges existing educational systems. Data science has emerged as a new discipline, and new data science courses, subjects, delivery methods, teaching and learning technology and relationships, and educational products are appearing as a result of this new science. In Chap. 11, the challenges, opportunities, and status of data science education are discussed.

As discussed in Sect. 2.7, data innovation is creating new forms and aspects of data products. These data products are driving the creation of a new form of economy, shifting from Internet and Web-focused economy to ubiquitous data economy in terms of new socio-economic ways of studying, working, entertaining, living, and traveling. They also represent the new generation of artificial intelligence systems.

3.4.2 Complexities in Data Science Systems

Complexity [294] is a term that is generally used to characterize systems that have many parts which interact with each other in multiple ways, "culminating in a higher order of emergence greater than the sum of its parts" [442].

While there is no absolute definition of complexity, we focus in this book on the sophisticated and difficult to manage characteristics, features and aspects of a data problem. Multiple aspects, multiple layers, mixed-relations, and mixed-structures are involved in building a system-centric characterization of complex data science problems. The study and quantification of these multiple interactive aspects, relations, and structures on various levels and scales are therefore the main goal and task of data system complexities and data science challenges.

When data science problems are treated as complex systems, *system complexities* [62, 335] can be explored to understand the characteristics, dynamics and challenges of data problems. In this way, a comprehensive overview of data problems can be created, and relevant solutions can be more fundamental, systematic and pragmatic.

The term *System complexities* describes the intrinsic characteristics, behaviors, interactions, dynamics, and consequences of a system. In general, complexities are embedded in complex systems, in which many constituents interact with each other and are integrated in various relationships, rules, norms, protocols, or agreements in organized, semi-organized and/or disorganized (self-organizing) ways on local and global levels.

For problem-solving complex systems, a bottom-up reductionist approach is often taken. *Bottom-up reductionism* assumes that a system can be decomposed into many layers and many local parts residing in a local subsystem, which interact with each other in multiple ways, and by following local rules and norms.

Reductionism uses system decomposition strategies which extract multiple layers of subsystems, and various interactive relationships within and between subsystems. The behaviors of individual subsystems can be more easily and deeply understood as a result.

To form a comprehensive picture of an underlying complex system, a *top-down holism* approach is necessary. *Top-down holism* combines subsystems into a single high level system by following composition strategies and rules to draw an overall picture from the sum of its parts. When each subsystem is clearly and precisely understood, subsystems are connected in terms of combination rules and norms.

In a complex data problem, data components (e.g., variables, subspaces, sample sets, or single data sources) interact in multiple ways and follow local rules, meaning that there is no reasonable higher instruction to define the various possible interactions. This may conform to the flat-structured and self-organizing nature of complex systems [62, 219, 294, 335].

Data science system complexities can be correspondingly explored from different perspectives, similar to the study of other complex systems. Our specific focus is on

- The system constituents and their characteristics in data problems: data and its internal granularities, scales, structures, relations, and dynamics, and its external environments, interactions, and dynamics;
- The processes for understanding and solving complex data problems: the technical processes for data extraction, processing, management, governance, and applications, and theoretical assurances such as building a theoretical understanding of motivation, foundation, proof, and performance.

3.4.3   The Framework for Data Science Thinking

In discussing such questions as "What makes data science a new science?", we believe exploring the view and structure of data science is a major issue. Accordingly, understanding the concept of "data science thinking" should include but also extend beyond relevant concepts such as "statistical thinking", "mathematical thinking", "creative thinking", "critical thinking", and more recently "data analytical thinking" [392].

As discussed in [23, 63–65, 205], a critical task in data science is to "think with data". To enable thinking with data and to support the process of thinking with data, we introduce the corresponding general problem-solving structure, components, skills and capabilities. These form the *data science overview and structure*, as shown in Fig. 3.5.

The overview and structure to support data science thinking consist of *four progression layers* with embedded thinking activities and skills that are required to handle data science: (1) the feed layer, (2) the mechanism design layer, (3) the deliverable layer, and (4) the assurance layer. There are also two *driving/enabling wings*: (1) the thought wing, and (2) the custody wing.

Let us start with the two *enabling wings*, since they are the drivers.

- *The thought wing*: This concerns thought leadership, in particular the thinking and intent that inspires, initiates and drives the data science research and practice

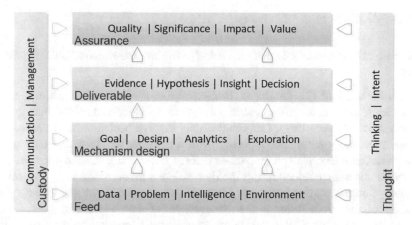

Fig. 3.5 Data science overview and structure

for a specific domain problem, the corresponding data understanding, mechanism designs, solutions, evaluation and refinement, and final data products.

- *The custody wing*: This concerns the support and assurance of the process, in particular communication and management, which enable the applications and business transformation of data science innovation and practice.

3.4.4   Data Science Thought

Two key enablers and drivers forming *the thought wing* are *thinking* and *intent*.

- *Thinking-driven*: The goals, tasks, milestones, delivery, validation and deployment of data-driven business understanding, analytics, and decision-support come from what we think about the given data and business. Differences in thinking result in setting different goals, tasks, milestones, deliverables, and outcomes. A better or deeper understanding of data requires creative, critical, systematic, problem-inspired, and balanced thinking (such as a tradeoff between local and global thinking, specific and comprehensive thinking, and subjective and objective thinking).
- *Intent and goal-steered*: Thinking generates intent; intent produces specific goals, tasks and milestone plans, which steer the subsequent data exploration and use. Generating appropriate intent and goals, or appropriately understanding intent and goals, is thus critical for data science research, innovation and applications.

3.4.5 Data Science Custody

Two core drivers and enablers in the *custody wing* are *communication* and *management*.

- *Communication*: This facilitates seamless and cost-effective interactions, collaborations, and coordination within a cross-disciplinary and mixed background-based data science team. Communications involve horizontal and vertical presentation, story-telling and reporting within the team and with reference groups and stakeholders, as well as back-and-forth reflection and refinement between layers and stakeholders during the data science process and in a data science team. Communication issues and skills for data science are available in Sect. 6.9.
- *Management*: In addition to the management of data, resources, roles, process and outcomes, particular efforts are necessary in relation to the governance, management of social and ethical issues and risk, and the transformation of data feed to deliverables to create high impact and decision value. Management issues and skills for data science are available in Sect. 6.8.

3.4.6 Data Science Feed

The *feed layer* in the progression consists of *problem*, *data*, *environment* and *intelligence*.

- *Problem-inspired*: Creative data science work needs to be problem-inspired. It needs to properly and deeply understand and address the nature, domain, challenges, complexities and differences of the underlying problems in a given domain. Even if the high level problem is obvious, e.g., a classification problem, it is still important to understand and capture the intrinsic and discriminative nature of the underlying problems, e.g., whether the classification is imbalanced or balanced, or consists of IID vs. non-IID data.
- *Data-driven*: It is data above all that enables the delivery of valuable data products. Being data-driven requires deeply and accurately understanding and representing specific characteristics (e.g., volume, variety, velocity, and veracity), data factors (e.g., structure, frequency and semantics of objects and attributes) [71], and X-complexities (e.g., data, domain, social, organizational and behavior complexities) [64] in a dataset. It requires allowing the data to tell the story, and using data to evidence recommendations and conclusions. Data-based approaches do not ignore domain knowledge. In fact, data is always from and about a particular domain, thus involving domain-driven factors is crucial for data-based exploration and decision-making.
- *Environment-specific*: A complex data world consists of a complex and specific environment which places constraints on the underlying problem and data. Involving environmental factors in goal setting, data design, analytics, and

exploration is necessary for context-aware and robust solutions. The complexities and intelligence associated with the data environment are discussed in Sect. 4.2.5 and Sect. 4.3.8 respectively.

- *Intelligence-enabled*: Data science has to effectively understand, represent, involve and utilize X-intelligence (e.g., data, human, domain, social, and meta-synthetic intelligence) [64]. Data science is intelligence science. Data scientists with different levels of intelligence and attitude will create a range of designs and produce different data products. See more discussion on various intelligences in data science in Sect. 4.3.

3.4.7 Mechanism Design for Data Science

The *mechanism design layer* in the progression consists of *goal*, *design*, *analytics* and *exploration* of learning and insight discovery, as described below.

- *Goal-motivated*: Data-based design and exploration for domain-specific problem-solving is motivated by goals, e.g., the generation of personalized solutions and insights. Although data analysis tends to produce general outcomes, these are often useless in deep analytics and are not actionable. Problem-specific and personalized data products, such as for personalized recommender systems, which are driven by specific thinking, intent and design and need to be aligned with specific domains, problems, data nature and deliverables, are demanding.
- *Design-guided*: Data and data problems are often invisible. Data scientists with wide-ranging experiences, intelligences and capabilities may devise very different designs, leading to alternative data exploration processes and diversified data products. Data science is an art. Quality data products originate in creative design by experienced designers. Data designers with creative thinking and clear intent may convert the invisible to visible and creative data artworks.
- *Analytics and exploration-empowered insights*: The transformation of data to insights represented by creative and innovative data artworks is empowered by iterative exploration, deep analytics and learning. In data artwork design, embedding data exploration and deep analytics and learning into the design and production stage is fundamental. This naturally involves the choice of method-ologies, methods and models, and how to undertake analytics and learning. X-analytics [64] may be used to enable the transformation (more discussion on X-analytics is available in Sect. 7.6).

3.4.8   Data Science Deliverables

The *deliverable layer* in the progression is composed of new *hypotheses*, *evidence*, *insights* and *decisions*.

- *Hypotheses and evidence-oriented research*: Data science differentiates itself from other sciences in several aspects; one is that data science is evidence-oriented and hypothesis-free. This is the expectation of data-driven discovery. It is not always easy or possible to generate and verify data-enabled evidence and new hypotheses, when we need to generate data-enabled indicators and hints.
- *Insight and decision-targeted findings*: Discovering new insights from data to enable better decisions is the pragmatic objective of data science. A data-enabled insight is an accurate and in-depth understanding of, and perspective on, an underlying intrinsic problem and the data complexities, nature and significance of that problem, which extend beyond observational, model-based, and hypothesis-driven perspectives and findings. With insight, we look for something new, novel, intrinsic, and ground-breaking, to enable smarter decisions to be made that are inspired and obtained by data and analytics.

3.4.9   Data Science Assurance

Four components, *quality*, *significance*, *impact* and *value*, form the key constituents of the *assurance layer* in the progression.

- *Quality and significance*: The data products resulting from the data design and analytics need to be verified in terms of their quality and significance, including technical (statistical) and business (business impact and value) significance. Only significant data products are valid and useful. Significance validation needs to be undertaken within certain domains, problems, data, design and analytical constraints and settings.
- *Impact and value*: In addition to quality and significance, a critical question to ask about a data product is *who cares* and *why they care*. This reflects the importance of demonstrating the impact of data insights and products, i.e., what difference they make. It is useful but often not easy to design appropriate measures and indicators to quantify impact, e.g., in terms of the effect on business. The business value of data is often focused within an industry or market, which emphasizes the benefits that may result from data-driven discovery and data products. Data science thinking requires a broader view of data value, such as enabling new data economy, theoretical breakthrough, and solving problems that were not previously solvable.

3.5 Critical Thinking in Data Science

This section discusses several perspectives on critical data science thinking. As discussed in Sect. 3.2.2.3, critical thinking is self-disciplined, domain-specific, and problem-oriented. Accordingly, critical thinking in data science may have many aspects, especially since data science is such a broad, cross-disciplinary and cross-domain topic.

3.5.1 Critical Thinking Perspectives

Given the many general aspects of critical thinking in data science and the questions related to specific areas of critical thinking in data science, we highlight only the following aspects in this section.

- *Data world complexities*: Gaps between knowledge levels and future world complexities, where we focus on the area of "we do not know what we do not know".
- *Data science paradigm*: Data-driven science constitutes the new paradigm in data science, in contrast to heuristics and experiment, hypothesis testing, logic and reasoning, computational intelligence, and simulation. Here we discuss the meaning of "data-driven".
- *The connections with conventional data research*: hypothesis testing, experiment, simulation, and model-based science are typical approaches in existing science. These are connected to data-driven discovery, and we thus discuss their relationships and connections.
- *Essential questions to ask in data science*: We discuss the general aspects that should be considered for critical thinking in data science and how they might drive critical thinking functions and habits in data science.

Our discussion in the following sections focuses on only some of the above-related perspectives.

3.5.2 We Do Not Know What We Do Not Know

In the era of data science and big data analytics, our research focus is on exploring the complexities and challenges we do not know well, or do not know at all, with the research objective of migrating from known to unknown spaces in data science.

As shown in Fig. 3.6 (which will be further discussed in Sect. 4.4.3), our current interest lies mainly in Space D; that is, to explore the unknown knowledge space we do not really understand, as discussed in Sect. 4.4.2. This is the stage referred to as "we do not know what we do not know".

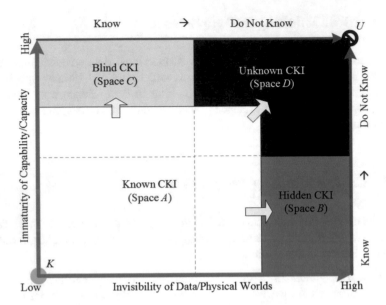

Fig. 3.6 Data science: known-to-unknown research evolution

More discussion on the evolution of data science research from known to unknown spaces is available in Sect. 4.4.3.

The stage "we do not know what we do not know" can be explained in terms of various unknown perspectives and scenarios. As shown in Fig. 3.7, the unknown world presents the unknown in terms of (1) problems, challenges, and complexities; (2) hierarchy, structures, distributions, relations, and heterogeneities, (3) capabilities, opportunities, and gaps, and (4) solutions.

First, there are unknown gaps between known capabilities and unknown challenges: This highlights the significant gaps between our current knowledge and intelligence capabilities and the invisible complexities and challenges in the hidden data world. There are many data science challenges and opportunities we are not yet aware of, and we do not even know how big the gaps in our knowledge might be.

Second, unknown challenges: While work has been conducted on many research issues, as outlined in many conference topics of interest, what are the new areas that have not been considered or discovered in our data science research? There has been increased discussion about the challenges in data science [64, 233, 252, 265], but what are the grand challenges that have not been defined and explored?

From the analytics and learning perspectives, we are familiar with several scenarios about unknown complexities. Unknown data types, structures, relations (note: the interchangeable term *relationship*), distributions, hierarchies, and heterogeneities are some aspects of unknown problems and unknown complexities. For example, we often talk about relation learning, but what do we mean by "relations"? Current research focus has been on such relations as dependency (including hidden relations modeled as dependency), association, correlation, linkage (including co-

Fig. 3.7 Data science: the unknown world

occurrence), but is this focus wide enough to capture all coupling relationships [61] in the data world? Are there new relations that have not been well defined and studied?

As an example, how can we model the relations in the following scenario. An American Indian data scientist speaks at a major Indian data science conference: Are dependency, correlation and/or association necessary and sufficient to represent why the Indian scientist addressed the Indian conference? How about other hidden factors, such as cultural or occasional reasons, or possibly the influence of a third party who might be the conference chair and know the speaker well? How do a mixture of relations interact with each other to form complex data systems? Are those relations explicit and/or implicit? Where in the system hierarchy is each relation applicable? Are these relations the same or similar? Do they play similar roles or contribute differently? There could be many questions about this simple but very common scenario.

Third, unknown opportunities: Data is becoming bigger, and more complex. The values are becoming richer as we usually expect, but they are often hidden. What are the new opportunities for developing breakthrough theories, technologies, data products, and data economy which we are as yet unaware of?

One example can be found in the "beyond IID" [33] initiatives. It is believed that existing IID assumption-based information theories are insufficient for managing complex data and problems. New opportunities are expected to eventuate from building new quantum information science and upgrading classic information theory (Shannon theory).

Another example is the exploration of non-occurring behaviors [76]. In general, people are only concerned about behaviors that have taken place or will occur, and little attention has been paid to behaviors that have not taken place. How to model, represent, analyze, learn, manage and control unexpected non-occurring behaviors is an open issue in behavioral science and data science. There could be enormous opportunities to develop non-occurring behavior informatics to address the above issues, especially from the behavioral data-driven perspective.

Lastly, unknown solutions: There are many known data science issues that have not been well addressed or have not even been studied. The most effective solutions for investigating these issues are unknown. There is an even greater case for seeing unknown solutions when unknown challenges are identified.

For example, in learning complex coupling relationships in the above Indian scientist example, what are the solutions for capturing explicit and implicit, known and unknown relations? When a data problem is too complex to be solved by one model, an ensemble of heterogeneous solutions is required to address different aspects and challenges. How can we ensure that each model will capture the respective challenges effectively, and how do the ensemble models complement a global understanding of the underlying problem as a system? How can we ensure the local model and global ensemble strategies are perfect and solid? Often, we do not know what the solutions are.

3.5.3   Data-Driven Scientific Discovery

3.5.3.1   What Is Data-Driven Discovery

It is possible that there are different understandings of what data-driven discovery is. In our definition, *data-driven discovery* is the methodology and process of conducting scientific discoveries primarily from data and data-based systems. Data-driven discovery is regarded as a unique new feature and a major transformation of existing scientific research, innovation and development paradigms.

In traditional scientific research, hypothesis testing and model-based design are the main approaches. As data is becoming increasingly intensive and valuable, and as data-driven discovery thinking is increasingly accepted, the roles and responsibilities of data-driven discovery and model-based design also become debatable.

The concept of *data-driven discovery* has the following meanings:

- Data is the main input of discovery: While there may be other input sources such as human and domain knowledge involvement, data is the major input and is the main resource to be relied on for discovery.
- Data drives the discovery process: The discovery process is not predefined, rather it is determined by the data, i.e., the data determines what the process looks like; for example, whether a learning problem is to cluster or classify objects is determined by the data and the nature of the problem.
- Data forms the discovery model: The model formation is not predefined, rather it is determined by data, i.e., data determines what the discovery model looks like. A data-driven discovery process customizes the most suitable model for discovering the underlying data and problem.
- Data dictates the discovery outcomes: The discovery outcomes (which could be models and/or data-analytical findings) are determined by the data, and the quality and impact of the outcomes are evaluated per the data.
- Perfect fitting: An ideal discovery is a process that perfectly fits the data observations to an online learned model and outcomes. It is not a bilateral process of fitting data to a given model or fitting a given model to data.

3.5.3.2 Data-Driven Discovery vs. Model-Based Design

Here we discuss the goodness of fit between data-driven discovery and model-based design, that is, how well a model fits data complexities.

There are five general points of view to take into account about the relationships and roles of data-driven discovery and model-based design when the latter are treated as two different types of scientific thinking and method. Figure 3.8 illustrates the various scenarios and roles played by data-driven discovery and model-based design in a data science problem in terms of the closeness of fit between the data challenges and model capabilities: perfect fitting, excessive data fitting, excessive model fitting, imperfect fitting, and non-fitting between data nature (D: intrinsic data complexities and characteristics) and model power (M: rich and deep model capabilities and performance).

The perfect fitting between data and model is illustrated in Fig. 3.8a. Data-driven discovery with perfect fitting is a one-directional process. *Perfect fitting* is a one way data-driven discovery, indicating that if the perfect model is learned, it can perfectly capture the challenges in data and there is no excessive or inadequate fitting of model or data. This reflects the best possible status of data-driven discovery, where the boundary of the data challenges matches that of the model capabilities, although it is often impossible for the discovery on complex data. Perfect fitting is also called *perfect data modeling*.

The excessive model fitting, between data and model is illustrated in Fig. 3.8b. *Excessive data fitting* in data-driven discovery means that data is more than sufficient while the model is inadequate for adequately capturing the data challenges. Excessive model fitting is also called *under-adequate data modeling*.

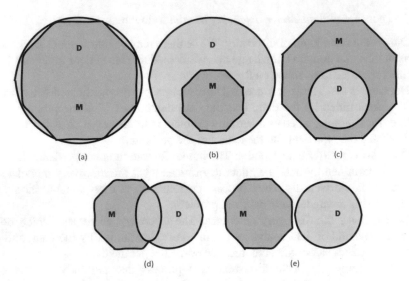

Fig. 3.8 Data-driven fitting vs. model-based discovery fitting. (**a**) Perfect fitting. (**b**) Inadequate data fitting. (**c**) Inadequate model fitting. (**d**) Limited fitting. (**e**) Failed fitting

The excessive data fitting between data-driven discovery and model-based design is illustrated in Fig. 3.8c. *Excessive data fitting* in data-driven discovery means that the data is limited while the model adequately represents all the data challenges and other aspects that do not exist in the data. Excessive data fitting is also called *over-adequate data modeling*.

The imperfect fitting between data and model is illustrated in Fig. 3.8d. *Imperfect fitting* in data-driven discovery refers to the scenario in which the data and model fit in some way, but neither fits the other perfectly. Some data challenges cannot be captured by the model, and the model also covers other data complexities that do not take place in the underlying data. Imperfect fitting is also called *limited data modeling*.

The non-fitting between data and model is illustrated in Fig. 3.8e. *Non-fitting* in data-driven discovery refers to the scenario in which neither the data fits a model nor a model fits the data. No data challenges can be captured by the model. Non-fitting is also called *failed data modeling*.

Excessive data fitting, excessive model fitting, imperfect fitting, and non-fitting are imperfect forms of data modeling. In data-driven discovery, we aim for perfect fitting.

Model-based design is often used in conventional data mining and machine learning research. A typical situation is that a model is accepted and modified to fit certain data characteristics, and a new version of the model is created. This reflects the model-driven discovery approach. *Model-driven discovery* (including hypothesis-driven discovery) refers to the testing, parameter tuning, or model tuning on a given model so that the revised model fits the underlying data better. This does not effectively utilize and capture data characteristics. In contrast, data-driven discovery takes the opposite approach; that is, it allows the data to tell the story (hypothesis), brings about a model that fits the data perfectly, and verifies the discovered model.

3.5.4 Data-Driven and Other Paradigms

In practice, data-driven discovery approaches often have to be synergized with other scientific paradigms, including experiments, simulation, and hypothesis-driven approaches. There are various reasons for this, such as an imperfect model which can be improved, a known data characteristic which fits a model or a model update, unknown data-driven solutions but trial and error problem-solving, and limitations in data-driven approaches.

3.5.4.1 Various Hybrid Paradigms

In addition to model-based design, which has largely been adopted in current scientific research in data science, the combination of data-driven approaches with other approaches may provide additional value in understanding complex data problems. We discuss a few of these combinations below.

First, *the synergy between data-driven discovery and hypothesis testing*: *Hypothesis testing* refers to the formal procedures used (typically by statisticians) to accept or reject certain (usually statistical) hypotheses. Hypothesis testing [271] has been the main scientific approach in classic scientific (especially statistical) research and has demonstrated strong value in current data-based system design. It has been increasingly used in complex web-based data problem understanding and solving, such as marketing campaign design in online business and interactive Web/user interface design in social media.

A very popular example is A/B testing [362]. *A/B testing*, which is also called *split testing* or *two-sample hypothesis testing*, compares two identical versions A and B except for one variation between A and B that may affect the consequences of A and B. If A is a version under control, then B is modified in some aspect to test the impact of the variation (or treatment). Hypothesis testing, like A/B testing, is widely used in business, especially in large-scale online services such as webpage design, to show which version of webpage design performs better. When unknown challenges, opportunities and solutions exist in a complex data problem, the testing of different hypotheses can help to understand the challenges progressively, before a clearer picture of the data challenges is visible.

Second, *the synergy between data-driven and experimental design methods*: *Experimental design* is an important area that refers to "a plan for assigning subjects to experimental conditions and the statistical analysis associated with the plan" [255]. It investigates scientific or research hypotheses by following a series of activities: the formulation of statistical hypotheses, the determination of the experimental conditions and the measurement to be used and extraneous conditions to be controlled, the specification of experimental units and population, the procedure for assigning subjects to experiments, and statistical analysis [255]. Experimental design ensures the sound process, solid specification, and valid verification of data experiments in data problem-solving, thus it is essential. The correct adherence to and execution of experimental design theories, specifications, procedures, and analysis will ensure that experiments and empirical analysis in data science are standardized, trustful and valid. This is particularly important for hypothesis-free data exploration.

Third, *human and nature-inspired intelligence representation and computing*: Human intelligence and natural intelligence are often integral in complex data systems specially in natural data systems and social data systems. Such data systems are often associated with sophisticated data challenges and system complexities, and the understanding, representation and computation of human intelligence and natural intelligence is imperative. Studies in human and natural intelligence provide complementary thinking, ideas, approaches and solutions for classic computing-based and other research approaches. These include *cognitive artificial intelligence* (AI that is analogous to human cognition and focuses on symbolic and conceptual systems) [366], *computational intelligence* (the nature-inspired design of intelligent systems, particularly by neural networks, fuzzy systems and evolutionary computation) [249], *bio-inspired computing* (the design of intelligent artifacts inspired by biology to be capable of autonomous operation in changing and often unknown environments) [165], and *human-machine interaction and cooperative systems* [43]. They can extend the methodologies and tools available and customizable for understanding and computing complex data science systems, and they complement existing computational, intelligence, and information theories and systems.

Lastly, *metasynthesis of various paradigms*: Different paradigms are suitable for specific settings and conditions, and are complementary to one another in addressing a significantly challenging scientific problem. When a data problem meets an extreme challenge (Sect. 4.9), and is associated with X-complexities (see detail in Sect. 4.2) and X-intelligence (see discussion in Sect. 4.3), it is often impossible and impractical to address the challenges within one paradigm and in one round of exploration. This triggers the need to explore new methodologies and cognitive methods, similar to the challenges seen in handling complex systems [62].

A reasonable methodology is probably the metasynthesis [62, 70, 114] of different relevant paradigms, and develops *metasynthetic analytics*. This *metasynthesis* may adhere to a principle such as the one that follows: (1) conducting high-level understanding of the nature, challenges and opportunities of the problem by unifying expertise and knowledge from a group of domain-relevant experts and human-centered computing, (2) exploring currently observable individual challenges by applying relevant paradigms familiar to the domain-specific experts and expert team, (3) synthesizing exploration methods and findings to deepen the understanding of the nature, challenges and opportunities of the problem, and (4) exploring other semi-transparent challenges by involving the findings obtained and new approaches. The metasynthesis of various paradigms is different from a simple integration of two or more approaches.

Metasynthetic analytics takes several iterations to deepen the data understanding and to achieve the level of expected understanding of the underlying problem. In each round, analytics are conducted to achieve a more detailed and deeper understanding of the problem complexities, and new open areas and issues are identified. New and additional analytical methods are discovered or invented to crack the newly identified data and problem complexities. The understanding process may be a qualitative-to-quantitative cognition process,[1] with increasing cognition and recognition of the problem nature and solutions. Both top-down and bottom-up thinking and exploration methods may be used in the process, with the aim of building a comprehensive, quantitative, and connected representation and solution of diversified complexities and challenges in the data problem.

3.5.4.2 Domain + Data-Driven Discovery

Data science has naturally built on several relevant data-analysis research areas and disciplines: statistics and mathematics, data mining (also called knowledge discovery), machine learning, signal processing, and pattern recognition. The core relevant areas are statistics, data mining and machine learning, which focus on and rely on data-centric analysis, and are often defined as being data-driven (namely, from data).

Literally, *data-driven* refers to the thinking, approach and process that are based on and driven by data. *Data* in this context refers to a collection of single pieces of information, such as facts, statistics, or code, that are organized or gathered. Typical features of *data-driven approaches* include:

- *Aim*: The aim of data-driven exploration is to discover and learn patterns and/or exceptions as well as broad intelligence and knowledge in data.
- *Input*: Input often only consists of data, although it is frequently necessary to involve other resources, such as domain knowledge. When different information resources are involved, data should play the main role in telling the stories.
- *Design*: Problem-solving design is often built on a given or selected hypothesis, model, or system. Typically, we fit the hypothesis, model, or system (referred to as 'model' for simplicity) in data and add additional considerations to capture additional or particular data complexities. This involves model tuning, adjustment, fitting, and optimization.
- *Process*: The process for achieving the aim is compelled by data rather than hypotheses, models, expert knowledge or intuition. Typically, a data-driven

---

[1] Also refer to Sect. 5.3.3 for more discussion on metasynthesis for complex data problems and broader discussion on data science methodologies in Sect. 5.3.

system explores data through a training and testing process to obtain a model that effectively and/or efficiently captures the target problems in data.

- *Output*: The output from data-driven explorations are the models, which represent a satisfactory design for achieving the aim from the input through the data exploration process.
- *Evaluation*: The evaluation of data-driven output is often dependent on the quality of the model in terms of its degree of fit with the ultimate groundtruth on data objects. The degree of fit is quantified in terms of learning performance measures such as accuracy.

Extensive discussion and a summary of the gaps and issues in this classic data-driven approach for actionable data-driven problem-solving are available in [54, 77]. Issues are related to the imbalance of models, output and decisions based on the output, weak dependability, repeatability, trust, explainability, deliverability and transferability, and the gaps between the models obtained and the underlying data complexities and problem challenges [77].

In contrast, *domain-driven* [54] refers to the thinking, approach and process that are based on and compelled by the domain. Here *domain* refers to the field in which the problem and associated data are embedded and the realm of field information, knowledge, etc. Domain-driven approaches highlight the involvement of general and domain-specific aspects, facts, data and objectives.

- *Aim*: The aim of domain-driven exploration is to understand, characterize and discover general and domain-specific intelligence and knowledge from the data-based domain.
- *Input*: The input consists of not only domain-specific data but also other relevant resources such as domain knowledge, human factors, organizational and social aspects, and environmental factors. In multi-source domain-based resources, the domain should play the main role in telling the stories.
- *Design*: The problem solving design is driven by the domain, that is, the X-complexities discussed in Sect. 4.2. New hypotheses, models, or systems are designed or obtained by understanding and solving domain-specific problems, understanding the relevant complexities and challenges embedded in the data-based domain. It is opposite to data-driven design in that it is inspired by the domain and the target data in the domain.
- *Process*: The process of achieving the aim is driven by the domain and its data without given or assumed hypotheses, models, expert knowledge or intuition. A domain-driven system still typically explores data through a training and testing process to obtain a model that effectively and/or efficiently represents and solves the target problems in the domain.
- *Output*: The output from domain-driven explorations is new knowledge and intelligence about the domain or its specific challenges.
- *Evaluation*: The evaluation of domain-driven output is the degree of fit between the discovered models and the intrinsic nature of the problem, data complexities and domain satisfaction, measured by the evaluation of the output; for example,

the actionability [59, 77] of the output for driving decision-making actions and its effect on addressing problem challenges.

In handling complex data problems, a reasonable thinking is to combine data-driven approaches and processes with domain-driven approaches and processes, to form *domain+data-driven discovery* or *data+domain-driven discovery*. *Domain+data-driven discovery* is the thinking, approach and process of discovering new knowledge, intelligence, insight and wisdom from data and its domain without prior assumptions. An example of this domain+data-driven discovery is the *domain-driven data mining* methodology [77]. *Domain-driven data mining* is a data mining methodology for discovering actionable knowledge and delivering actionable insights from complex data and behaviors in a complex environment. It studies the corresponding foundations, frameworks, algorithms, models, architectures, and evaluation systems for actionable knowledge discovery by involving and synthesizing relevant X-complexities and X-intelligences.

The migration from data-driven discovery to domain+data-driven discovery should address the aspects featured in Table 3.1.

Table 3.1 Data-driven to data+domain-driven discovery

Dimensions	Data-driven discovery	Data+domain-driven discovery
Aims	Discover and learn patterns and/or exceptions and broad intelligence and knowledge in data	Understand, characterize and discover general and domain-specific intelligence and knowledge from the data and data-based domain
Drivers	Data	Data+domain
Thinking	Methodology, approach and process are based on and driven by data	Methodology, approach and process are based on and driven by data and the domain
Input	Often consists of only data, although other resources such as domain knowledge may also be involved	Domain-specific data but also other relevant resources such as domain knowledge, human factors, organizational and social aspects, and environmental factors
Design	Often built on a given or selected hypothesis, model, or system	Understand and quantify the relevant complexities and challenges embedded in a data-based domain
Process	Driven by data, rather than hypotheses, models, expert knowledge or intuition	Driven by the domain and its data without given or assumed hypotheses, models, expert knowledge or intuition
Output	Models fit data	New knowledge and intelligence about the domain or its specific challenges
Evaluation	The quality of a model and its its degree of fit with the ultimate groundtruth on data objects	Satisfaction of discovered models capturing the intrinsic problem nature, data complexities and domain value

3.5.5 Essential Questions to Ask in Data Science

Asking essential questions in data science is very challenging and risky, as the domain is broad, developing, and uncertain at this stage. In this section, building on the above discussion about scientific thinking, data science view and structure, and specific points about viewing data science in terms of complex systems and critical thinking, we list a few such questions on objective, input, design, method and process, outcome, and evaluation.

A. Essential questions about data science objectives:

- What is the nature of the problem?
- What outcomes are expected from the studies of data?
- What are the opportunities for knowledge advancement?
- What outcomes can we obtain from the data rather than anything else (such as expert knowledge)?
- How will the data be used?
- How will the outcomes be used?
- Who cares about the problem, data and outcomes?

B. Essential questions about data science input:

- What relevant sources of data are required?
- What is the nature of the data?
- What data characteristics are in the data?
- What data challenges are in the data?
- What are the common characteristics in the data?
- What makes this data different or unique?

C. Essential questions about data science design, method and process:

- How do the design, method and process advance the knowledge?
- How do the design, method and process capture the intrinsic data characteristics?
- How do the design, method and process capture the intrinsic data complexities?
- How sensitive are the design, method and process to the data?
- How irrelevant are the design, method and process to the data?
- How perfectly do the design, method and process fit the data?
- How deeply do the design, method and process fit the data?
- How broadly do the design, method and process fit the data?
- How rational are the design, method and process?
- How creative are the design, method and process?
- How reproducible are the design, method and process?
- What are the alternative designs, methods and processes?
- Are there any other possibilities in the design, method and process?
- What might the design, method and process be if some data are not involved?
- What might the design, method and process be if some data characteristics change?

- What are the critical aspects in the design, method and process?
- What are the good and bad in the design, method and process?
- What are the constraints and assumptions in the design, method and process?
- Are there any errors in the design, method and process?

  D. Essential questions about data science output:

- How surprising is the output?
- What is new (novel) in the output?
- Is the output accurate?
- Is the output solid?
- Is the output fair?
- Is the output testable?
- Is the output constant?
- Is there evidence to support the output?
- What are the conclusions that can be drawn from the output?
- What are the intellectual outcomes?
- What are the socio-economic and other benefits?

  E. Essential questions about data science evaluation:

- What are the good reasons for the design and output?
- What makes the design and output solid, rational, and reproducible?
- What are the solid connections between the input, design and output?
- How general are the design and output?
- How explainable are the design and output?
- What impact and value can the output bring to business?
- What in the data have not been sufficiently captured by the design?
- What in the data have not been correctly captured by the design?
- What could be the better ways for the design?
- What could be the better output?

3.6 Summary

This chapter addresses a very important issue: data science thinking. Effective data science cannot exist without reasonable data science thinking. The discussion in this chapter aims to elevate data science from a merely technical level to a philosophical level, and to build high level thinking and a considered view of what makes data science a new science, as well as which thinking skills and mental traits are required to conduct data science activities. What makes data science thinking different? This is a very challenging and open problem. It is understandable that different perspectives of thinking may be targeted, and what should be included in data science thinking is highly debatable. On one hand, this chapter attempts to explore data science thinking in terms of conventional frameworks of thinking, namely scientific thinking, creative thinking, logical thinking, and critical thinking.

On the other hand, we put forward our view of data science thinking in terms of data system complexities, problem-solving frameworks and processes, and the roles of relevant scientific paradigms.

The chapter makes several important observations:

- Complex data science problems are complex systems;
- A systematic view is appropriate and essential for understanding data science complexities;
- High-level thought structures, mental habits and traits are required in data science;
- While existing scientific paradigms are necessary and useful in data science, data+domain-driven discovery reflects the nature of data science problems;
- Conducting data science raises many essential questions to which answers are sought.

In practice, the quality of data science work is highly dependent on the quality of data science thinking. It is data science thinking that differentiates a good data scientist from an average one, and a quality data science output from a bad or average one. Given the same data, data scientists with more powerful data science thinking may deliver much better outcomes than data scientists who have limited data science thinking capabilities.

Many important concepts and aspects have been raised in this chapter with only limited discussion or explanation. They include X-complexities, X-intelligence, unknown worlds and challenges. We will explain some of these in Chap. 4, and further discussion is also available in other chapters, such as Chap. 5.

Part II
Challenges and Foundations

Chapter 4
Data Science Challenges

4.1 Introduction

What are the greatest challenges of big data and data science? This question itself is problematic as data science is at a very early stage and has been built on existing disciplines. This chapter explores this important issue.

Rather than focusing on specific research issues from a disciplinary perspective, such as complex relation representation or learning in hybrid networks, this chapter discusses those challenges that are

- driven by low-level factors and aspects, especially the complexities and intelligence embedded in the data, behavior and environment of a data science problem;
- caused by system complexities, as a result of which a data science problem is viewed as a complex system;
- embedded in data quality, which determines what and how much value we can obtain through data science;
- complicated by the social and ethical issues surrounding data science problems.

We discuss the intrinsic X-complexities of data science problems from a low-level perspective and the associated X-intelligence that is emerging as a result of processing the X-complexities. This is followed by discussion about the corresponding research challenges and directions encountered in addressing data science complexities and characteristics. Our observations particularly draw on the viewpoint that data science problems are complex systems [62, 294], and we will focus on issues from knowledge and intelligence perspectives.

In addition, to illustrate the conceptual discussion about various types of data science complexities and corresponding intelligences, we will consider the following challenges.

- Non-IIDness in data science problems: the concepts of non-IIDness and non-IID learning when data and problems are treated as being non-independent and identically distributed (non-IID);
- Human-like machine intelligence: the evolution of machine intelligence to human-like intelligence and the driving roles played by data science in obtaining such human-like intelligence in machines;
- Data quality and ethical issues: issues of data quality and data ethics extend the discussion to the broader nature of data science, i.e., where social, legal and ethical matters are integral components.

To conclude this chapter, we discuss the scenario of the extreme data challenge, which embraces and combines various X-complexities and X-intelligence in one problem, typically in highly complex data science problems.

4.2   X-Complexities in Data Science

In Sect. 1.5.1, we briefly introduced the concept of X-complexities. In this section, we elaborate on the complexities and challenges of data problems and data systems.

A core objective of data science research and innovation is to effectively explore the sophisticated and comprehensive complexities [294] trapped in data, business, data science tasks and problem-solving processes and practices, which form a complex system [62]. A data science problem is a complex system in which comprehensive system complexities are embedded. These complexities are explored from the following specific aspects:

- Data (characteristics) complexity
- Behavior complexity
- Domain complexity
- Social complexity
- Environment complexity
- Human complexity
- Learning (process) complexity
- Deliverable complexity

4.2.1   Data Complexity

Data complexity is concerned with the sophisticated data circumstances and characteristics of a data problem or system. Typical data characteristics of concern are largeness of scale, high dimensionality, extreme imbalance, online and real-time engagement, cross-media applications, mixed sources, strong dynamics, high frequency, uncertainty, noise mixed with valuable data, unclear structures, unclear hierarchy, heterogeneous or unclear distribution, strong sparsity, and unclear availability of specific data.

A very important issue concerns the *complex coupling relationships* [60] hidden in data and business, which form a key component of data characteristics and are critical in properly understanding the hidden driving forces in data and business. Complex couplings may consist of comprehensive relations that may not be describable by existing association, correlation, dependence and causality theories and systems. Learning mixed explicit and implicit couplings, structural relations, nonstructural relations, semantic relations, hierarchical and vertical relations, relation evolution and reasoning are critical and challenging.

4.2.2 Behavior Complexity

Behavior complexity becomes increasingly visible in understanding what actually takes place in business and a data science system. This is because behaviors carry the semantics and processes of behavioral objects and subjects in the *physical world* that are often ignored or largely simplified in the transformed *data world* following the physical-to-data conversion undertaken by existing data management systems. Hence, *behavior complexity* refers to characteristics and issues in the behaviors of the actual things in the physical world which may be overlooked or represented in terms of specific symbols in a data problem or data system.

Behavior complexities are embodied in such aspects as coupled individual and group behaviors [55], behavior networking and interactions [61, 74], collective behaviors, behavior divergence and convergence, non-occurring behaviors [76], behavior network evolution, and group behavior reasoning. Other issues concern the insights [31], impact, utility and effect of behaviors, the recovery of what has happened, is happening, or will happen in the physical world from the highly deformed (and disconnected, scattered) information collected in the data world, and the emergence of behavior intelligence.

4.2.3 Domain Complexity

Domain complexity has become increasingly recognized [77] as a critical aspect for deeply and genuinely discovering data characteristics, value and actionable insights. *Domain complexity* refers to the characteristics and issues in the domain of a data science problem and in the interactions between data, the underlying domain, and domain-specific data products.

Domain complexities are reflected in such aspects as domain factors, domain processes, norms, policies, qualitative to quantitative domain knowledge, expert knowledge, hypotheses, meta-knowledge, the involvement of and interaction with domain experts and professionals [335], multiple and cross-domain interactions, experience acquisition, human-machine synthesis, roles and leadership in the domain.

As data science problems are often domain-specific, unique challenges sometimes emerge from particular domain problems and domain complexities. This may be the mixture of domain complexities with data and other aspects of complexity, or the embedding of domain characteristics into data. An example is evident in the spatio-temporal data complexities in environmental data systems, in which the significant environmental change and disappearance of regular environmental phenomena are embodied in spatial and temporal data characteristics in environmental entities and their relations.

4.2.4 Social Complexity

Social complexity is embedded in business and data, and its existence is inevitable in data and business understanding. *Social complexities* refer to social issues embedded in or surrounding a data science problem and its resolution. Social complexities are integral components of data science problems, since a data science problem sits in a social environment, thus the creation, acquisition, use, dissemination, and curation of data and data products naturally involve social or even legal and ethical issues.

Social complexities may be embedded in such aspects as

- complexities in social data science problems, such as social networking, community emergence, social dynamics, social media networking, group formation and evolution, group interactions and collaborations;
- complexities imposed on data science problems from social and ethical perspectives, such as social conventions, social contexts, social cognition, social intelligence, economic and cultural factors, social norms, emotion, sentiment and opinion spreading and influence processes, and impact evolution.

In fact, social issues and ethical issues in data science emerge as new research areas in social science, psychological science, and ethics research. Social and ethical issues including security, privacy, trust, risk and accountability in social contexts are challenging and integral in data science. More discussion on data social and ethical matters is available in Sect. 4.8.

4.2.5 Environment Complexity

Environment complexity plays an important role in complex data and business understanding. *Environment complexity*, also called *context complexity*, concerns characteristics and complexities in the context of a data problem or system, and the interactions between data and context.

Environment complexities are reflected in environmental factors, relevant contexts, context dynamics, adaptive engagement of contexts, complex contextual interactions between environment and data systems, significant changes in environment, and variations and uncertainty in the interactions between data and environment.

4.2.6   Human-Machine-Cooperation Complexity

Complex data science problem-solving has to involve domain experts and expert knowledge to form human-machine cooperative data science solutions and systems, which results in human-machine-cooperation complexities. *Human-machine-cooperation complexities* refer to issues and problems that need to be addressed to enable the involvement of humans and human knowledge in data science problem-solving or human-centered data science systems.

Human-machine-cooperation complexities may be embodied in multiple ways, depending on the level and manner of the human and human intelligence involvement in data science. Issues to be considered may include, but not be limited to:

- determining the specific roles, types of domain experts, and levels of human and/or human knowledge involvement are needed in a data science problem-solving system,
- determining how to incorporate human and human knowledge into data science systems, e.g., how to engage domain experts in a data science team, and how to support a domain expert's interactions during an analytics and learning process,
- determining how to model and dynamically learn user behaviors when users are constituents of a data system,
- determining how to mimic human critical and creative thinking, and non-logical thinking such as imaginary thinking in data science thinking and problem-solving, and
- determining where and how to involve human intelligence and expert knowledge that cannot be replaced by data-driven discovery systems.

4.2.7   Learning Complexity

Learning (process) complexity has to be properly addressed to achieve the goal of data analytics. *Learning complexities* refer to issues in the process and relevant actions of understanding data, context and solutions, and in designing, implementing and evaluating learning systems.

Typical challenges include developing effective methodologies, common task frameworks and learning paradigms to handle various aspects of data, domain, behavioral, social and environmental complexities in a learning system. For example, there are additional challenges in

- learning multiple sources and inputs, parallel and distributed inputs, heterogeneous inputs, and dynamics in real time;
- supporting on-the-fly active and adaptive learning, as well as ensemble learning, while considering the relations and interactions between ensembles;
- supporting hierarchical learning across significantly different inputs;
- enabling combined learning across multiple learning objectives, sources, feature sets, analytical methods, frameworks and outcomes;
- accommodating non-IID learning [60] by mixing couplings with heterogeneity; and
- conducting meta-analysis [326].

Other elements include the appropriate design of experiments and mechanisms.

4.2.8   Deliverable Complexity

Deliverable complexity refers to the challenges of creating and converting data science findings that will best fulfill the requirements and expectations of stakeholders. Deliverable complexity particularly becomes an issue when actionable insights and actionable knowledge discovery [77] are the focus of data science.

This necessitates the identification and evaluation of outcomes that satisfy technical significance and have high business value from both objective and subjective perspectives. The challenges are also embedded in designing the appropriate evaluation, presentation, visualization, refinement and prescription of learning outcomes and deliverables to satisfy diversified business needs, stakeholders, and decision purposes.

As an example, *complex patterns* [59] are deliverables that are presented as patterns but capture complex relations between patterns or pattern components which disclose complex interactions between multiple sources of data and can inform actionable decision-making. Such patterns can be identified by knowledge and ontology-based pattern discovery, tree-based pattern learning, graph-based pattern learning, image/video-based pattern learning, combined pattern learning, cross-media pattern generation, framework and prototype-based pattern learning, indirect pattern discovery through reasoning and similarity learning, and automated pattern discovery.

In general, deliverables to business are expected to be easy to understand and interpretable from the non-professional perspective, disclosing and presenting valuable but hidden insights that can inform and enable decision-making actions and possibly have a transformative effect on business processes and problem-solving.

An in-depth exploration and understanding of the above various complexities and their respective issues and challenges in each type of complexity will disclose general and/or unique opportunities for data science research and innovation. These may fall into the family of research issues commonly and currently explored in the relevant communities, but it is likely that new perspectives and opportunities will be disclosed by the intrinsic nature of the problem.

4.3 X-Intelligence in Data Science

The nature of data science is the drive to achieve a successful transformation from data to knowledge, intelligence and wisdom [342]. During this transformation process, comprehensive intelligence is often involved in complex data science problems, from data to domain, organizational, social and human aspects, and the representation and synthesis of these aspects.

In Sect. 1.5.2, we briefly introduced the nature of the intelligence surrounding data science problem, process and solution, which is called "X-intelligence". Here, we explore the various aspects, challenges and opportunities embedded in the comprehensive intelligence [62] embedded in a domain-specific data science problem. This consists of data intelligence, behavior intelligence, domain intelligence, human intelligence, network intelligence, organizational intelligence, social intelligence, and environmental intelligence.

The term *X-intelligence* refers to the various abilities and knowledge that inform or support the integrative comprehension, representation and problem-solving of underlying complexities and challenges, but also the specific, deeper, more structured and organized aspect of those complexities. X-intelligence clearly emerges from the understanding, analysis and processing of X-complexities. The undertaking of each analysis of X-complexities corresponds to the respective form of X-intelligence.

Below, we discuss the X-intelligence associated with the different types of complexities discussed in Sect. 4.2.

4.3.1 Data Intelligence

Data intelligence refers to the intelligence embedded in data. It highlights the interesting information and stories about the formation of business problems or driving forces and their reflection in the corresponding data. Intelligence hidden in data is obtained through understanding specific data characteristics and complexities.

Apart from the usual focus on exploring the complexities in data structures, distribution, quantity, speed, and quality issues from the individual data object perspective, our focus is on the intelligence hidden in the unknown space D in Fig. 4.1. For example, to understand unknown couplings and heterogeneities at different levels from value, feature, object to data source and learning outcomes, non-IID learning has to be considered (see Sect. 4.5). The success criteria for learning data intelligence are how much and to what extent we can understand and capture data characteristics and complexities.

4.3.2   Behavior Intelligence

Behavior intelligence refers to the intelligence embedded in the behaviors of a data problem or data system. It is discovered by understanding the activities, processes, dynamics and impact of the individual and group actors who are the data quantifiers, owners and users in the physical world.

Behaviors sit between the scattered *data world* and its corresponding original *physical world*, and bridge the gaps between the data world and the physical world by connecting what has happened, is happening, and will happen to the formation and dynamics of the real world problem.

Exploring behavior intelligence and insights [31] is critical for a complete and genuine understanding of data and problems, and for a full picture and genuine understanding of business development. Behavior actors, coupling relationships and impact are key aspects to be explored through building the theory of *behavior informatics* and *behavior computing* [72].

4.3.3   Domain Intelligence

Domain intelligence is concerned with the intelligence embedded in the domain of a data problem or data system. It emerges from properly involving relevant domain factors, knowledge and meta-knowledge, and other domain-specific resources that not only wrap a problem and its target data but also assist in problem understanding and the development of problem-solving solutions.

Involving qualitative and quantitative domain intelligence can inform and enable a deep understanding of domain complexities and their critical roles in discovering unknown knowledge and actionable insights. Accordingly, domain-specific data science problems and data science research and practices, such as bioinformatics and social computing, take place. In addition, domain intelligence is embedded in data systems for a better understanding of data formation, characteristics, and processing. Another aspect is to explore data problems across different domains and to transfer data science solutions from one domain to another. This is the basis of *cross-domain data science* problems and *transfer data science* research.

4.3.4   Human Intelligence

Human intelligence refers to the intelligence contributed by humans to a data problem or data system. In complex data science problem-solving, human intelligence plays a critical role in understanding the underlying data and its contexts, the design and implementation of data solutions, and the discovery, evolution, application and use of data values.

Human intelligence may be embodied in

- the explicit or direct involvement of human empirical knowledge, belief, intention, expectation, run-time supervision, and evaluation in complex data science problems and systems by individuals or expert groups.
- the implicit or indirect involvement of human intelligence as imaginary thinking, emotional intelligence, inspiration, brainstorming, reasoning inputs and embodied cognition, such as convergent thinking through interaction with other members in the process of data science problem-solving.

Here, human intelligence focuses on the intelligence associated with human individuals. We refer to the intelligence from a group or society of humans as *human social intelligence*, to be discussed in Sect. 4.3.7.

Depending on the level of complexity and associated requirements, human intelligence may play a role in data science in different modes; for example, in a data-centric system, i.e., human as a system module in the data system; in human-data interaction, i.e., a human-data cooperative system; and in a human-centric data system, i.e., human-driven data system.

4.3.5 Network Intelligence

Network intelligence refers to the intelligence embedded in the networks that encompass a data problem or data system.

Network intelligence emerges from either Web intelligence or broad-based networking activities and resources, especially social media networks, mobile services and networked systems and organisms. Networking-based intelligence is embodied in such aspects as information and resource distribution, linkages between distributed objects and systems, relationships and interactions between networked nodes, hidden communities and groups, information and resources from networks, and, in particular, the Web, distributed and cloud infrastructure and computing facilities, information retrieval, searching, and structuralization from distributed repositories and the environment.

The information and networking facilities in the networks that support and surround a target data problem either serve as the problem constituents or contribute to useful information for complex data science problem solving. Examples of the former are social networks and social media, and of the latter is cyber security in distributed networks.

4.3.6 Organization Intelligence

Organization intelligence refers to intelligence embedded in and emerging from organizations that own a data problem or data system. Organization intelligence plays an important role in forming the context of a data problem or data system, or driving the resolution of data problems and the establishment and operations of data systems.

Organization intelligence emerges from the involvement of organization-oriented factors and resources in data science problem-solving. Incorporating organization intelligence into data science may require the understanding and involvement of organizational goals, actors and roles, as well as organizational structures, behaviors, evolution and dynamics, regulation and convention, process and workflow in data science systems.

In data science, organization intelligence is also embodied in the organizational competency of data science. This involves the level and extent of knowledge, capability and capacity within an organization for conducting data science. More discussion is available in Sect. 10.5.

4.3.7 Social Intelligence

Social intelligence refers to the intelligence associated with social aspects (and other non-technical aspects, such as economic, cultural, legal, or ethical) related to a data problem or data system. Social intelligence is important for

- involving social factors such as social relationships and ethical regulation in data science,
- coping with social data science problems, such as cultural data analysis and data-driven cultural modeling, and
- involving data science in social problem solving, for example, incorporating data science in smart city planning, design and optimization.

Social intelligence may be extracted and learned from group interactions, social networking, social factors, cultural factors, economic factors, and corresponding social norms and regulation.

Social intelligence covers both human social intelligence and animated intelligence. *Human social intelligence* is related to such aspects as social interactions, group goals and intentions, social cognition, emotional intelligence, consensus construction, and group decision-making, taking place in a group or society of humans. Animated intelligence emerges from animated systems, such as artificial life, collective and swarm intelligence, and intelligent robotics.

Social intelligence is often associated with *social network intelligence* and *collective interactions*, as well as the business rules, law, trust and reputation for governing the emergence and use of social intelligence.

4.3.8 Environment Intelligence

Environment intelligence refers to the intelligence hidden in the environment of a data science problem. Environmental factors may thus be described in terms of, and be broadly connected to, the underlying domain, organizational, social, human and network intelligence and complexities.

In complex systems, environments may be open or closed. An *open environment* has interactions with its underlying systems in terms of information, energy or material. In contrast, a *closed environment* or semi-closed environment may exist in data systems in which only a limited number of interactions, or no interactions, take place.

Environmental factors in a data problem or data system are domain specific and problem specific. In addition to domain-specific contextual factors, common environmental issues are the interactions and relationships within the context, the interactions and relationships between the context and the data system, the impact of contextual factors and contextual relationships and interactions on the data system, and the dynamics and evolution of an environment and its impact on a data problem. These factors are important aspects of data understanding and manipulation.

4.4 Known-to-Unknown Data-Capability-Knowledge Cognitive Path

4.4.1 The Data Science Cognitive Path

The data science journey of understanding X-complexities and learning X-intelligence is a known-to-unknown exploration process which achieves the transformation from data to knowledge by applying certain capabilities.

Figure 4.1 (repeating Fig 3.6 for reading convenience) shows a known-to-unknown data science exploration map. The goal and task of data science is to reduce the *capability immaturity* (*y*-axis) to better understand *data invisibility* (*x*-axis) from the 100% known state K to the 100% unknown state U. When data invisibility increases, the capability immaturity also grows.

The known and unknown states are associated with the level and extent of our transparent, genuine and complete understanding and control of X-complexities and X-intelligence in a data science system. When the relevant complexities and intelligence become visible and manageable in data systems, they become known to us. The unknown space D represents the ultimate challenge and complexity we face in X-complexities and X-intelligence in highly complex data science problems.

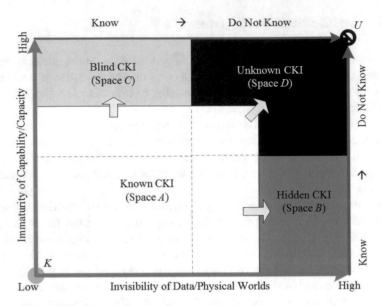

Fig. 4.1 Data science: known-to-unknown exploration

4.4.2 Four Knowledge Spaces in Data Science

According to the status and levels of data visibility and capability maturity, our knowledge about a data science problem can be categorized into four elements:

- Space A representing the *known knowledge space* i.e., *I know* (my mature capability) *what I know* (about visible data). This is similar to the ability of sighted people to recognize an elephant by seeing the whole animal, whereas non-sighted people might only get to identify part of the animal by touch. The knowledge in the visible data is known to people with mature capability (the level of capability maturity is sufficient to understand the level of data invisibility). Typically, work on profiling and descriptive analysis fall into this space.
- Space B representing the *hidden knowledge space* i.e., *I know what I do not know* (about the invisible data). Although the capability is mature, knowledge in the invisible data is hidden from the low level of capability maturity. Forecasting, prediction and exploratory analysis may be undertaken to understand the invisible data.
- Space C representing the *blind knowledge space* i.e., *I do not know* (my immature capability) *what I know*, which is similar to the ability of blind people to recognize an elephant, as described above; although the data is visible, the immaturity of the capability makes the knowledge blind. Descriptive analytics and hypothesis testing may be exercised to improve capability maturity.

- Space D representing the *unknown knowledge space* i.e., *I do not know what I do not know*, knowledge in the invisible data is unknown as a result of immature capability. Future research and discovery are focused in this area.

4.4.3 Data Science Known-to-Unknown Evolution

The diagram in Fig. 4.1 presents insights into, and an explanation of, the evolution of data science in terms of the following perspectives:

- The data science evolution from Spaces A to D corresponds to the paradigm shift from descriptive analytics to deep analytics, forming a known-to-unknown journey of exploration;
- From the time perspective, the space migration from A to B and D corresponds to analytics on historical to future data;
- With the transformation of capability maturity, our knowledge awareness migrates from known (A) to hidden (C) areas in data;
- The current focus in data science research is on Space D, i.e., to invent and significantly enhance our capability maturity about invisible data and highly improbable [379] insights.

A typical example to demonstrate the above evolution, transformation and paradigm shift is the analytical and learning theory shift from IID to non-IID [60, 175] data understanding. In classic theories and systems, mature capability is still required to understand the invisibility of IID data, as shown by the currently blind and hidden knowledge in IID cases, illustrated by Spaces B and C. The current critical challenge in learning and analytics is to invent non-IID capabilities to explore the non-IID data invisibility in Space D.

4.4.4 Opportunities for Significant Original Invention

With the understanding and recognition of significant gaps between the X-complexities and the availability of X-intelligences, the gaps between the valuable but invisible X-opportunities and current limited capability to take advantage of them, there are great opportunities for us to explore significant explicit and/or implicit occurrences and emergent opportunities that were not previously possible.

As a result, many original invention opportunities and unexpected outcomes may emerge by managing these gaps, for example:

- Research that previously could not be carried out, such as conducting statistical data analysis by involving large scale of mixed data and deep analytics and learning of such data, becomes feasible;

- Achieving business goals that were previously impossible, such as building data-driven global counter-terrorism systems that collect, detect, and analyze data and information on a massive scale from mobile networks, social networks, social media, traditional media and intelligence systems, to recommend and implement instant actions to prevent or intervene in possible terrorist activities;
- Knowledge and insights that were previously hidden may become clear and obvious through the analysis of large scale data for unknown knowledge and insights, thus reducing the blind and hidden spaces,
- Trials and hypotheses that were originally non-standardized and unacceptable as routine processes may become more standardized and informative policy-guided practices, such as identifying new medical and health hypotheses from large medical and health data which can be verified by domain experts and converted to new domain-oriented protocols,
- Previous myths and pitfalls in data-based applications and practices may be verified and then corrected and adjusted towards more appropriate thinking, designs and implementations, probably within certain constraints, and
- Data-driven scientific approaches may significantly supplement and reform classic scientific research thinking, theories and methods; for example, data-driven financial and economic research may change traditional finance and economics research from model application-based story-telling to data-driven story telling.

The additional discussion on X-complexities and X-intelligences in relation to data science thinking in Chap. 3, data science as a discipline in Chap. 5, and the relevant foundations in Chap. 6, indicate the unlimited opportunities for theoretical breakthroughs, cutting-edge new technologies, and innovative practices.

4.5 Non-IIDness in Data Science Problems

To illustrate the challenges, concepts and potential of exploring data complexities in complex data science problems, we introduce the non-IIDness of complex data in this section. Many other data characteristics and complexities discussed in Sect. 4.2.1 could be explored further in the context of big data and complex data, although it may be difficult to gain a deep and sound understanding of some of them.

4.5.1 IIDness vs. Non-IIDness

Complex interactive and heterogeneous data and applications present major challenges to current data processing, analytic and learning theories and systems. Big data, in particular, presents specific complexities of weakly structured and

Fig. 4.2 IIDness vs. non-IIDness in data science problems. (**a**) Learning problem. (**b**) IID learning. (**c**) Non-IID learning

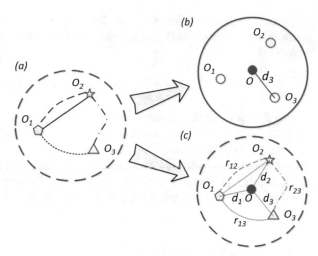

unstructured data distribution, dynamics, relations, interactions and structures, which challenge the existing theoretical and commercial systems in mathematics, statistics and computer science. The latter have been mainly designed on the assumption that data is independent and identically distributed (IID), whereas big/complex data is essentially non-IID. This raises a critical challenge in data science and analytics, namely, to understand, compute and analyze non-IID data and the *non-IIDness* in data and business [60, 175].

Non-IIDness (see Fig. 4.2c) caters for the *couplings* and *heterogeneities* in a data problem (see Fig. 4.2a), while *IIDness* ignores or simplifies heterogeneities and coupling relationships, as shown in Fig. 4.2b. Here *couplings* refer to any relationships (for instance, co-occurrence, neighborhood, dependency, linkage, correlation, or causality). *Heterogeneity* refers to the diversities within and between two or more aspects, such as entity, entity class, entity property (variable), process, fact and state of affairs, or other types of entities or properties (such as learners and learned results) appearing or produced prior to, during and after a target process (such as a learning task).

At a high level of abstraction, IID learning, i.e., learning IID data, shown in Fig. 4.2b, can be described as solving the objective function below to represent and learn the original data problem in Fig. 4.2a:

$$d_3 = ||O_3 - O|| \qquad (4.1)$$

where d_3 represents the distance (or similarity) between object O_3 and baseline O, and O_3 and O also represent the function (e.g., measures) calculated on objects O_3 and baseline O based on the same approach. In this case, we assume object O_3 follows the same distribution and is homogeneous to objects O_1 and O_2, and distance d_3 follows the same distribution as distances d_1 and d_2.

In contrast, non-IID learning, i.e., learning non-IID data by taking the non-IID assumption as shown in Fig. 4.2c, takes the following high level view of learning objective:

$$d_3 = ||O_3(r_{13}, r_{23}) - O(d_1, d_2)|| \qquad (4.2)$$

where d_3 represents the distance (or similarity) between object O_3 and baseline O, and the function O_3 is obtained on object O_3 by involving the influence of objects and their functions O_1 and O_2, while the baseline function O is also affected by its relationships d_1 and d_2 between objects O_1 and O_2 and the baseline O. In this case, object representations O_1, O_2 and O_3 may follow different distributions, and the same is true for distance representations d_1, d_2 and d_3.

4.5.2   Non-IID Challenges

Major challenges appear in handling complex non-IIDness hidden in *complex data* (here *complex data* particularly refers to data that has the comprehensive complexities discussed in Sect. 4.2, including complex coupling relationships and/or mixed distributions, formats, types and variables, and unstructured and weakly structured data). Learning visible and especially invisible non-IIDness can complement and assist the understanding of weakly structured and unstructured data.

In many cases, locally visible but globally invisible (or vice versa) non-IIDness are presented in a range of forms, structures, and layers and on diverse entities. Often, individual learners cannot tell the whole story due to their inability to identify the comprehensive complex non-IIDness. Effectively learning the widespread, various, visible and invisible non-IIDness is thus crucial for obtaining the truth and a complete picture of the underlying problem.

Computing non-IIDness refers to understanding, formalizing and quantifying the non-IID aspects, entities, interactions, layers, forms and strength. This includes extracting, discovering and estimating the interactions and heterogeneities between learning components, including the method, objective, task, level, dimension, process, measure and outcome, especially when the learning involves multiples of the above components, as in multi-methods or multi-tasks. We are concerned about understanding non-IIDness at a range of levels from values, attributes, objects, methods and measures to processing outcomes (such as mined patterns). Such non-IIDness is both comprehensive and complex.

We frequently only focus on the explicit non-IIDness which is visible to us and easy to learn. Work in hybridized multiple methods and the combination of multiple sources of data into a big table for analysis typically fall into this category.

In a learning system, however, explicit couplings and implicit couplings (as well as heterogeneities) may exist within and/or between aspects, such as entities (objects, object class, instance, or group/community) and their properties (variables), context (environment) and its constraints, interactions (exchange of

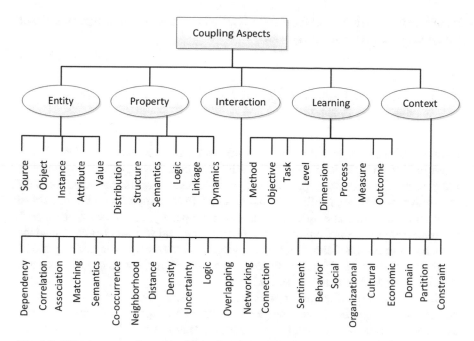

Fig. 4.3 Ubiquitous coupling relationships in coupling learning

information, material or energy) between entities or between an entity and its environment, learning objectives (targets, such as risk level or fraud), the corresponding learning methods (models, algorithms or systems) and resultant outcomes (such as patterns or clusters). This is illustrated in Fig. 4.3.

Coupling as a concept is much richer than existing terms such as dependence [307, 419], correlation [171, 232], and association [86, 201] would suggest, and involves similarity learning [146]. Classic research on dependence, correlation and association are much more specific, descriptive and explicit, often having specific meaning and ways of being measured.

More information about the definition, types, forms, and hierarchy of coupling, as well as its distinction from dependence, correlation, association and uncertainty can be found in Sect. 4.5 and in [59, 61].

4.6 Human-Like Machine Intelligence Revolution

Artificial Intelligence (AI) is experiencing yet another of its many boom cycles. The current AI boom is driven by different factors to those in previous AI history, and is more intense. Today's AI is predominantly data science-empowered. This section discusses the evolution of machine intelligence towards human-like machine intelligence through mimicking human intelligence, and the fundamental roles of data science in this evolutionary wave.

4.6.1   Next-Generation Artificial Intelligence: Human-Like Machine Intelligence

Human intelligence is embodied in the cognitive ability to comprehend ideas, understand and create logic, learn, invent and drive new concepts, reason and synthesize, imagine and think critically, process and produce new thoughts and ideas, and infer and derive knowledge, intelligence and wisdom from complicated and networked environments. These intelligent activities involve perception, consciousness, self-awareness, inference, reasoning, reflection, thinking, and volition. Humans have the capacity to create and learn language, engage in culture and societal interactions, remember and recall things, recognize patterns and exceptions, plan, explore, communicate, socialize, and make decisions.

Not all of these abilities and capabilities are possessed by, or can be implemented in, existing artificial intelligence systems. Artificial intelligence has until now been essentially driven by informatics, computing, learning and interaction theories and systems, including computational logic, planning, reinforcement learning, statistical/probabilistic representation, inference and learning, case-based reasoning, evolutionary computing, and deep search. These elements essentially cover a large proportion of existing popular research areas and targeted artificial intelligence, and can be summarized as

- *memory emulation*: including information extraction, encoding, storage, indexing, search and retrieval;
- *analytics and learning*: including statistics, machine learning, data mining, and pattern recognition;
- *interactions and recognition*: including speech, language, voice, image, vision and multimedia recognition;
- *simulation and optimization*: including neural computing, fuzzy and evolutionary computing, experimental design, simulation, and optimization.

The human-like machine intelligence revolution relies on building new mechanisms in machines or the fundamental revolution of existing machine intelligence. Core human intelligence that has not been well simulated in machines but which needs to be incorporated in next-generation artificial intelligence includes intuition, enthusiasm, curiosity, imaginary thinking, inspiration, creative thinking, hierarchical complexities and intelligence, and connectivity.

First, a critical capability of humans, starting from childhood, is to be *curious*. We often naturally and eagerly want to know what, how and why. *Curiosity* connects other cognitive activities, in particular, imagination, reasoning, aggregation, creativity and enthusiasm, which then often results in new ideas, observations, concepts, knowledge, and decisions. During this process, human intelligence is augmented. A critical task is to enable machines to generate and retain curiosity through learning inquisitively from data and generating curiosity in data.

Second, human *imaginary thinking* differentiates humans from machines, which have sense-effect working mechanisms. Human *imagination* is creative, evolving, and sometimes uncertain, and it cannot be generated by following patterns and pre-defined sense-effect mechanisms. The machine intelligence revolution needs to simulate human imagination processes and mechanisms before creative machines can be said to exist.

Third, existing knowledge representation, reasoning and aggregation and computational logics, reasoning and logic thinking incorporated into machines do not support curiosity and imagination, and machines are not *creative*. Giving machines the ability to think creatively is a crucial task, and existing computer theories, operating systems, system architectures and infrastructures, computing languages, and data management thus need to be fundamentally reformed. To create creative machines, one way is to simulate, learn, reason and synthesize from data and engage other intelligence in a non-predefined way, in contrast to existing simulation, learning and computation, which are largely predefined by default.

Further, the exploration of X-intelligence in complex data problems requires the *micro-meso-societal level* of hierarchical complexities and intelligence to be learned. A major future direction is to develop and enable *human-like* and "non-traditional" machine intelligence, progressing toward imaginary thinking (i.e., non-logical thinking) and new (networked) mechanisms for invisibility learning, knowledge creation, and complex reasoning and consensus building through connecting heterogeneous and relevant data and intelligence.

Lastly, *data-science thinking* (see more discussion in Chap. 3) needs to be built into data science systems and developed in data professionals. *Data-science thinking* is not only explicit, descriptive and predictive, but is also implicit and prescriptive. Data science thinking mimics human thinking by involving and synthesizing comprehensive data, information, knowledge and intelligence through various cognitive processing methods and activities. This reflects the requirements of developing thinking-based data science and engineering, although this has rarely been recognized in the data science community.

In pursuing the next-generation of artificial intelligence, our goal is to mimic human intelligence. The key challenge facing current artificial intelligence research is how to create a human mind [263] and implement human thought in machines.

4.6.2   Data Science-Enabled Human-Like Machine Intelligence

Artificial Intelligence (AI) is experiencing a significant revolution which is fundamentally driven by data science. The life of AI has been quite volatile, and AI is again becoming popular. This wave of AI recognition relies significantly on the research and development of a data-driven engine for more intelligent systems and decision-making. A typical example is the use of deep learning [123] for much better representation and learning of natural language processing and computer vision, and the application of reinforcement learning in DeepMind and AlphaGo [189]. The data-driven analytical and learning engine plays the core role in creating more intelligent systems and decision-making.

It is often debated whether machines could replace humans [375]. While it may not be possible to build *intelligent thinking machines* that have identical abilities to humans, big data analytics and data science are driving the revolution from logical thinking-centered machine intelligence to imaginary thinking-oriented "non-traditional" machine intelligence. This may be evidenced by the success of Google AlphaGo in beating Lee Sedol and Jie Ke [189], and the Facebook emotion experiment [261]. However, these examples essentially reflect the power of large scale data-driven deep representation, inference and search, none of which exhibits actual human-like imaginary thinking.

This transformation from learning, logic and inference-based machine thinking to human-like machine thinking, if the latter is able to mimic the above human intelligence well, may reform machine intelligence and significantly or even fundamentally change the current man-machine role and segmentation of responsibilities. One way to enable this transformation is to build a data-driven human-like machine by developing data science capabilities and collaboration with human beings to

- simulate the working mechanism of the human brain, in particular, human imaginary thinking and processing;
- learn and absorb societal and human intelligence hidden in data and business during the problem-solving process;
- understand unstructured and mixed structured data and intelligence, to enable extraction of these structures, and the conversion of unstructured data and intelligence to structured representation and models;
- understand qualitative problems and factors, and quantify qualitative factors and problems to form quantitative representations and models;
- observe, measure and learn human behaviors and societal activities, and evaluate and select the preferred behaviors and activities to undertake;
- synthesize collective intelligence to solve problems that cannot be handled by individuals;
- generate knowledge of knowledge (abstraction and summarization) and derive new knowledge based on implicit and networked connections in existing data, knowledge and processing;
- ask questions actively and be motivated by online learning and inspiration from certain learning procedures, objects and groups;
- be capable of creating and discovering new knowledge adaptively and online;
- gain insights and optimal solutions to address grand problems on a web-scale or global scale through experiments on massively possible hypotheses, scenarios and trials;

- provide personalized and evolving services and decisions based on an intrinsic understanding of personal characteristics, behaviors, needs, emotion and changes in circumstance.

Data science must play the fundamental role in involving, translating, and empowering the comprehensive X-complexities and X-intelligence into artificial intelligence devices and artificial life systems [328] before the existing design principles and working mechanism of computers and computing theories can be fundamentally replaced by another suite of computing devices and theories that extend beyond the von Neumann computer model and architecture [306].

4.7 Data Quality

Data quality is another important aspect of data complexities that plays a critical role in data problem-solving. Data quality [22, 251, 338] can be categorized from a number of perspectives, for example, in terms of the quality of the data source, data analytics, or data product. In general, data quality refers to the quality of data sources. In this section, we discuss the quality of data sources, data analytics and manipulation, and data systems and data products.

4.7.1 Data Quality Issues

Data science tasks involve roles and follow processes that differ from more generalized and standardized IT projects, because data science and analytics works tend to be data-driven, which goes beyond pre-defined processes, modeling, design and execution. Data science works are more creative, intelligent, exploratory, nonstandard, unautomated, and personalized. They have the objective of discovering evidence and indicators for decision-making actions. Since data science is data-driven, it inevitably involves data quality [208, 355, 443, 463, 464] issues. Data quality ultimately determines the quality of data analytics and manipulation, and the values of data science outputs and outcomes.

Data quality refers to the condition of the data, i.e., whether, how, and to what extent the data fit specific or general purposes, fulfill requirements, or portray actual scenarios, often in a given context. In general, higher data quality indicates better fitness for serving certain purposes, better satisfaction of requirements, or excellent portrayal of realities. Data quality may be assessed in terms of fitness of purpose for understanding business problems and improving operations, optimization, planning, and decision-support.

A *data quality issue* refers to a scenario or condition that affects or violates the direct and normal understanding and/or use of data. Data quality issues may appear in data, data analytics and manipulation, and data products.

Given a data science problem, we should not assume that

- The data available is perfect;
- The data always generates good outcomes;
- The outputs (findings) generated are always good and meaningful; or that
- The outcomes can always inform better decisions.

These assumption myths concern the quality of the data (input), the model, and the outcomes (output).

There are many different perspectives from which data quality issues can be discussed, for example, data validity, veracity, variability, and reliability, and social issues such as privacy, security, accountability, and trust, which must also be considered in data science and analytics. In so-called big data analytics, data validity, veracity, variability, reliability, objectivity, integrity, generalizability, relevance, and utility are particularly important. We briefly discuss these aspects below.

Validity refers to the closeness between the destination and target. *Data validity* and *analytics validity* determine whether a data model, concept, conclusion, or measurement is well-founded and corresponds accurately to the data characteristics and real-world facts, making it capable of giving the right answer.

Similarly, *data veracity* and *analytics veracity* determine the correctness and accuracy of data and analytics outcomes. Both validity and veracity must be checked from the perspectives of data content, representation, design, modeling, experiments, and evaluation.

Data variability and *analytics variability* are determined by the changing and uncertain nature of data, reflecting business dynamics (including the problem context and problem-solving purposes), and thus require the corresponding analytics to adapt to the dynamic nature of data. Because of the changing nature of data, the need to check the validity, veracity, and reliability of the data used and analytics undertaken is very important.

Data reliability and *analytics reliability* refer to the reproducibility, repeatability, and trust properties of the data used, the analytic models generated, and the outcomes delivered on the data. Reliable data and analytics are not necessarily static. Making data analytics adaptive to the evolving, streaming, and dynamic nature of data, business, and decision requests is a critical challenge in data science and analytics.

In addition, *data integrity* refers to the consistency across data and the accuracy obtainable across data analytics. *Data objectivity* is the need for analytical results and conclusions to be based on statistically sound models and valid data. *Data generalizability* means that data samples correctly and sufficiently represent the characteristics of the population, and that the analytics/manipulation models and results are widely applicable to various data, contexts and purposes. *Data relevance* is concerned with the degree to which the sampled data fits the learning purposes, requirements and actual situations. Lastly, *data utility* is concerned with the value, impact and benefit of data and analytics/manipulation; for example, the difference made in business revenue.

A number of different issues may be identified in any data science problem. Data quality issues are problem, data, analytics, and purpose specific. Data quality issues may appear along the lifecycle of data creation, processing, analysis, deployment, and curation. They may also occur in the lifelong period of data formation, analytical process, or data product building.

4.7.2   Data Quality Metrics

Data quality and data quality issues are judged in terms of qualitative and quantitative measures or metrics. *Data quality measurement* enables the perception, representation, quantification and assessment of the data's fitness to serve certain purposes in a given context, fulfill specific requirements, or align with actual scenarios.

Data quality may be measured by many different data quality indicators, metrics and measures according to the context and intended use of the data. A *data quality indicator* (or *data quality measure* or *data quality metric*) characterizes a specific condition of data fitness.

A large number of data quality indicators are involved in qualitatively or quantitatively characterizing data quality from different aspects, in various forms, for diverse purposes, and even according to specific meanings. Typical data quality indicators are accessibility, availability, validity, redundancy, accuracy or correctness, reasonableness or soundness, currency, relevance, completeness or totality, consistency, disagreement, timeliness, accuracy, actionablility, coherence, interpretability, and conformity. These indicators do not necessarily converge to a commonly agreed understanding of meaning, aspect, scope, goal, or criteria.

Examples of applying the above data quality indicators are the validity of data values, samples and attributes, the redundancy of attributes, the accuracy of data products such as identified findings, the relevance of data and features to a learning objective, the completeness or totality of data samples and features fitting a requirement, the appearance of missing and noisy values, the consistency or disagreement between different time periods of data distributions, the timeliness of the data value decay rate or various timestamps and the positions of those timestamps, the reasonableness and soundness of analytical results conforming to assumed logic, hypothesis and theoretical guarantee, the conformity and coherence of experimental results with theoretical analysis or of data and results to business requirements, the interpretability of analytical results, the accessibility of high dimensional or latent data complexities and characteristics such as relationships.

Typically, *data quality metrics* are categorized in terms of data source quality, data analytics quality, and data product quality.

Data source quality may be measured in terms of such indicators and metrics as noise rate, redundancy rate, completeness rate, consistency rate, decay rate, and relevance rate.

Data analytics quality may be measured in terms of (1) *learning performance* such as accuracy, precision (or positive predictive value), recall (or hit rate, true positive rate or sensitivity), and specificity (true negative rate), F1 score (the harmonic mean of precision and recall), and Matthews correlation coefficient [440]; (2) *theoretical performance* in reproducibility, robustness, and convergence; and (3) computational performance such as running time, scalability, and memory cost.

Data product quality is measured on application, software or system quality in terms of intrinsic data product, customer reflection, in-process quality, and maintenance quality [243]. Data product quality can be estimated by such measures as (1) intrinsic product reliability (e.g., functional defect density rate, mean time to failure, lines of code, or life of product), and (2) customer problems (e.g., problems per user month, number of license-months) and customer satisfaction (e.g., the IBM categories: capability, documentation/information, functionality, installability, maintainability, usability, performance, reliability, service, and overall; and the HP categories: functionality, usability, reliability, performance, and service; or satisfaction levels: very satisfied, satisfied, neutral, dissatisfied, and very dissatisfied).

4.7.3   Data Quality Assurance and Control

Data quality assurance ensures that the data fit expected purposes, satisfy certain condition, fulfill specific requirements, and/or portray actual scenarios. Data quality assurance is undertaken through certain methodologies, processes, and tools. In assuring data quality, *data quality control* refers to the process and operations of ensuring data quality satisfaction and the achievement of certain target data quality metrics.

Data quality assurance requires the implementation of technical tests, checks, verification and possible guarantee/proof of data authenticity, accountability, privacy, security, and trust. Quality assurance and enhancement relies on the quality of the assurance process, the policies for data, repository and code review, and if necessary, human judgment.

To assure and control data quality, the following aspects need to be addressed: *data quality technical assurance*, *data quality social resolution*, and *data quality ethical assurance* in relation to data repositories, data analytics and manipulation, and data product. Here we focus on data quality technical assurance, while leaving the discussion on data quality social resolution and data quality ethical assurance to Sect. 4.8.3.

Data quality technical assurance requires appropriate methodologies, methods, tools for testing, exception reporting, checking, monitoring, alerting, case management, and intervention in data quality issues.

Methodologies ensure that relevant, often domain and/or objective-specific policies, conceptual systems, code of conduct, and governance structure are created, applied, reviewed, and adjusted in a data ecosystem and in the life of the data and data product. Similar methodologies applied to software quality assurance [90, 173, 351] could be used as a basis for forming methodologies for data product quality assurance. However, alternative strategies, policies and processes are required for testing, reviewing, verifying, managing and intervention issues in (1) data repositories and data characteristics, including data validity, veracity, variability, and reliability; (2) data analytics and manipulation quality, including authenticity, soundness, transparency, reproducibility; and (3) data product quality, including reproducibility, immutation, usefulness, and actionability.

Methods are the appropriate and relevant processes, framework, models, and strategies for ensuring the proper implementation of data quality assurance methodologies. They may involve alternative perspectives and approaches; for example, lifelong process-based sequential test and review, outlier/anomaly-based identification and detection for exception discovery, or rule-based risk scenario checking and management.

The implementation of methodologies and methods for data quality technical assurance requires the development of corresponding tools, including systems, platforms, and application programming interfaces (API), as well as mobile applications and online services, for conducting, reporting, and data quality assurance intervention. Existing tools and systems for data cleaning, transformation and processing can play a role in assuring and enhancing data quality, typically in relation to missing values and noise in data, but additional efforts are needed to address other issues, and for data analytics and manipulation, and data products, as well as the provision of data quality assurance services.

Tools are required to address other data quality issues, including validity, veracity, variability, and reliability, especially in big data applications, such as multinational data applications and services, and in mobile services and social media applications.

Tools for resolving issues associated with data analytics and manipulation may have the functions of (1) providing open and shared mobile and online facilities for open analytics and open manipulation with sound and publicly available functions and methods embedded and changes tracked in the platform, (2) recording and reporting specific settings and parameter tuning in data analysis and manipulation to align with the explanation of outputs, (3) outputting models with relevant accountability, input and settings, and (4) identifying possible inconsistencies, conflict and exceptions with publicly available functions, methods and regulation rules.

For data quality social and ethical issue assurance, data owners, end-users and regulators (curators) need to jointly form a data governance team, process, compliance checklist, and mitigation strategies for the regulation, compliance, management, intervention, and enhancement of data quality. The data governance team looks after the definition, quantification, prioritization, identification, tracking, reporting and resolution of data quality issues.

4.7.4 Data Quality Analytics

Data quality analytics is a promising direction for characterizing, detecting and predicting issues in data. Data quality analytics may be undertaken with basic tools and tasks including data profiling, data standardization, data matching, and data monitoring, and more advanced tools and tasks such as data consistency test, data contrast analysis, data anomaly detection, and data change detection in the data lifecycle, specific phase or periodic procedure.

Data profiling characterizes, analyzes, and identifies important features, characteristics, relationships, structures, and issues in a dataset. Data profiling may be conducted by applying statistical/descriptive analytics methodologies, methods and tools, such as trend analysis, factor analysis, and distribution analysis. More advanced analytics may also be involved to profile data characteristics and complexities, such as clustering. Data profiling produces a standard (or regular) understanding and description of data characteristics and complexities, which is usually quantified in terms of data factors and indicators.

Data standardization builds benchmarks for measuring data quality, and ensures that data conforms to data quality specifications and rules. *Data matching* connects and compares multiple sources or editions of data to find inconsistencies across data inputs based on identity matching. *Data monitoring* keeps track of issues that are often known or visible over time, based on a pre-defined checklist (rules), and identifies and reports variations.

The *data consistency test* checks the consistency of data over time in terms of data, analytics, results and data systems. *Data consistency* may refer to the following conditions: (1) two or more data snapshots along different timestamps can be perfectly matched, which is also called *point-in-time data consistency*; (2) data analytics and manipulation models and results are simultaneously visible and constant to all parties attempting to access the results, also called *analytics consistency*; (3) data products and systems driven by the same data and same data analytics and manipulation methods are the same or constant.

Contrast analysis can be used to detect, analyze and quantify data quality issues. *Data contrast analysis* may be conducted on two or multiple data sample sets, data manipulation instances, or data systems to verify their consistent conformation to data quality metrics.

Data anomaly detection identifies exceptions, significant inconsistencies, and anomalies in data, analytics and/or systems. *Data anomalies* are the significant inconsistencies from the baseline or conditions satisfied by the majority of the data. Data anomalies may appear in attribute values, sampling methods, sample type, structure or distribution, analytical design or results, or data products. Data anomalies are measured by specific data outlierness measures in terms of attribute values, attributes and objects.

Data evolves and thus may undergo significant unexpected change. *Data change detection* seeks to identify significant dynamics that violate the assumed conditions in the data, analytics and/or data system, such as time, range, structure, relation, density, and distribution. The change detection of data quality identifies significant inconsistencies, alterations, or mutations measured by significant variations in the data quality metrics.

Data factors and *data indicators* need to be defined to quantify the measurement of data quality and describe the quality of the data. They are used in quality check and control methods, and include profiling, standardization, matching, monitoring, consistency testing, contrast analysis, anomaly detection, and change detection. A *data quality indicator* may quantify a data condition in terms of data types, distribution, attribute types, attribute lengths, attribute cardinality, data granularity, attribute value statistics, data format patterns, data content distributions and trend, factor distribution, the relationships between attributes, data sources, and domains, and the types, forms, and cardinality of those relationships.

4.7.5   Data Quality Checklist

A data quality compliance and assessment checklist consists of selected data quality metrics or indicators and their applicable methods, resolutions, conditions, contexts and scenarios. A checklist is specified by a data science team for a given data problem or data system, with the participation of business and domain experts. Each issue should be defined in terms of the nature of the problem, its context, its resolution, how that resolution is to be implemented and with what level of priority. Checklists have been developed for software quality assessment [293], but very limited information is available on data quality assessment [148, 404].

Sample data quality checklists are given below.

A. Data source quality checklist:

1. Is the data valid?
2. Is the data complete?
3. Is the data relevant?
4. Is the data consistent?
5. Is the data timely?
6. Is the data publicly accessible?
7. Are there any missing values?
8. Are there any noisy values and features?
9. Are there any redundant features?
10. What data characteristics have been captured and measured?
11. What data complexities have been quantified and addressed?
12. What are the data factors and indicators for describing and representing data characteristics and complexities?
13. Are the data collection methods sound?
14. Are the data sampling methods biased?
15. What are the main limitations in the data?

16. What are the main limitations in the data collection?
17. What are the main limitations in the data sampling?

B. Data analytics and manipulation quality checklist:

1. Are the results valid?
2. Are the results accurate?
3. Are the results reasonable?
4. Are the results statistically confident?
5. Are the results robust?
6. Are the results reproducible?
7. Are the results interpretable?
8. Are the results actionable?
9. Are the results surprising?
10. Are the results biased or misleading?
11. Do the results incur false positive or false negative prediction?
12. Is the analytics/manipulation technically sound?
13. Is the analytics/manipulation scalable?
14. Is the analytics/manipulation efficient?
15. Is the analytics/manipulation convergent to optimal outputs?
16. Does the analytics/manipulation fit its assumption and hypothesis?
17. Does the analytics/manipulation fit the data sampled?
18. Why does the analytics/manipulation work well?
19. Why do the results make sense of the data?
20. What are the main limitations in the fitness between data and data analytics/manipulation?
21. What are the main limitations in the data analytics/manipulation?

C. Data product quality checklist:

1. Does the data product satisfy general software quality requirements?
2. Does the data product satisfy general application quality requirements?
3. Does the data product satisfy general software test requirements?
4. How many functional defects are identified?
5. What is the mean time to failure?
6. How many lines of code are there in a month?
7. What is the life of the data product?
8. How many problems are identified per customer month?
9. How usable is the data product in relation to customer profile?
10. How reliable is the data product?
11. What is the customer satisfaction rate of the data product?
12. What is the unresolved customer dissatisfaction rate of the data product?

4.8 Data Social and Ethical Issues

Technical professionals naturally pay attention to data quality issues; however social scientists and ethics professionals often identify social and ethical issues when they look at data. This section thus discusses data social issues, data ethical issues, and data quality social resolution and data quality ethical assurance.

4.8.1 Data Social Issues

Domain-specific data and business are embedded in social contexts and integrated with social issues. Data social issues do not refer to social problems themselves, such as social media or welfare problems, but rather the societal aspects related to or embedded in data and data ecosystems.

Data social issues refer to the social concerns that need to be considered in data, data analytics and manipulation, and data products. Data social issues can be categorized in terms of data ownership, data sharing and use, and data values.

In terms of data ownership, data science tasks typically involve such social issues as *ownership*, and *openness*. The essential features of data sharing and use are *privacy*, *security*, *accountability*, and *trust* of data, modeling, and products. *Data values* are concerned with socio-economic benefits, utility, and impact.

Data ownership refers to legal rights and control over data creation, acquisition, editing, sharing, usage, distribution and governance. Four roles often play a part in data creation, acquisition, use, dissemination and governance: data producers (data creators, data suppliers), data service providers (data compilers, data managers, data acquirers, data processors, data delivery agents, data packagers), data consumers (end-users), and data governors (data curators). Data service providers often collect, acquire and own data, such as in medical services and social media services, and people who produce (as in social media) or contribute (as in health and medical testing) data may not own their data. This is quite different from classic scenarios in which data creators own their data; for example, book authors usually retain copyright of their written content, although they often license this copyright to publishers.

Typical ownership may reside in data creators, data consumers, data service providers, enterprise, funding organizations, data decoders, data packers, purchasers and licensers. The ownership paradigm [277] may also include everyone as owner.

Two other important concepts are data sovereignty and data residency. *Data sovereignty* means that data is subject to the laws and regulations of the country in which the data is located. This is particularly relevant in data business (data economy) conducted via globalized cloud computing, multinational online data storage and online services, in which geopolitical boundaries are broken down. Typical legislative matters concerning data sovereignty and data residency relate to which country's legal system applies to privacy obligations, data security, and data breaches.

Countries have differing regulations and policies on data protection, privacy, safeguards, and breaches, thus it is important to understand this before using data storage services in a particular country. General data protection regulation can be found in [435] and a list of national data protection authorities are in [438].

As discussed in Sect. 6.7.1.3, data openness is an important societal feature of the data world and data economy. *Data openness* differentiates the data world and data economy from classic and conventional systems and economic forms and enables free access and transparency with limited or no restrictions. Openness may cause social issues (as well as legal and ethical issues) when it is not properly applied or governed by effective principles and regulations, such as breaches of data privacy, lack of accountability, and misinformation, which damages credibility and trustfulness in data acquisition, sharing, distribution and use. Below, we discuss these important concepts: privacy, security, accountability, and trust.

Data privacy addresses the challenge of collecting, analyzing, disseminating, and sharing data and analytics while protecting personally identifiable or other sensitive information and analytics from improper disclosure. Protection technology, regulation, and policies are required to balance protection and appropriate disclosure in the process of data manipulation.

Data security protects target objects from both destructive forces and the actions of unauthorized users, such as improper use or disclosure. It addresses not only privacy issues but also such aspects as software and hardware backup and recovery. Data and analytics security also involves the development of political or legal regulation mechanisms and systems to address such issues.

Data accountability refers to an obligation to comply with data privacy and security legislation and to report, explain, trace, and identify the data manipulated and analytics conducted to maintain the transparency, traceability, liability, and warranty of both the measurement and results, as well as the efficacy and verifiability of the analytics and protection.

Data trust is the belief in the reliability, truth, or ability of data and analytics to achieve the relevant goals. This involves the development of appropriate technology, social norms, ethical rules, or legislation to ensure, measure, and protect trust in the data and analytics used and to engender confidence in the corresponding outcomes and evaluation of analytics.

In addition, *data values* are the benefit, impact, and utility derived from analyzing and manipulating data that contributes to problem solving. Data values may be evaluated from intellectual, economic, social, cultural, or environmental perspectives. Qualitative and/or quantitative measures can be generated to estimate or sum such values.

Both direct and indirect data values exist. *Direct data values* refer to visible and observable benefit, impact and utility. Examples of direct data values are the market value of a social media website like Twitter or Facebook, or the dollar savings achieved by applying a fraud detection model in online banking risk management. *Indirect data values* refer to hidden or third-party benefit, impact and utility; for example, insights discovered from data might be indirect but could still be very valuable. Both direct and indirect data values can be estimated by developing corresponding valuation methods, although it is often more difficult to accurately estimate indirect values.

4.8.2 Data Science Ethics

Ethics has been a fundamental and integral part of the data science journey. *Data science ethics* concern the code of ethics and codes of conduct in relation to data and data-centric behaviors, products and applications.

Data ethical issues are both technically and socially important in the data world, and can especially be seen in open data, open access, open innovation and open science. The term *data science ethics* refers to ethical, or moral, norms and rules, as well as the codes of conduct applicable to data and data-centric analytics and manipulation and products during the process of their creation, acquisition, review, sharing, distribution, application, reproduction and curation.

Data ethics [132, 167, 273, 296, 315, 378] is a new branch of ethics that builds on computer and information ethics [49, 166, 291]. Compared to information ethics and computer ethics, data ethics address ethical issues at an even lower level of the data, information, knowledge, intelligence and wisdom (DIKIW) framework discussed in Sect. 2.3 and on aspects from computer and information to even the original low-level data.

Data ethics complement the handling of data legal and social issues. Data legal issues in data ecosystems are addressed by laws, regulations, governance policies, and sovereignty-related legal systems. Data ethical issues are not governed by data legal systems but by moral mechanisms.

Data ethical issues are broad, ad hoc, domain and scenario-specific, and they are often complex as they are frequently integrated with legal and social matters. Discussion on data ethics must take account of the various perspectives on different ethical aspects and their diverse backgrounds. In [167], data ethics are mapped within three axes of conceptual space: the ethics of data, the ethics of algorithms, and the ethics of practices. In this work, we discuss various ethical matters in relation to three layers and perspectives of data science: data, data analytics and manipulation, and data products. Privacy, Identity, Security, Trust (reputation), Responsibilities (accountability), and Ownership (PISTRO) are major data science ethics issues. The components of PISTRO in the data science sense are explained below.

- *Privacy* is the state of maintaining seclusion without being maliciously concealed.
- *Identity* is the state, fact or condition of having unique qualities or remaining the same.
- *Security* is the freedom and protection from harm.

- *Trust* is the confidence in data and its manipulation.
- *Responsibility* is the state or fact of being answerable or accountable.
- *Ownership* is the state, fact or right of possession.

Data ethics discussed from a broad perspective relate to many extended aspects of the above important matters. Examples of typical data ethical issues are

- the integrity, reproducibility and reliability of data, methods and processes, data philanthropy, and validity including an informed representation of possible limitations, defects or bias and the adverse consequences of failure;
- the trust, transparency, fairness, and appropriateness of data, processes, methods and results, and the resolution of conflicts of interest;
- anonymity, confidentiality, accountability, sensitivity, consent, user privacy, secondary use, antisocial content disclosure, possible illegal identification of individuals, improper disclosure of personal data;
- the moral responsibilities to science, the public, clients and funding bodies; respect of human rights, social acceptability and social preferability;
- negative targeting, including malicious test of security and privacy, subjective human judgment including reputational damage, and negative public sentiment about branding;
- discrimination in collecting, processing and using sensitive data and decisional autonomy;
- group discrimination, and group-targeted violence.

4.8.3   Data Ethics Assurance

The achievement of data science values for public and social good relies on the assurance of data science ethics.

Assuring data ethics is a double-edged tool which has to tackle potential conflict and requires a balance and tradeoff between harnessing data values and enforcing ethical compliance. Both the oversight and violation of data ethical norms and the overemphasis and empowerment of rigid data ethics assurance could have a negative impact on harnessing the values and development of data science. A suitable balance requires an appropriate tradeoff between the professional conduct and over- or under-conduct on data. *Professional conduct* reflects the most appropriate respect of data ethics while maximizing data values, whereas *data under-conduct* does not sufficiently respect or follow the data code of ethics, which may cause breaches of the code of conduct. *Data over-conduct*, in contrast, applies the code of data ethics too rigidly, which may eventually downgrade the values of data.

Data ethical assurance requires a holistic ethical conceptual space, a comprehensive and consistent and inclusive ethical framework, balanced enforcement, and a balanced auditing process and system. These are created, implemented and audited in line with individual, group, area-specific, and organizational scenarios, purposes, objectives, requirements, and outcomes. Such assurance plans and actions should also consider ethical issues that may be associated with stakeholders (such as clients, customers, and sponsors), stakeholders' associates, or their data.

The above data ethical assurance systems are usually instantiated into codes of conduct and codes of ethics for specific or all data science roles (as discussed in Sect. 10.3), processes (e.g., analytics), and/or aspects (e.g., data quality) to address social and ethical issues, such as data quality social resolution and data quality ethical assurance.

Examples of codes of conduct/ethics for data and analytics, or broadly information-related ethics, are

- the Ethical Guidelines for Statistical Practice of the American Statistical Association [13], which focuses on the responsibilities of different roles and stakeholders;
- the Data Science Code of Professional Conduct of the Data Science Association [134], which highlights the responsibilities of data scientists in handling client relationships, and
- the Code of Ethics for Certified Analytics Professionals of INFORMS [228], which emphasizes the promotion of professional standards and integrity.

The following are examples of misconduct [134] (1) commit a criminal act related to the data scientist's professional services; (2) engage in data science involving dishonesty, fraud, deceit or misrepresentation; (3) engage in conduct that is prejudicial to methods of science; and (4) misuse data science results to communicate a false reality or promote an illusion of understanding.

In summary, the above exemplars of codes of ethics cover the following key aspects:

- the integrity of data and related methods and practices;
- the duties and responsibilities in different data roles (especially analysts and data scientists) in coping with client and stakeholder relationships;
- the quality of data, data products and practices; and
- possible misconduct and conflict in the executing the codes of conduct for data.

4.9 The Extreme Data Challenge

Different types and levels of analytical problems trouble the existing knowledge base, and we are especially challenged by the problems in complex data and complex environments. Our focus on data science research and innovation concerns what we call an *extreme data challenge* in data science and analytics. The *extreme data challenge*, as illustrated in Fig. 4.4, seeks to discover and deliver complex knowledge in complex data, taking into account complex behaviors within a complex environment to achieve actionable insights which can inform and enable decision action-taking in complex business problems that cannot be better handled by other methods. In this context, *extreme* refers to

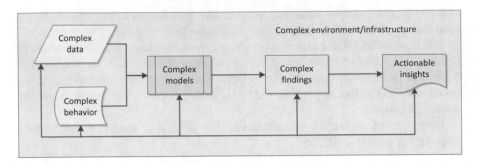

Fig. 4.4 A general framework of the extreme data challenge

(1) the essential combination of many aspects of complexity: data, behavior, environment, model, findings, and insights, which exceed existing hybridization theories and methods;
(2) the critical degree of complexity in each aspect that exceeds existing capabilities and capacities; and
(3) the level of complexity that consists of combining multiple relevant aspects that extend beyond existing data theories and methods.

This general extreme data challenge also has the following complex nature in terms of the highly challenging aspects of a complex data science problem. This type of complex data science problem represents the critical future direction of data science research and innovation.

- *Complex data* Data consists of or is embedded with complex data and associated characteristics and complexities (as discussed in Sect. 4.2 on X-complexities and Sect. 4.3 on X-intelligence; for more information, see also [64] and [62]);
- *Complex behaviors* Hierarchical data entities are associated with complex activities, relationships and dynamics (as discussed above in relation to behavior complexity and intelligence; for more information about complex behaviors, see discussion in [60, 74, 416]);
- *Complex environments* The complex data and behaviors concerned are embedded with, and interact with, sophisticated contexts which contain domain-specific, organizational, social, environmental and human-related complexities and intelligence;
- *Complex models* The models (methods) for understanding and addressing the data, behavior and complexities, and for involving and learning the various intelligences in a complex environment, must be complex (see the relevant discussion on learning complexities and decision complexities). This requires the

involvement, representation, modeling and integration of relevant complexities and intelligences;[1]

- *Complex findings* The data-driven findings uncover hidden but technically interesting and commercially-influential observations, indicators or evidence, hypotheses, statements, or presentations; and
- *Actionable insights*: The insights identified indicate the next best or worst situation and inform the corresponding next best actions and strategies for executing effective business decision-making (see more discussion on actionability in [54, 77].

Many real-life problems fall into this level of complexities and challenges, as the extreme data challenge shows, and they have thus far not been addressed well. For example,

- understanding the financial manipulation behaviors of large numbers of investors from different countries, acting in different markets, and working for different stakeholders, where multiple intentions and objectives, markets, manipulation strategies and methods may be involved. Complex interactions, relationships and conflict-resolving negotiations exist between markets, investors and stakeholders. Within this pool of large-scale cross-capital markets and internationally collaborative investors [74], each of the manipulators plays a role by connecting information from the underlying markets, social media, other financial markets, socio-economic data, and policies.
- predicting local climate change and effect by connecting local, regional, and global climate data, geographical, vegetational and agricultural data, and other relevant information [157].
- online, dynamic and proactive prediction of organized cyber-security attacks across multiple networks and media channels that are manipulated by groups of hackers, strategists, intelligence forces, and media resources. In this case, it is necessary to collect and select multiple sources of information from various channels, determine the roles and responsibilities to be shared in the campaign, and understand sophisticated workload sharing and cooperation.

Many global and cross-country social, economic and environmental problems fall into this level of complexities. They involve complexities related to relevant data, behavior, and environment/context.

4.10   Summary

We live in the era of data science and analytics. The possible disciplinary transformation and revolution of data science creates unique opportunities for breakthrough research, cutting-edge technological innovation, and significant new challenges.

---

[1]More discussion on involving and synthesizing various complexities and intelligences is available in Sects. 5.3.2 and 5.3.3.

These need to be explored from the perspective of the intrinsic complexities and nature of data science problems, with the aim of exploring original challenges and issues in data science research, innovation and practices.

This chapter has outlined some of the major challenges and directions in complex data science problems from both data-driven and system complexity perspectives. Our discussion has extended beyond data alone to explore an overall picture of complex data problems by considering them as complex systems. It is evident from this discussion that data science is driving a significant paradigm shift in scientific research and innovation, and the future and impact of data science may be unpredictable.

The new science known as data science seeks to address these challenges, and for this, a data science discipline must be established. In Chap. 5, we discuss the disciplinary gaps in existing sciences for handling X-complexities and X-intelligence in data science problems, and the methodologies, disciplinary framework and some major research issues that apply to the field of data science.

Chapter 5
Data Science Discipline

5.1 Introduction

What forms the data science discipline?

This chapter presents a comprehensive view of this important and challenging question, mainly from the research and disciplinary development perspective. The initial discussion is motivated by the disciplinary gaps between data potential and the limited existing disciplinary capabilities. Three philosophical methodologies: reductionism, holism and systematism, are then discussed, which are useful for guiding the thinking and problem-solving process in handling complex data science problems. A research map is presented to summarize the main research directions and specific issues in the data science family, and we look at how systematic research approaches may be helpful in addressing key research issues.

In addition, we highlight several fundamental areas which we believe are critical components of data science research.

5.2 Data-Capability Disciplinary Gaps

The discussion in Chap. 4 on handling the X-complexities in sophisticated data problems by involving X-intelligence to achieve X-opportunities pave a pathway toward the disciplinary development of data science. There are significant disciplinary gaps between addressing the X-complexities and achieving the X-opportunities however. The rapid increase in data and data-based applications and services is widening the already significant gaps between what is in the data and how much we can understand.

Figure 5.1 empirically illustrates the *growing disciplinary gaps* between the growth of the X-complexities and *data potential* and the development of state-of-the-art *disciplinary capabilities*. Such gaps have increased in the past 10 years with

© Springer International Publishing AG, part of Springer Nature 2018
L. Cao, *Data Science Thinking*, Data Analytics,
https://doi.org/10.1007/978-3-319-95092-1_5

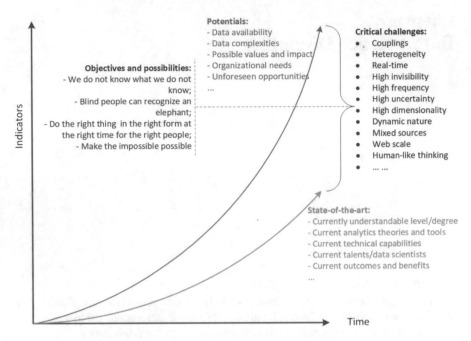

Fig. 5.1 Growing disciplinary gaps between data potential and disciplinary capabilities

the significant increase of data quantification and applications and the exponential increase in the imbalance between data potential and state-of-the-art capabilities.

Data potential is indicated by key factors and aspects including potential values and their impact, data availability, unforeseen X-complexities, X-intelligence and X-opportunities from data, and the possibility of satisfying organizational needs. *Data disciplinary capabilities* refer to currently understandable levels of addressing X-complexities and X-intelligence, current data theories and methods, as well as technical capabilities, the skills of existing data talent and data scientists, and the outcomes and benefits achievable from data by existing theories and methods.

The gaps between data potential and data disciplinary capabilities may be evident in a variety of scenarios, for example,

- between data availability and the currently understandable level, scale, and degree of the data;
- between data complexities and the currently available analytics theories and tools;
- between data complexities and the currently available technical capabilities;
- between possible data values and data impact and the currently achievable outcomes and benefits;
- between organizational needs and the currently available talent (that is, data scientists); and

- between potential opportunities and the outcomes and benefits currently achievable.

These gaps are widening with the rapid growth of data potential and the slower development of data disciplinary capabilities, contributing to the critical challenges facing existing disciplinary capabilities; for example, the development of human-like thinking in data products, understanding subjective data, handling complex hierarchical explicit and implicit coupling relationships, modeling hierarchical diverse heterogeneities, tackling real-time data and satisfying real-time business decision needs, and modeling high invisibility in X-complexities and X-intelligences. Many other challenges currently confront the relevant communities, such as the high levels of frequency and uncertainty of data, extremely high dimensionality, the dynamic nature and mix of complex data sources, and handling Web scale of data.

There is a shortage of effective theories and tools for handling these critical challenges. For example, a typical challenge in complex data concerns intrinsic complex coupling relationships and heterogeneity, forming data that are not independent and identically distributed (non-IID) [60], which cannot be simplified in a way that can be handled by classic IID learning theories and systems. Other examples include the real-time learning of large-scale online data, such as detecting online shopping manipulation and making real-time recommendations on high frequency data in the "11-11" shopping seasons launched by Alibaba, or identifying suspects in an imbalanced and multisource data environment such as fraud detection in high-frequency trading in stock markets.

To manage the gaps, it is necessary in the development of data science discipline to address the following high-level disciplinary objectives:

- Understanding "we do not know what we do not know" in data;
- Enabling the scenario "blind people can recognize an elephant";
- Achieving the scenario "doing the right thing in the right form at the right time for the right people"; and
- Making the impossible possible.

5.3 Methodologies for Complex Data Science Problems

The discussion thus far illustrates that the problem-solving system of a complex data science problem is a complex intelligent system [62, 335]. Building such a complex intelligent system requires effective methodologies to understand, specify, quantify, manipulate and utilize the underlying X-complexities and X-intelligence. In this section, we discuss appropriate methodologies for understanding and building complex data science systems.

5.3.1   From Reductionism and Holism to Systematism

In this section, three major scientific philosophies are briefly discussed: reductionism, holism, and systematism. They represent different ways of thinking in understanding and handling complex problems. When a data science problem is complex, neither reductionism nor holism is sufficient; systematism is essential. This drives the need for the appropriate selection of research methodologies in data science.

5.3.1.1   Bottom-Up Reductionism

Reductionism is a methodology that believes a system is equivalent to the sum of its parts, and that a system can be equivalently understood by exploring its individual constituents. In this methodology, the complexity of a data system is decomposed into smaller complexities in data subsystems or smaller (even atomic) building blocks in a data system.

Classic data analysis and engineering follow this methodology. The characteristics and complexities in a data set are modeled as the smallest possible entities or particles, namely data points or data objects and their attributes. Reductionism targets the understanding of object behaviors at a very low level and then sums the findings from each object to build a complete understanding of the higher-level data subsets.

Reductionism may be more applicable for simple data systems, since such systems experience no significant changes, and the interactions and coupling relationships between system constituents may be ignored; that is, data objects and attributes are assumed to be independent. As this assumption does not fit the reality of data, the summation of individual object behaviors cannot form a genuine picture of a data set.

Real-life data problems cannot be effectively understood by reductionism alone. Real-life data problems often consist of large scale system constituents, heterogeneous components, and various types and levels of coupling relationships between system constituents and components, as well as dynamic evolution driven by nonlinear and uncertain factors.

Data problems related to natural systems, ecosystems, and human society typically consist of such phenomena as emergence, interaction, and feedback loops across data subsystems and data samples. For example, a dynamically evolving data system will see additional objects, relations, structures and distributions that do not exist in the original system. Modeling and integrating current data object relations cannot represent the system's status in its next moment.

5.3.1.2 Top-Down Holism

Holism is the belief that neither the behavior nor the properties of a data system can be deduced from its constituents (subsets, objects, and attributes) alone. Holism holds that a system functions as a whole and its properties cannot be viewed as merely a collection of component parts.

Holism has been particularly appreciated by researchers in social science and systems science, who adopt strong systems thinking and derivatives, such as emergent behaviors, chaos theory and system complexities.

Unexpected data characteristics and behaviors in a data system may have surprising effects on the system whole and on the collection of parts. Accordingly, the characteristics and complexities of a data system as a whole cannot be modeled or predicted from an understanding, however good, of all its constituents. A holistic approach such as *complexity theory* [62, 294, 442] perceives the need to explore unpredictable characteristics and behaviors that emerge from individual data subsets and objects in a data system.

When a data system has data subsystems or objects on a large scale, even weak couplings and interactions can evolve into significant unexpected emergent behaviors. Complex data systems often have other typical system complexities including openness, human involvement, hierarchy, nonlinearity, and uncertainty. These may further impact on the hierarchical couplings and interactions and make a system more unpredictable.

Modeling the emergence of surprising behaviors or phenomena in complex data problems is challenging. It is necessary to model the sophisticated coupling relationships hierarchically embedded in data entities as a whole rather than individually. This is a typical feature of complex data systems, in which visible and invisible couplings and interactions connect data subsystems, groups of objects, and single objects to form an organic whole.

5.3.1.3 Integrative Systematism

Systematism embodies the concept that (1) a good understanding of system members not only contributes to forming an overall picture of the whole, but also reduces the complexity of understanding and makes a complex problem comprehensible or solvable; and (2) a complex system is more than the sum of its parts as a result of its system complexities and unexpected and unpredictable system characteristics, thus special theories and tools need to be developed to address such complexities and characteristics. Although the sum of parts is not equal to the whole, collections of parts do contribute to the whole.

Systematism aims to integrate the valuable qualities of holism and reductionism and to form a loop for systematic thinking and exploration by integrating bottom-up and top-down approaches. It is not possible to obtain a complete understanding of a complex problem in a one-off loop. Hence, the loop may appear as an iterative and qualitative-to-quantitative progression.

The systematological process may start from a top-down understanding of major characteristics, often in a qualitative way, by which the main challenges, core functions and functioning components may be grasped. Once a basic understanding (which may be wrong or biased) of high-level system characteristics and complexities is achieved, the focus will move to quantitatively understanding those major (or likely to be major) components (subsystems) through a reductionist approach.

Understanding the major components helps to deepen the comprehension of system complexities and to adjust the original "estimation" (a guess which may or may not be correct) to achieve a more appropriate understanding of the whole and parts and their interactions and coupling relationships. Several rounds of these deepening and adjusting actions may be conducted before a confident picture about the whole and parts and their relations is extracted. Importantly, findings in each step are accumulated and synthesized, with conflicts or diffused cognition mitigated or managed to build a common consensus.

5.3.1.4 Appropriate Methodological Adoption

The respective focuses and advantages of top-down-oriented holism and bottom-up-based reductionism only address certain challenges in complex systems. Both holism and reductionism thus have only limited advantages and tools for understanding and explaining partial system complexities.

Reductionism is a very mature methodology, particularly in scientific and social areas, with many effective tools and theories as well as successful case studies. By focusing on decomposing a complex whole into parts, reductionism has the advantage of reducing system complexity and understanding the functioning of individual parts, in contrast to the focus of holism. The strengths of reductionism are very necessary for analyzing, designing, and implementing complex systems, as most existing engineering technologies, tools, and systems have been built on reductionism.

Holism, however, is good at creating an overall view of the whole rather than its parts, and centers on system-level complexity, providing the means to capture system-level characteristics such as emergence and nonlinearity, as well as the underlying drivers, which include interactions and coupling relationships between parts. These features are not central in reductionism. The challenges faced by holism may lie in its limitations in understanding local functioning and composition mechanisms and its simplified way of capturing concrete system particles and predictable behaviors. Such needs are widely apparent in simple systems and in the subsystems of complex systems.

Due to the limitations of each philosophy, the outcomes from the applied methods may be misleading, incomplete or even incorrect. Assuming that the result produced by reductionism-based analytics and learning is correct may overlook the whole picture of system characteristics and complexities in a data science problem. Adopting only holism-based research may neglect low-level features that are often important in data-driven discovery.

Whether a holistic or a reductionist approach is appropriate is very much dependent on the status, characteristics, scale, and other system complexities in a specific data system or at a specific stage. When a data problem is very complex and has a number of levels and aspects of invisible characteristics and complexities, it is essential to combine the strengths and advantageous tools of holism and reductionism. The iterative process of qualitative-to-quantitative systematism is thus essential for complex data problems.

5.3.2 Synthesizing X-Intelligence

Successfully cracking the extreme data challenge discussed in Sect. 4.9 relies on the effective transformation of the comprehensive complexities highlighted in Sect. 4.2 and the X-intelligence examined in Sect. 4.3. Here we discuss how X-intelligence can be used and synthesized for complex data science problem-solving.

The use of X-intelligence may follow one of two paths: *single intelligence engagement* or *multi-aspect intelligence engagement*. An example of single intelligence engagement is the involvement of domain knowledge in data mining and the consideration of user preferences in recommender systems. This also applies to simple data science problem solving and systems.

In general, multi-aspect X-intelligence is used for complex data science problems, which are usually very challenging but inevitable in complex data and analytics. As shown in Fig. 5.2, it is necessary to integrate every type of intelligence in a complex data system, despite the challenges of involving and modeling the X-intelligence.

The performance of a data science problem-solving system is highly dependent on the effective recognition, acquisition, representation and integration of relevant intelligence and indicative human, domain, organization and society, network and web factors. For this, new methodologies and techniques need to be developed. The theory of *metasynthetic engineering* [62, 333] and the approach to the *integration of ubiquitous intelligence* may provide useful methodologies and techniques for synthesizing X-intelligence in complex data and analytics.

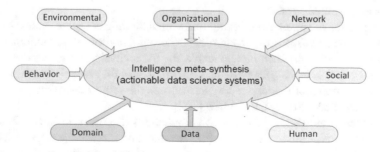

Fig. 5.2 Synthesizing X-intelligence in data science

From a high level perspective, the principle of intelligence meta-synthesis [62, 335] is to involve, synthesize and use ubiquitous intelligence in complex data and its environment to discover actionable knowledge and insights [77].

As an example of integrating diversified intelligence for complex knowledge discovery problems, *domain driven data mining* [77] has been proposed for introducing ubiquitous intelligence into data mining. It advocates a comprehensive process of interaction and integration between multiple kinds of intelligence, as well as encouraging the emergence of intelligence to deliver actionable knowledge. Domain driven data mining achieves this goal by:

- Properly understanding data characteristics as the most important task in analytics,
- Acquiring and representing unstructured, ill-structured and uncertain domain and human knowledge,
- Supporting the dynamic involvement of business experts and their knowledge and intelligence in the analytics process,
- Acquiring and representing expert thinking, such as imaginary thinking and creative thinking, in group heuristic discussions during data understanding and analytics,
- Acquiring and representing group and collective interaction behaviors and their impact, and
- Building infrastructure that supports the involvement and synthesis of ubiquitous intelligence.

5.3.3   Qualitative-to-Quantitative Metasynthesis

Systematology methodology relies on forming a qualitative-to-quantitative cognitive process, which can be achieved according to the theory of *metasynthesis* of X-intelligence [62, 335].

The process of solving complex data science problems through intelligence metasynthesis involves complex system engineering. Often, several aspects of complexities and intelligence are embedded in the data, environment and problem-solving process, therefore it is preferable to handle a data science problem-solving system as a complex system.

The theories of *system complexities* and corresponding complex system methodologies of *systematism* (or *systematology*, the combination of reductionism with holism) [62, 335] may then be applicable for the analysis, design and evaluation of complex data science problems. This also indicates that complex data problems may not be sufficiently addressed by existing engineering methodologies such as the analytics methodology CRISP-DM [104].

For those complicated data science complexities discussed in Sect. 4.2, simply using the *reductionism* methodology [62] for data and knowledge exploration may not work well. This is because the problem may not initially be clear, certain, specific and quantitative, thus it cannot be effectively decomposed and analyzed. Further, the analysis of the whole does not equal the sum of the analysis of the parts, which is the common challenge of complex systems [335]. An appropriate problem understanding and solving process is thus required to comprehend the problem from the system complexity perspective, before a more specific, certain and quantitative view can be achieved.

When a data science problem involves large scale objects, multiple levels of sub-tasks or objects, multiple sources and types of data objects from online, business, mobile or social networks, complicated contexts, human involvement and domain constraints, it presents the characteristics of an open complex system [62, 335]. It is likely to present typical *system complexities*, including openness, large or giant scale, hierarchy, human involvement, societal characteristics, dynamic characteristics, uncertainty and imprecision [62, 294, 335].

Typically, a big data analytical task satisfies most if not all of the above system complexities. The complexities discussed in Sect. 4.2 and the X-intelligence discussed in Sect. 4.3 in major data science and analytics tasks render a complex data project equivalent to an open complex intelligent system [62]. To address such problems, one methodology that may be effective is *qualitative-to-quantitative metasynthesis* [62, 335], which was initially proposed to guide the engineering of open complex giant systems [335]. Qualitative-to-quantitative metasynthesis supports the exploration of open complex systems by engaging various intelligences. In implementing this methodology to engineer open complex intelligent systems, the metasynthetic computing and engineering (MCE) approach [62] provides a systematic computing and engineering guide and a suite of tools to build the framework, processes, analysis and design tools for engineering and computing complex systems.

Figure 5.3 illustrates the process of applying the qualitative-to-quantitative metasynthesis methodology to address a complex analytical problem. For a complex analytics task, the MCE approach supports an iterative and hierarchical problem-solving process, which starts by incorporating the corresponding input, including data, information, domain knowledge, initial hypothesis and underlying environmental factors. Motivations are set for analytics goals and tasks to be explored on the data and environment. With preliminary observations from both the domain and experience, hypotheses and estimations are identified and verified, guiding the development of the modeling and analytics method. Findings are then evaluated and simulated, and are fed back to the corresponding procedures for refinement, optimization and adjustment, for the purpose of achieving new goals, tasks, hypotheses, models and parameters. Following these iterative and hierarchical explorations of qualitative-to-quantitative intelligence, quantitative and actionable knowledge is identified and delivered to address data complexities and analytical goals.

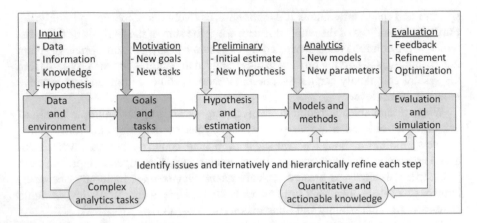

Fig. 5.3 Methodology for complex data science problems: Qualitative-to-quantitative metasynthesis

5.4   Data Science Disciplinary Framework

This section presents a disciplinary framework for data science, using the methodologies for complex data science problems. This framework is described in terms of (1) interdisciplinary fusion for data science, (2) the research map for data science, and (3) the data A-Z dictionary covering disciplinary activities.

5.4.1   Interdisciplinary Fusion for Data Science

As discussed in Chap. 6, data science involves knowledge and capability components from many disciplines. These components comprise quantitative and computational fields: from pattern recognition, knowledge discovery and machine learning to statistics and mathematics; qualitative and heuristic sciences and approaches: from cognitive science to social science, management science, decision science, and communication studies.

Figure 5.4 shows a human-like structure of the interdisciplinary fusion that forms the data science discipline. If we regard a complex data science problem as being like a human body, which is of course also very complex, then cognitive science and communication studies would be the head, intelligence science and information science would be the two feet, management & decision and social science would be the hands, and mathematics, statistics, and the core subjects of pattern recognition, knowledge discovery and machine learning would form the body.

The roles played by the respective disciplines and the interactions between those disciplines and data science are discussed in Chap. 6. The roles of the respective

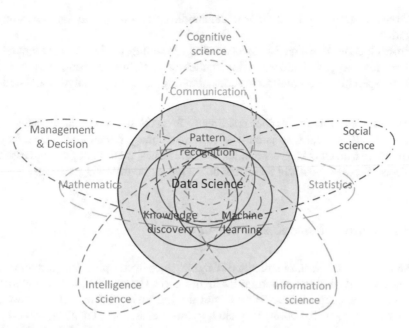

Fig. 5.4 Interdisciplinary fusion for data science

disciplines in fusing and forming the human-like data science disciplinary system are discussed below.

- Motivating discipline of data science: serves as the head discipline of data science and consists of cognitive science and communication studies, which motivate the formation and development of data science as a new scientific discipline;
- Foundational disciplines for data science: contribute to the main body of data science, consisting of statistics and mathematics that build the foundation for data science and its principal disciplines and areas;
- Principal disciplines/areas of data science: serve as the heart of data science, and include pattern recognition, data mining/knowledge discovery, and machine learning, which conduct the core tasks of data science;
- Essential disciplines for data science: serve as the foot disciplines of data science, consisting of information science and intelligence science which contribute to the essential knowledge and capabilities required to implement data science;
- Enabling disciplines for data science: serve as the hand disciplines of data science, consisting of social science and management/decision science, which speed up the formation, development and revolution of data science.

While there are different views about disciplinary roles and interactions in the data science fusion, each discipline contributes complementary knowledge and capabilities to the data science knowledge map. The synthesized complementary knowledge and capabilities effect the transformation:

- from qualitative to quantitative understanding, modeling, analytics, and computation,
- from objective to subjective understanding, modeling, analytics and computation,
- from local to global understanding, modeling, analytics and computation, and
- from specific to comprehensive understanding, modeling, analytics and computation.

Nevertheless, this fusion does not directly form the science of data. Data science as a new independent field is built on interdisciplinary fusion and creates its own disciplinary framework, research map, research issues, methodologies, and essential subjects. These aspects will be further discussed in the following sections.

5.4.2 Data Science Research Map

In addressing the X-complexities and X-intelligence in complex data problems, data science not only delivers opportunities for theoretical breakthroughs but also enables the mining of deep business values from the increasing volume of complex data in many data intensive domains, including those of finance, business, science, the public sector and online/social services. This presents research challenges from both theoretical and practical perspectives, driving the development of a data science research map.

Figure 5.5 illustrates some of the major challenges facing the data science community given the perspective that a data science problem is a complex system. We categorize the challenges facing domain-specific data applications and problems in five major areas and are interested in the gaps that cannot be managed by existing methodologies, theories and systems:

- *Data/business understanding challenges*: To identify, specify, represent and quantify comprehensive complexities, known as X-complexities [62, 64]) and intelligence, known as X-intelligence [62, 64]). Such X-complexities and X-intelligence cannot be managed well by existing theories and techniques. However, they nonetheless exist and are embedded in domain-specific data and business problems. The issue is to understand in what form, at what level, and to what extent they exist, and to understand how the respective complexities and intelligence interact and integrate with one another. An in-depth understanding of X-complexities and X-intelligence will ultimately result in the design of effective methodologies and technologies for incorporating them into data science tasks and processes.
- *Mathematical and statistical foundation challenges*: To discover and explore whether, how and why existing theoretical foundations are insufficient, absent, or problematic in disclosing, describing, representing, and capturing the above complexities and intelligence and obtaining actionable insights. Existing theories may need to be extended or substantially redeveloped to cater for the complexities in complex data and business, for example, supporting multiple,

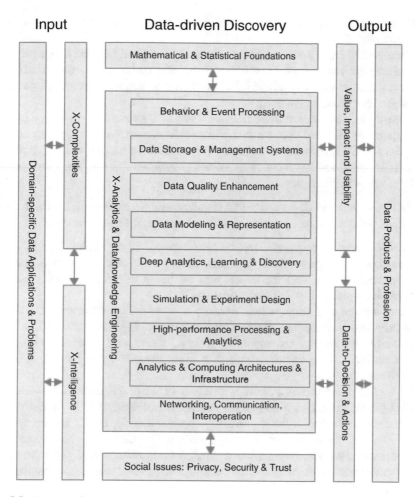

Fig. 5.5 A systematic structure of data science

heterogeneous and large scale hypothesis testing, learning inconsistency, change and uncertainty across multiple data sources, enabling large scale fine-grained personalized predictions, supporting non-IID data analysis, and creating scalable, transparent, flexible, interpretable, personalized and parameter-free models.

- *X-analytics and data/knowledge engineering challenges*: To develop domain-specific analytic theories, tools and systems that are not yet available in the body of knowledge. They will represent, discover, implement and manage the relevant and resultant data, knowledge and intelligence, and support the engineering of big data storage and management, behavior and event processing.
- *Quality and social issues challenges*: To identify, specify and respect data quality and social issues related to domain-specific data and business understanding and data science processes, including processing and protecting privacy, security and

trust and enabling social issues-based data science tasks, which have not so far been handled well.

- *Data value, impact and utility challenges*: To identify, specify, quantify and evaluate the value, impact and utility associated with domain-specific data that cannot be addressed by existing theories and systems, from technical, business, subjective and objective perspectives.
- *Data-to-decision and action-taking challenges*: The need to develop decision-support theories and systems that will enable data-driven decision generation, insight-to-decision transformation, as well as decision-making action generation, and data-driven decision management and governance. These cannot be managed by existing technologies.

The challenges in X-analytics and data/knowledge engineering involve many research issues that have not been properly addressed, for example:

- *Behavior and event processing*: how to capture, store, model, match, query, visualize and manage behaviors and events and their properties, behavior sequences/streams, and the impact and evolution of behaviors and events of individuals and groups in the physical world.
- *Data storage and management systems*: how to design effective and efficient storage and management systems that can handle big data with high volume, velocity and variety, and support real-time, online, and on-the-fly processing and analytics; how to house such data in an Internet-based (including cloud) environment.
- *Data quality enhancement*: how to handle both existing data quality issues, such as noise, uncertainty, missing values and imbalance which may be present at very different levels due to the significantly increased scale, extent and complexity of data. At the same time, how to handle new data quality issues emerging in the big data and Internet-based data/business environment, such as cross-organizational, cross-media, cross-cultural, and cross-economic mechanism data science problems.
- *Data modeling, learning and mining*: how to model, learn, analyze and mine data that is embedded with comprehensive complexity and intelligence.
- *Deep analytics, learning and discovery*: how to discover unknown knowledge and intelligence hidden in the space D in Fig. 4.1 (unknown complexities, knowledge and intelligence, see Cao [64]) through inventing new theories and algorithms for implicit and deep analytics that cannot be handled by existing latent learning and descriptive and predictive analytics. Also, how to integrate data-driven and model-based problem-solving which balances common learning models/frameworks and domain-specific data complexity and intelligence-driven evidence learning.
- *Simulation and experimental design*: how to simulate the complexity and intelligence, working mechanisms, processes, dynamics and evolution in data and business, and how to design experiments and explore the subsequent impact if certain data-driven decisions and actions are undertaken in a business.

- *High-performance processing and analytics*: how to support large scale, real-time, online, high frequency, Internet-based (including cloud-based), multi-national and cross-organizational data processing and analytics while balancing local and global resource involvement and objectives. This requires new batch, array, memory, disk storage and processing technologies and systems, and massive parallel processing and distributed/parallel and high-performance processing infrastructure, as well as cloud-based processing and storage. It also requires large and complex matrix calculation, mixed data structures and management systems, and data-to-knowledge management.
- *Analytics and computing architectures and infrastructure*: how to facilitate the above tasks and processes by inventing efficient analytics and computing architectures and infrastructure based on memory, disk, cloud and Internet-based resources and facilities.
- *Networking, communication and interoperation*: how to support the networking, communication and inter-operation between the various data science roles in a distributed data science team and during whole-of-cycle data science problem-solving. This requires the distributed cooperative management of projects, data, goals, tasks, models, outcomes, workflows, task scheduling, version control, reporting and governance.

5.4.3 Systematic Research Approaches

There are two ways to explore major research challenges: to summarize what concerns the relevant communities, and to scrutinize the potential issues arising from the intrinsic complexities and nature of data science problems as complex systems [62, 62].

With the first approach, we can grasp the main research challenges by summarizing the main topics and issues in the statistics communities [87, 145, 467], informatics and computing communities [64, 343], vendors [374], government initiatives [101, 196, 395, 399, 407] and research institutions [302, 409] that focus on data science and analytics. The second approach is much more challenging, because we must explore the unknown space of the complexities and comprehensive intelligence in complex data problems.

In either case, exploration of the fundamental challenges and innovative opportunities facing big data and data science needs to encompass domain problem-, goal-, and data-driven discovery.

- *Problem-driven discovery* requires understanding the problem's intrinsic nature, characteristics, complexities, and boundaries and then analyzing the gaps between the problem complexities and the existing capability set. This gap analysis is critical for original research and breakthrough scientific discovery.
- *Goal-driven discovery* requires understanding the business, technical, and decision goals to be achieved by understanding the problem and then conducting gap

analysis of what has been implemented and achieved and what is expected to be achieved.

- *Data-driven discovery* requires understanding the data characteristics, complexities, and challenges in data and the gaps between the nature of a problem and the data capabilities. Because of the limitations of existing data systems, projection from the underlying physical world where the problem sits to the data world where the problem is datafied can be biased, dishonest, or manipulated. As a result, the data does not completely capture the problem and thus cannot create a full picture of it through any type of data exploration.

The exploration of the research issues in data science and analytics and domain problem-driven, goal-oriented data discovery require systematic and interdisciplinary approaches. This may require synergy between many related research areas, including data representation, preparation and preprocessing, distributed systems and information processing, parallel computing, high performance computing, cloud computing, data management, fuzzy systems, neural networks, evolutionary computation, systems architecture, enterprise infrastructure, network and communication, interoperation, data modeling, data analytics, data mining, machine learning, cloud computing, service computing, simulation, evaluation, business process management, industry transformation, project management, enterprise information systems, privacy processing, information security, trust and reputation, business intelligence, business value, business impact modeling, and the utility of data and services.

It is therefore necessary to establish inter-disciplinary subjects and initiatives to bridge the gaps between the respective disciplines, and to create opportunities to develop and invent new technologies, theories and tools to address critical complexities in complex data science problems that cannot be addressed by singular disciplinary efforts. For example, new learning theories and algorithms are expected to explore non-IID data through collaboration between statistics, data mining and machine learning, and new data structures and detection algorithms are required to handle high frequency real-time risk analytics issues in extremely large online businesses, such as online shopping and cross-market trading.

5.4.4 Data A-Z for Data Science

The big data community often uses multiple "V"s to describe *what constitutes big data*, i.e., the characteristics, challenges, and opportunities of big data. These comprise volume (size), velocity (speed), variety (diversity), veracity (quality and trust), value (insight), visualization, and variability (formality).

In fact, these terms cannot completely describe big data and the field of data science. It is therefore very valuable to build a *Data A-Z Dictionary* to capture and represent the intrinsic comprehensive but diverse aspects, characteristics, challenges, domains, tasks, processes, purposes, applications and outcomes of data.

To this end, we list a sample sequence of data science keywords below.

$$Actionability/Adaptation, Behavior/Boosting,$$
$$Causality/Change, Dimensionality/Divergence,$$
$$Embedding/Ethics, Fusion/Forecasting,$$
$$Governance/Generalization, Hashing/$$
$$Heterogeneity, Integrity/Inference, Join/Jungle,$$
$$Kernelization/Knowledge, Linkage/Learning,$$
$$Metrology/Migration, Normalization/Novelty,$$
$$Optimization/Outlier, Privacy/Provenance,$$
$$Quality/Quantity, Relation/Regularization,$$
$$Scalability/Sparsity, Transformation/Transfer,$$
$$Utility/Uncertainty, Variation/Visibility,$$
$$Wrangling/Weighting, X - analytics/$$
$$X - informatics, Yield/Yipit, Zettabyte/Zenit. \qquad (5.1)$$

It is notable that such a Data A-Z ontology probably covers most of the topics of interest to major data science communities. The exercise of constructing a Data A-Z can substantially deepen and broaden the understanding of intrinsic data characteristics, complexities, challenges, prospects, and opportunities [177].

In addition, we list many relevant keywords and terms in the data science information portal: www.datasciences.info, which covers a very wide spectrum of the data science knowledge base.

5.5   Some Essential Data Science Research Areas

In this section, we discuss several research areas that are intrinsic to the data science discipline: developing data science thinking, understanding data characteristics and complexities, discovering deep behavior insights, fusing data science with social and management science, and developing autonomous algorithms and automated data systems. While there are many other areas and issues that could be discussed here, these areas have been selected because they are fundamental to the discipline of data science yet have not been well addressed in the literature or relevant communities.

5.5.1   Developing Data Science Thinking

A fundamental disciplinary, research and course component in data science is data science thinking.[1] In this section, we discuss the development and training of data science thinking.

5.5.1.1   Aspects of Data Science Thinking

The purpose of this component is to build the mindset, thinking structure and habits, views, conceptual framework, and processes for data scientists and data professionals undertaking data-related tasks and projects. Researchers, practitioners and students need to address such issues and aspects related to data science thinking as

- What makes data science a new science, or what differentiates data science from classic data analysis, statistics, informatics, computing and related areas?
- What are the paradigms that drive data science and differentiate data science from existing science?
- What are the methodologies for data science research, development, and innovation? What differentiates reductionism, holism and systematism in conducting complex data science tasks?
- What are the views, structures, and systems of data science? What differentiates them from existing statistics and analysis?
- What forms scientific thinking and critical thinking in data science? What aspects are unscientific and uncritical for data science?
- What differentiates data science thinking from statistical thinking [42, 44, 97, 203, 271, 467] and data analytical thinking [23, 149, 392]?
- What are the conceptual frameworks and processes for understanding complex data science problems using effective and well-grounded arguments?
- What does data-driven mean and how should data-driven scientific discovery be conducted?
- What are the essential questions to ask in conducting data science research, innovation and development?

In conducting data science work, data scientists need to select, define, analyze, debate and evaluate a set of issues and topics that either (1) address fundamental research problems and issues in existing theories and systems, or (2) tackle real-life business and natural problems and challenges that cannot be handled well by other approaches. The emphasis will be on a data-driven reasoned approach with sound, reproducible and compelling arguments to evidence that approach.

---

[1]Refer to Chap. 3 for more discussion about data science thinking.

Cognitive thinking, psychological judgment and socio-economic decision-making are involved and evaluated in the process.

5.5.1.2 Tasks in Developing Data Science Thinking

A range of tasks using data-driven methodologies and processes may be undertaken to develop and train data science thinking through data science problem solving. For example, a project may involve the following series of basic processes and building blocks:

- problem selection, structuring, formalization and formulation, to pose the right questions, frame unstructured problems, and formulate the data science questions;
- proposals of hypotheses and arguments, to put forward hypotheses and arguments and identify assertions that may relate to the data problems;
- data acquisition and quantification, to identify and quantify data that is relevant;
- feature engineering, to select, analyze, construct and refine features that are relevant and discriminative but are not noisy, redundant or irrelevant;
- definition and quantification of data complexities, to select, define, quantify and evaluate data characteristics in the data;
- structuring of deductive, reductive and abductive reasoning, to support different ways of thinking and reasoning about data and findings from data;
- identification of solid and sound evidence and indicators, to detect indicators and evidence from data to support or reject the proposed arguments, when often only empirical evidence and analysis is available;
- development and evaluation of alternatives, to recognize the potential limitations of identified arguments and developed approaches and to perceive other possible arguments and approaches;
- justification and evaluation of arguments and outcomes with theoretical analysis, to identify and evaluate, when possible, the assumptions and outcomes in terms of theoretical soundness and support;
- evaluation of possible misunderstanding and incomplete understanding of data complexities, to critically evaluate the understanding and quantification of data complexities;
- evaluation of improper data modeling and learning, to provide well-reasoned arguments for the modeling and learning, and to identify/justify possible incorrectness and the reasons for it;
- justification of possible bias and mistakes in the data, modeling and evaluation, to check the soundness of the assumptions and to identify and explain possible cognitive and analytical bias and error; and
- analysis of the constraints on outcomes and new hypotheses, to impose conditions for the proper understanding and application of the identified outcomes and hypotheses.

By completing this exercise, data scientists should achieve a comprehensive grasp of data science thinking and gain the capacity to establish data science thinking habits, structures, logics, processes and questioning. Attention is paid to both positive and possible negative aspects in data science research, innovation and development. As it becomes increasingly difficult to address complex data problems using to well-developed existing theories and systems, it is important to be aware of the possible constraints and conditions imposed on data, modeling and outcomes, and the possible existence of bias and error.

5.5.2 *Understanding Data Characteristics and Complexities*

We believe that the values, complexities in data modeling and quality of data-driven discovery are driven by the characteristics and complexities of the data. Accordingly, this section discusses the definition, identification, characterization, and evaluation of data characteristics.

The term *data characteristics* refers to the profile and complexity of data (in general, a data set), which can be described from multiple aspects. *Data complexities* refers to the aspects of data characteristics that are difficult to understand or engage with, such as form, structure, relation level, state or quality. Examples of data characteristics and their expressions of complexity are

1. data types, e.g., textual data, video data, image data, or transactional data;
2. attribute types, dimensionality of attributes, attribute length, structure, cardinality, granularity, value statistics, and distribution;
3. data granularity and hierarchy;
4. data distribution type, structure and balance;
5. data format patterns, trends, and exceptions;
6. data scale and volume;
7. data timeliness, frequency and speed (velocity);
8. data invisibility including implicit structures, hierarchies, distributions, and relations;
9. data heterogeneities of attributes, objects, sources and domains—their heterogeneous structure, distribution and relations, and their type, form, and degree of heterogeneity;
10. data relationships between attribute values, attributes, data objects, data sources and domains, and the type, form, and cardinality of those relationships; and
11. other data aspects such as data dynamics, uncertainty, and sparsity.

Understanding data characteristics is a fundamental research issue in data science for many good reasons, including the fact that

- Data characteristics describe the profile and complexity of a data point or data set.

- Data characteristics determine means to appropriately understand a data set and its complexity.
- Data characteristics indicate the potential value and significance of a data set.

Unfortunately, a very limited number of research outcomes and systematic approaches and tools are available in the literature. Understanding data characteristics addresses the following issues:

- what data characteristics are, namely, how to define data characteristics;
- how to represent and model data characteristics, namely, how to quantify the different aspects of data characteristics;
- how to conduct data characteristics-driven data understanding, analysis, learning and management; and
- how to evaluate the quality of data understanding, analysis, learning and management in terms of data characteristics.

Having a genuine and complete estimation of the data characteristics and complexities in a data set is very challenging, as it is often unclear what they are. Different estimation methods lead to inconsistent fitting outcomes, as only certain aspects of the data characteristics and complexities may be captured. Performance is also highly dependent on the aspect, level, order, degree and completeness of information captured.

- Aspect: As discussed in the above definition of data characteristics, many possible aspects may be used to observe the data characteristics and complexities of a data set.
- Level: Level refers to granularity, hierarchy or other tier-related information; for example, value to attribute. An important consideration is whether a data factor captures the aspect from a very low level perspective or a high level perspective, e.g., based on keywords in documents vs. extracted abstract topics.
- Order: Order refers to the power of information, e.g., higher-order statistics that use the third or higher power of a sample, compared to lower-order statistics that rely on constant (zeroth power), linear (first power) or quadratic (second power) terms. A higher-order power may be more expressive and indicate stronger semantics, while possibly incurring less well-behaved properties (e.g., being less robust).
- Rank: Rank refers to the dimension of the vector space generated to represent a data characteristic. In matrix-based modeling, a matrix's column and row ranks refer to the dimensions of its column and row spaces.
- Degree: Degree refers to the extent to which the interested data characteristic or complexity of interest is captured. For example, in imbalanced data, the distribution function may determine the degree of tailed distribution captured.
- Completeness: Completeness refers to the richness of information captured by a data factor, which may be quantified by the combination of other indicators, such as aspects, level, order, rank, or degree. For example, in matrix-based modeling, this may relate to whether low or high rank is involved; in order-

based information representation, whether low-order or high-order information is captured.

In [67], the above issues, the definition, characterization, representation and evaluation of data characteristics and complexities are discussed in Chap. 2: data characteristics and complexities.

5.5.3   Discovering Deep Behavior Insight

Behaviors in natural and artificial systems are ubiquitous, thus discovering deep behavior insights is critical for deeply understanding and modeling principles, dynamics and changes, and exceptional and unexpected behaviors. Behaviors do not exist in data management systems, however, thus the direct analysis of behaviors cannot be conducted on data systems. This section discusses how to discover behavior insights.

5.5.3.1   Physical World–Data World–Behavior World

A critical problem with data is its failure to capture behaviors that have taken place in the physical world, in which behaviors are fundamental components. This is because current data management systems only capture entities and entity relationships but overlook the behaviors of entities.

Discovering deep *behavior insights* is important for many reasons.

- In physical and virtual worlds, behaviors are everywhere and are common and direct representations of worldly activities, processes, states, and dynamics.
- In the data world, behaviors are often excluded or decomposed, since data typically only captures segmented information about behaviors from entities and entity relationships, which are the foundation of existing database and storage systems. Behavior-related activities, semantics, processes, states, and dynamics are consequently often not captured in data, and there are big gaps between what has been captured by the data world and the realities of the physical world.
- Due to the gaps between the data world and the *behavior world*, a direct understanding of the collected data cannot reflect the reality of the physical world, and insights about physical behaviors cannot be directly discovered from the data-based findings, hence issues resulting from physical behaviors cannot be fully addressed.

A fundamental task is therefore to recreate and represent behaviors from the collected data when the underlying physical behaviors are unknown. This *behavior construction* is to transform transactional data and data collected and managed by existing data management systems into *behavioral data* according to certain behavior models, logics, relationships, and semantics that may be gleaned from the domain knowledge and domain intelligence. The intention is for the converted behavioral data to describe behaviors in the corresponding physical world.

The conversion results in the *behavior world*, which is embodied by the behaviors of subjects and interactions between subjects and objects in the corresponding physical world. The behavior world is typically constructed by following a behavior model and representing it from the quantified data for a particular domain.

To enable the discovery of deep behavior insights, we need to

- deeply understand the transformation from the physical world to the data world using existing data engineering and information management theories and systems;
- effectively translate the data world to the behavior world so that behaviors in the physical world are reconstructed and recovered, and can be analyzed and managed through developing behavior informatics.

5.5.3.2   Behavior Informatics for Discovering Behavior Insights

To understand behaviors, it is necessary to build behavior models that can capture behavioral actions and the properties and other aspects of those actions. By aligning with a specific domain, a *behavior model* describes what a behavior is in terms of behavioral properties, including behavior subject, object, action, time, place, and the goals and beliefs that drive a behavior or a behavior sequence. It should also describe the impact, cost and utility of a behavior or behavior sequence, and the relationships between multiple behaviors. Behavior models guide the construction or transformation of behavioral data.

Once behavioral data has been formed, the behavior insights in the transformed behavior data can be analyzed and extracted to discover behavior intelligence that reflects behavior insights in the corresponding physical world.

The discovery of behavior insight relies on theories, systems and tools from the field of *behavior informatics* [55] to represent, analyze, evaluate and manage behaviors. *Behavior informatics* refers to the field that handles the behavior processing and engineering of behavior systems. It consists of main elements: behavior representation, behavior reasoning, behavior analytics, and behavior management. Figure 5.6 illustrates the main process and components for discovering behavior insights using behavior informatics.

Behavior representation, also known as *behavior modeling*, represents, models, reasons about and checks behaviors in business problems. A top-down approach is often taken in conducting behavior representation. The objective is to understand behavioral properties, relationships, constraints, interactions, context and dynamics, and to develop visual and formal representation theories and tools to represent behaviors, behavior sequences, behavior interaction relationships and forms. Additionally, the purpose is to invent representation rules for constraint reduction, relation reasoning, reduction and aggregation, and to build theories and tools to check the validity, soundness, issues and performance of behavioral models.

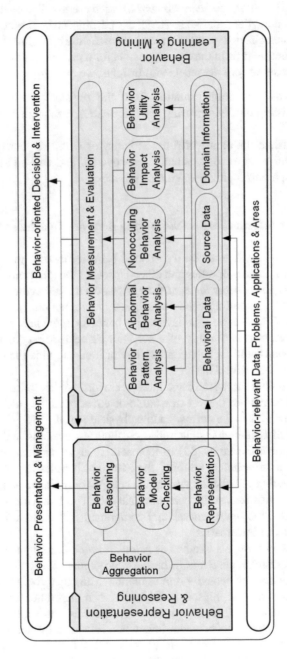

Fig. 5.6 Discovering behavior insights with behavior informatics

The second objective is to analyze and learn behaviors. This includes the development of algorithms, methods, and measurements for analyzing and learning behavioral patterns, outliers, impact, utility, and so on. Analytical methods and models are required to understand both individual and group behaviors. The behaviors of a group of actors are often conjoined, forming coupled group behaviors. In *coupled group behaviors*, the behaviors of one actor and the behaviors of other actors are coupled in various relationships [74]. Analyzing this coupled behavior [74] is critical for obtaining a picture of the intrinsic problems, including how a system behaves, what drives the system dynamics, and how actors interact. In practice, we are interested in behaviors that are associated with a particular impact or utility; for example, trading behaviors that are generating high profit or disturbing market efficiency and fairness.

Although *occurring behaviors* [79] are more visible and easily understandable, non-occurring behaviors [76] are sometimes critical but rarely explored. For example, the failure of a cancer patient to take prescribed medical treatments may result in their condition becoming worse and worse; in high frequency trading, a trading strategy that fails to follow financial risk monitoring protocols may be exposed to abnormal market dynamics. Analyzing non-occurring behaviors is challenging and has rarely been explored; there is very little literature on the topic. The relevant research issues include the anomaly detection of non-occurring behaviors, the discovery of non-occurring behavior patterns, and the clustering and classification of non-occurring behaviors.

Lastly, it is necessary to develop techniques and systems for behavior management to achieve business goals. For example, in intelligent transport systems, behavior analytics can identify and predict vehicle transport behavior patterns and anomalies, which can then be used to interrupt or prevent abnormal transport activities and driving behaviors. The identified behavior indicators, patterns and anomalies could be converted to behavior sequences to monitor behavior development and determine the actions to be taken, based on already occurring behaviors.

5.5.4 Fusing Data Science with Social and Management Science

A pivotal feature differentiating data science from classic statistics, information science and intelligence science[2] is the fusion between data science and social and management science. This section discusses the complementarity between data science and social and management science, and the opportunities that emerge from their fusion.

---

[2]Refer to Chap. 6 for more discussion about the relationship between data science and these disciplines.

5.5.4.1 Complementarity of Data Science and Social and Management Science

The fusion of data science with social and management science is driven by bilateral interactions and the complementary elements of data science and social and management science.

Data science, data systems and data problem-solving involve social, management, legal, economic and communication-related issues, which may be described as

(1) data-related social and ethical issues (see more discussion in Sect. 4.8) such as appropriate use or misuse and sharing, privacy, security, ownership, openness, sovereignty, accountability, and trust;
(2) the governance, assurance, integrity and regulation (including possible policies and law) of data, modeling, outcomes and applications;
(3) the evaluation of data values, social and environmental benefits, and impact on business of data and data-based outcomes;
(4) effective and efficient communications with relevant stakeholders;

On the other hand, social science and management science suffer from the over-accumulation of internal and external data, and need data science to provide more quantitative and evidence-based understanding, analysis, and optimization of management and decision-making problems. This may be achieved by

(1) enabling data-driven management [274], business and governance [384] through the integration of management, data information systems, knowledge management, organization theories, and data analytics. This will develop data-driven innovative solutions and evidence-based management and governance for managing information-driven organizations and decision-making within their operations.
(2) supporting data-driven social problem understanding, optimization, and solving through the development of social informatics and social computing theories and technologies [163], which will model, represent, analyze, learn, and evaluate social problems and solutions by incorporating and inventing new informatics and computing theories and systems for specific social issues, such as social media analytics [174] and social network analysis [353].
(3) creating new innovation, business strategies and entrepreneurship through inventing data-driven social and management systems and business.

5.5.4.2 Interdisciplinary Areas and Capability Set

The interactions between, and fusion of, data science and social and management science create unique and unlimited prospects through complementary studies and interdisciplinary research. This transcends the respective disciplinary boundaries and will create new theories and strategies for solving challenging problems that arise in social science and management practice. It takes advantage of the unique capabilities that integrate data manipulation and analytics with empirical, conceptual, qualitative analysis and leadership. This is critical for data-intensive organizations and the transformation of management in complex, innovation-intensive, data-driven environments. Data science enables the explosion of very large volumes of a range of internal and external data to be made available to business managers. The data analytics of private and public business and management-related data can disclose, inform and better assist in the identification, formulation, optimization and solving of complex social and management problems.

Several interdisciplinary areas and interdisciplinary capability sets need to be combined to enable the above bilateral interactions and interdisciplinary fusion. They include but are not limited to

(1) social and management science-related knowledge, such as economics, political economy, finance, marketing, business administration, law, accounting management, quantitative methods, survey modeling and simulation, and decision modeling;
(2) data analytics-related knowledge, such as computer science (in particular, data management, business intelligence, data mining, machine learning, network theory, and visualization), statistics (descriptive analytics) and applied mathematics (stochastic systems), and even natural sciences; and
(3) interdisciplinary knowledge, such as business psychology, econometrics, supply chain management, enterprise resource planning, operations management, game theories, systems modeling and optimization, and system engineering.

5.5.4.3 Creating New Interdisciplinary Areas and Professionals

The fusion of social science and informatics with data science will also create new interdisciplinary areas and knowledge, such as

- new informatics for social problems, such as behavior informatics, business informatics, and social informatics (see more discussion in Sect. 6.4.2);
- new analytics for managerial and business problems, such as business analytics, social analytics, behavior analytics, and marketing analytics (see more discussion about X-analytics in Sect. 7.6);
- social data science (definitions and discussion on social data science are available in Sect. 6.7.1.2); and
- management analytics (definitions and discussion about management analytics are available in Sect. 6.8.3).

These new areas will explore social and managerial issues and problems in a data-driven, informatics and analytics-oriented way by translating business problems into analytical problems, which

- identify, define, quantify, represent and formulate social problems, management issues, and organizational challenges for quantitative understanding;

- apply and invent statistical, analytical, and learning theories, systems and tools to deeply analyze the identified and formulated social, managerial and organizational problems;
- provide rigorous theoretical and empirical analysis, experiments and evaluation to underpin the social, managerial and organizational problem representation and data analysis;
- create new values, benefits and business impact and improve performance and decision-making that cannot be achieved by social and management methods only; and
- deliver solutions across business functional areas, disclose insights for managerial decision-making, and communicate results to business stakeholders and decision-making managers.

The above interdisciplinary areas will train and foster a new professional elite who can

- invent theories and advanced tools to analyze, visualize, interpret, and manage a multitude of complex data in economic, social, managerial, organizational, scientific, technological and cultural domains;
- understand how organizations currently use data and data science and identify gaps and optimal strategies to improve business production and operations, create new value and better performance;
- apply sound data science theories, systems and tools to real-life problems using sound understanding and interpretation skills;
- characterize and assess the quality of the data and its limitations in supporting data-driven evidence extraction and decision-making;
- motivate the development of new data science and social/management theories and techniques from practical problem solving; and
- make informed decisions and employ evidence-based strategies to improve decision quality and fulfill business objectives.

5.5.5 Developing Analytics Repositories and Autonomous Data Systems

Computer technologies have been extensively accepted in almost every scientific area and business domain. There are clearly many good reasons for this, but a key factor is the ease of accessing, programming, and applying computer systems, and the wide availability of a variety of toolboxes. Over the last few decades, computer science has been rapidly adopted by classic scientific fields and engineering operations. Data science sets a high bar for itself, to differentiate it from existing disciplines, including statistics and intelligence science. This is not helpful for promoting its development and broad acceptance by various communities, even though data science itself demonstrates strong cross-disciplinary and practical merits. This section summarizes the benchmarks of existing analysis systems, and discusses the development of autonomous analytics and learning systems.

5.5.5.1   Benchmarks of Existing Analytics Systems

Most existing data analytical systems, including those that are widely used commercially and those that are open sourced, usually provide workflow-based interfaces and dashboards to enable users to interact with systems to undertake data analytics tasks. Typical analytical development and programming systems, such as RapidMiner [337] and SAS Enterprise Miner [346], are interactive. They have built the following benchmarks for analysts and engineers: *predefined modeling blocks*, *multi-mode interactions*, *workflow building*, *visualization and presentation*, *programming language support*, *project management*, and *importing and exporting*. Their production systems usually wrap up established models at the backend as blackbox and show dashboard at the frontend for presentation and user interactions.

The functionalities of such interactive analytical systems are summarized below.

- Predefined modeling blocks: Many modeling blocks are hard coded into the analytical platform and categorized according to their focus, methods, and analytical purpose. They are employed as building blocks to construct more complicated analytical workflows, projects and tasks.
- Multi-mode Interaction: These systems usually support different ways for modelers and engineers to interact with the system, such as command line window, workflow/process, or programming environment.
- Workflow building: This enables analysts and data modelers to create an analysis process or workflow by dragging and dropping modeling blocks (methods or operators) from the relevant model base or operator library, adding local or external repositories, and connecting the selected operators and methods or sub-workflows (sub-processes). The engine checks the connection validity to form valid workflows and processes.
- Visualization and presentation: Graphs and special views are usually supported in addition to tables and charts.
- Programming language support: Analytics systems usually support existing programming language, typically R, Python and Java, or provide vendor-specific programming languages such as SAS so that professional analysts and data modelers can code analytical algorithms themselves.
- Project management: An analytics process is managed as a project. A project may consist of multiple processes and their input data and results.
- Importing and exporting: Importing projects, processes and data sets are usually supported. Developers and users can export projects, processes, and results to files or other formats. Importing and exporting are frequently supported by wizards, in which dialogs guide users step by step through the importing and exporting process.

- Frontend dashboard: When the modeling has been completed, a prototype or production version of the analytics can be generated. The prototype and production versions are usually presented as dashboards, which report at-a-glance views of all key indicators, results, and dynamics, and support user interactions (e.g., drill up or down, queries, configure indicators).
- Backend blackbox: When a data analysis system is compiled, its core processes and models are usually converted into a blackbox, while a frontend dashboard is released to interact with the backend. Modeling processes and analytical logic are fixed in the configuration, with limited adjustment and input access.

5.5.5.2   Towards Autonomous Analytics and Learning Systems

While interactive analytical and learning systems are mostly convenient and useful for analytical and learning professionals, they are not feasible for non-professional business analysts and operators or ordinary business people. The high standard and capability requirements demanded of data scientists scare non-professionals and prevent the wide access and application of data science enjoyed by social media and mobile applications.

To enable people to accept and apply data science in their daily business at a grassroots level, there are many important considerations on a data scientist's agenda. One option is to develop autonomous analytics, autonomous learning agents and multi-agent learning systems which differ from existing professional interactive and programming-based analytics and learning systems. This may require the creation of (1) analytics and learning repositories, (2) autonomous data modeling agents, and (3) autonomous data modeling multi-agent systems. Here, *data modeling* refers to the processing, analysis, learning of data, and the discovery of knowledge and insights from data.

Analytics and learning repositories are factories in which base analytics and learning modules are produced, configured or standardized. Each base module fulfills one low-level, simple and particular analytic/learning job as a method or function. All modules form the basic modeling blocks that can be called to form specific and sophisticated analytical and learning functions.

Open source efforts significantly push the sharing of knowledge and usually provide complete solutions for problems, but they still set a high qualification bar for the general population. Analytics and learning repositories may be open sourced or licensed to end users, as already seen in the activity of producers of images and freelance commentary and review contributors.

In fact, many open sourced software and toolboxes have already been made available, such as OpenCV [317], OpenNLP [8], TensorFlow [41], Weka [390], Apache Spark MLlib [9], Shogun [368], Azure [289] and many other tools shared through platforms such as Github, although they usually offer complete solutions rather than focus on base blocks.

Autonomous data modeling agents are independent data modelers that undertake specific data modeling tasks. Their role is to check and establish preconditions and postconditions for task execution. Such executions are autonomous, with no human or external system intervention.

For data modelers to be autonomous, they usually need to incorporate self-contained and self-organizing capabilities and business logic, to analyze data and sense the data-based domain, to react to the preconditions and sensed input, and to execute the corresponding analytical and learning processes, logic and models. Traditionally, such agents need to be imbued with sensing/reaction and planning capabilities, and logic-based reasoning and reinforcement learning are often empowered. Today's agents can collect and analyze data and behaviors about themselves and the environment, thus functionalities like self-learning, self-training, self-organization and self-decision-making should be enforced. In this sense, data modelers are intelligent agents, and agents are learners and modelers on the fly. The concept of *agent mining* [53, 73, 75] was proposed to synergize agent intelligence with analytical and learning intelligence to create super-intelligent agent learners.

Autonomous data modeling multi-agent systems are data systems composed of autonomous data modeling agents. Many autonomous data modeling agents may take different roles and responsibilities in a large-scale data analytics and learning project, work on different phases of an analytical and learning process, undertake specific analytical and learning jobs, communicate with each other for optimal collaboration and cooperation on multiple complex tasks and objectives, acquire and analyze the running process-based data, system status and progress, and make adjustments and choices to achieve optimal results.

Autonomous data modeling multi-agent systems are a combination of (1) autonomous intelligent agents with data modeling, (2) design-time modeling and system settings with run-time modeling and reconfiguration, (3) modeling both external input data and internal system data, and (4) various aspects of intelligence from data, domain, and environment.

Both autonomous data modeling agents and autonomous data modeling multi-agent systems are intelligent systems. Their design and building rely on interdisciplinary technologies, such as

(1) classic artificial intelligence and intelligent systems theories and technologies for pattern recognition, computational intelligence, knowledge management, planning, and logic-based reasoning;
(2) data science and analytics for understanding, analyzing, learning, inferring, reasoning about and evaluating data and data-driven knowledge and insight, usually from multiple modals, sources and systems, media channels, and networks and for multiple tasks and objectives;
(3) automated system design and technologies for making data systems autonomous and automated; and
(4) control systems and cybernetics for data system communication, control, planning, scheduling, conflict resolution, risk management, and optimization.

Building autonomous data systems is increasingly important for developing next-generation artificial intelligence and intelligent systems. These systems will effectively widen the application of data science and analytics to the life and environment of ordinary people, and will enhance the real-time and online analysis, detection, discovery, prediction, intervention and action capabilities and performance of existing automated systems such as driverless cars, unmanned aerial vehicles, and autonomous military systems.

5.6 Summary

This chapter has explored a collection of relevant aspects and issues to answer the question "What forms the data science discipline?" In the process, we have looked at the disciplinary gaps, the methodologies, the disciplinary framework, and several other essential areas. This new field of data science has its foundations—and will continue to be built on—many related disciplines and their disciplinary and transdisciplinary developments. This forms the basis for establishing data science, which will be discussed in Chap. 6.

There is little discussion in this chapter about the central point of data science: analytics, learning and optimization. These topics will be covered separately in Chap. 7 and in book [67] by the same author.

We have not covered the specific subjects and knowledge components of a data science course either. Data science course-related matters are included in Chap. 11.

Readers may need to synthesize the information in this and the above chapters, as well as Chap. 6 on the foundations of data science, in order to build a comprehensive understanding of data science research and education.

Chapter 6
Data Science Foundations

This chapter addresses the fundamental question: what lays the foundations for data science as a new science? Several relevant disciplines and areas are included: cognitive science, statistics, information science, intelligence science, computing, social science, management, and communication studies.

We discuss the roles and responsibilities, connections, and opportunities for disciplines to contribute to data science formation and address specific issues. The possible expansion of the disciplines driven by data science challenges is also discussed. The framework and subject structure of data science as a new discipline is not discussed here, but in Chap. 5.

In addition, this chapter discusses other relevant areas and issues, including broad business, management and social areas, domain and expert knowledge, and invention, innovation and practice.

6.1 Introduction

While the field of statistics plays a critical role in building data science foundation and theories, it is not the only discipline that contributes to data science. On one hand, data science is a disciplinary fusion of many disciplines, in particular, cognitive science, statistics, information science, intelligence science, computing, social science, management, and communication studies. On the other hand, the reach of data science extends beyond any individual discipline and the trans-disciplinary fusion of existing disciplines.

The interdisciplinary fusion of data science involves multiple diversified scientific paradigms and research methods, including human cognition and thinking, mathematical and statistical quantification and modeling, informatics and computing, logic and reasoning (including symbolism and reinforcement learning), computational intelligence (including neural network, fuzzy logic and evolutionary

© Springer International Publishing AG, part of Springer Nature 2018
L. Cao, *Data Science Thinking*, Data Analytics,
https://doi.org/10.1007/978-3-319-95092-1_6

computing), data mining, analysis and learning, and experiments and simulation. These are respectively defined in each of the above relevant disciplines, and contribute to data science.

Many disciplines and areas have the capacity to be involved or contribute to data science, especially for domain-specific data science problems, and the involvement and fusion of some disciplines are ad hoc and on demand, while others form the core foundations of data science. Accordingly, we group relevant disciplines to two categories: *hard foundations* and *soft foundations*.

- Hard foundations of data science include those disciplines and areas that are at the core of data science, are domain-independent and problem-independent, and provide knowledge, methods and tools for general data science problems and systems. They include cognitive science, statistics, information science, intelligence science, and computing.
- Soft foundations of data science consist of areas and knowledge that are elective and problem-specific, which may be required in some stages of data science or domain-specific data science problems. "Soft" does not mean they are not important: in fact, they sometimes play a critical role. They are social science, management, communication studies, domain and expert knowledge, and innovation and associated practices.

Figure 6.1 outlines these hard and soft foundations of data science.

Fig. 6.1 Hard foundations and soft foundations of data science

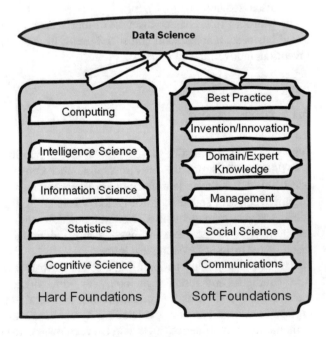

6.2 Cognitive Science and Brain Science for Data Science

Cognitive science and brain science are critical for data science. *Cognitive science* is an interdisciplinary science that has been built on several areas: neuroscience, philosophy, anthropology, psychology, linguistics and artificial intelligence. These are also relevant and contribute to data science formation and development.

Cognitive science provides fundamental theories, mechanisms, functionalities, and methods for data science, including but not limited to:

- thinking methods, cognitive functionalities, cognitive activity types, cognitive working mechanisms, and positive mental habits and traits such as imagination for building data science thinking (see more in Chap. 3);
- mechanisms of brain memorization, imaging, conceptualization, perception, cognition, recognition, representation and processing for informing and designing original and powerful human-like data science algorithms, models, and systems; and
- mechanisms and methods of brain computation, reasoning, learning, and intelligence, and decision formation of complex problems for informing and designing intelligent data science activities, especially data-driven machine recognition, mining, analytics and learning complex data and problems.

The progress made in cognitive science and brain science has significantly benefited the development and evolution of artificial intelligence and machine intelligence. Typical examples are knowledge representation and reasoning, connectionism-based shallow neural networks and recently deep architectures and deep networks-based deep learning [123], and reinforcement learning for Deepmind [189]. Complex data science problem-solving will require significant progress to be made on such areas as

- computational and representational understanding and models of the human mind in terms of advanced thinking, representation, learning, reasoning, cognition, and intelligence;
- new and powerful computational architectures for perceiving, imaging, memorizing, storing, and processing large-scale data and information with enormous entities and complex relationships, structures, hierarchies and granularities;
- new and powerful human-like computational models and methods for deep representation, analysis, reasoning, inference, optimization and processing of complex data problems;
- effective actions of the human brain in selecting, retrieving, synthesizing, and applying diverse resources, knowledge, skills, and intelligence; and
- positive mental traits and habits in handling complexities, conflicts, errors and refinement that may inform better training of data science thinking and better ideas and strategies for tackling X-complexities.

Figure 6.2 outlines the main functionalities of cognitive science and data science, and the connections between them. This fusion enables the creation and involvement

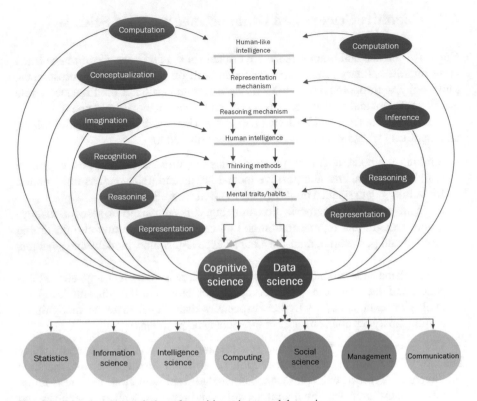

Fig. 6.2 Interdisciplinary fusion of cognitive science and data science

of positive mental traits and habits, thinking methods, human intelligence, reasoning and inference mechanisms, and representation mechanisms in data science, as well as general human-like intelligence in data products.

6.3 Statistics and Data Science

The term "data science" was originally proposed by statisticians, and the relationships between statistics and data science are tight and bilateral. On one hand, statistics drives the emergence of data science. On the other hand, data science is redefined by interdisciplinary forces and thus challenges statistics and drives its evolution. In this section, the bilateral relationships between statistics and data science are discussed.

6.3.1 Statistics for Data Science

Statistics, as well as mathematics, are seen as being increasingly important for forming the core theoretical foundation and modeling techniques of data exploration in data science.

The role played by statistics in data science is arguably mainly in the area of data analysis and analytics. Statistics provides quantitative theories, techniques, and tools for data description, sampling, manipulation, wrangling, munging, modeling, experimental design, hypothesis testing, and inference. It also builds a foundation for measurement, probability theories, error and confidence theories, and tools for testing and validation.

As statistics has evolved, statisticians have increasingly realized the need for, and challenges of,

1. handling increasingly bigger, faster and more complex data (e.g., sparser, mixed data),
2. making invisible and uncertain characteristics and complexities visible and certain, and
3. enabling hypothesis-free data exploration, i.e., data-driven exploration, for the discovery of new hypotheses.

Recognizing the need for more data-driven analysis, data science has been viewed as the future and next-generation of statistics for roughly the last 50 years. Today, the need to upgrade statistics to data science is driven by the *larger, more, longer, sparser, faster* and *deeper* analysis of big data, forming new-generation statistics.

- "Larger" corresponds to the volume of data and refers to the handling of larger scale data, or the full dataset rather than sampling-based exploration.
- "More" corresponds to the variety of data and refers to the processing of diversified or mixed data from multiple sources or modalities.
- "Longer" refers to the greater length of time required to analyze bigger data, or the more stable models required for managing evolving data and its context.
- "Sparser" refers to data that are not balanced and uniformly distributed but heavy-tailed (heavy-tailed distributions are probability distributions whose tails are not exponentially bounded). This may mean that the data follow long-tailed, fat-tailed, or subexponential distribution. Existing theories and estimations established for balanced and normally distributed data will not work.
- "Faster" refers to the velocity of data, in relation to the more efficient processing of data or the real-time analysis of data streams and high frequency data.
- "Deeper" refers to the exploration of invisibility and the hidden nature of data, in relation to the deep analytics, modeling and learning of sophisticated data characteristics and complexities in data and its context.

6.3.2   Data Science for Statistics

The statistical goals outlined above, driven by big data, significantly challenge existing statistical systems, which often handle small-scale samples of static data on the assumption of independent and identical distributed (IID) variables and samples. To address the above requirements, data science is driving a revolution in statistics to invent more advanced statistical and mathematical theories, methods, models, language, sampling and testing methods to address these significant challenges.

Data science presents unlimited opportunities for major innovation and revolution in statistics that will lead to a new generation of statistical analysis which is

- *Scalable*: statistical models are required to be scale-free and applicable to any scale of data samples without significant output difference. Sampling techniques should ensure the learning outcome consistency between a sampled set and the full dataset.
- *Heterogeneous*: statistical models have to tackle increasingly diversified data for enterprise and cross-source analytical objectives. Typical cases consist of mixed data types, multi-view, multi-source and cross-domain applications, multi-task and multi-objective analysis. Samples and variables in such cases have different statistical distributions and/or probabilistic density functions.
- *Adaptive*: statistical models have to address changing data and changing contexts and thus need to be relatively stable and robust at certain stages, while remaining active and self-adaptive to other levels of change, and capable of automated re-learning when changes are significant. Changes in data and context can take place in data value (new value or value range), type (new variable type), scale (from finite to infinite), distribution (new distribution or distribution deviation), structure (hierarchy change, presentation change; e.g., tree to star model), and relation (new relations, such as association and dependency, new qualitative relations).
- *Coupled* (or *relational*): in addition to mathematically quantified coupling relationships such as association, correlation and dependency, new statistic quantification and formalization models and tools are necessary for representing other types of coupling relationships, such as social network connections formed as a result of cultural and socio-economic relationships [61]
- *Efficient*: many statistical (probabilistic) models are suitable for small samples and cannot adapt to high frequency and high speed data. Faster and larger data analysis requires efficient statistical models, sampling, inference, and parameter tuning support.
- *Uncertain*: there is often uncertainty about the characteristics and complexities of data (e.g., distribution, structure and relations), and the full picture or ground truth. A critical yet difficult question is to what extent the uncertainty might become certain. This requires statistical models to understand and represent the various aspects of uncertainty, to make the uncertainty visible.

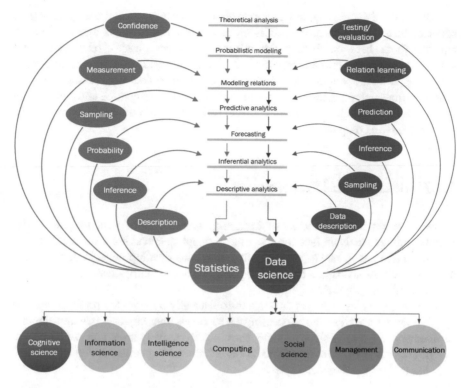

Fig. 6.3 Interdisciplinary fusion of statistics and data science

Figure 6.3 summarizes the main bilateral connections and complementarity between statistics and data science in their interdisciplinary fusion and evolution. This fusion enables fundamental data science works including descriptive mathematical modeling, analytics, inferential analytics, forecasting, predictive analytics, modeling relations, probabilistic modeling, and creating the foundation for theoretical analysis.

6.4   Information Science Meets Data Science

Information science is an interdisciplinary field that is concerned with information access, collection, storage, retrieval, processing, analysis, protection, presentation and publishing.

The roles and relationships of information science and data science are quite sophisticated. Information scientists often pose the question: "information science has existed for a long time and does all the jobs of data science—why do we need data science?" to data science advocates.

Putting the debate aside, information science certainly forms a core part of data science. Information science itself is an amalgamation of areas and disciplines, and we focus here on three major subareas that are directly relevant to data science: analysis and processing, informatics, and general information technologies including software engineering, information systems and networking for data science. Other potential, knowledge components such as computing and data mining and machine learning, are separately discussed in Sect. 6.5 on intelligence science and Sect. 6.6 on computing.

6.4.1   Analysis and Processing

Analysis refers to the process of breaking a whole into parts to gain better understanding. Analysis is a long lasting concept that has been formalized as a technique and has grown to jointly evolve alongside developments in mathematics and statistics. Analysis is a core technique in classic scientific research areas such as chemistry and physics.

In recent decades, analysis progressed to formally become the professional and specific act of analysis and processing (AP) of various object types and domain problems. Analysis has since become instrumental in many areas and domains with the development of a range of media-based, domain-specific, and goal-oriented analysis and processing techniques.

- *Media-based analysis and processing*: such as signal analysis and processing, audio and speech analysis and processing, language analysis and processing, video analysis and processing, image analysis and processing, and historical event analysis;
- *Domain-specific analysis and processing*: such as business analysis, econometrics, linguistics, financial analysis, technical analysis in finance, and quantitative finance, engineering analysis, management analysis, accounting analysis, auditing analysis, literature analysis, Web analysis, medical analysis, policy analysis, psychoanalysis, and philosophical analysis,
- *Goal-oriented analysis and processing*: such as requirement analysis, semantic analysis, syntax analysis, structural analysis, marketing analysis, and risk analysis.

Classic analysis and processing mainly relies on mathematical and statistical concepts, methods and tools. These include

- *basic mathematical building blocks* such as real numbers, complex variables, functions, vectors, matrix, tensor, algorithms, and infinity, and
- *basic statistical concepts* including factor, multivariate, variance, deviation, regression, principal component, scale, time-series, sequence, spatial attributes, probability, conditional probability, sensitivity, correlation, inference, forecasting, prediction, and clustering.

These methods and tools have further synthesized into many focused analysis and processing methods which form a core foundational unit in data science. Information analysis and information processing have evolved and expanded to a new generation, containing typical subareas such as data mining, data analytics, and machine learning. These are included under another increasingly recognized concept: intelligence science, to be discussed in Sect. 6.5.

6.4.2 Informatics for Data Science

Informatics is critical for data science. *Informatics* is the science of information and information systems, which handles the information processing and engineering of information systems. It directly contributes theories, tools and systems for data-to-information and information processing and engineering.

The specific role of informatics in data science is somewhat subjective in relation to the differences between informatics and data science and between information science and data science. In spite of conflicting views on these differences, a fundamental perspective can be gained by understanding the concepts and dissimilarities between the four progressive layers: data, information, knowledge and wisdom. The discussion on these differences can be found in Sect. 2.3.

Contradictory views have been expressed in the literature by researchers and practitioners. Some believe that informatics forms a core part of data science, while others argue that data science fits the broad scope of informatics or information science. The conceptualization of data and information indicates that data is much richer and broader than information in such aspects as the evolution of domains and sources, the level of cognition and intelligence as shown in Fig. 2.1, and the richness and size of content.

Accordingly, we believe that data science is an extension of information science. Data science absorbs and processes the lowest level of resources (i.e., data), provides the broadest opportunity to convert low-level data to intelligence and wisdom, and generates the greatest potential for innovation and added value.

As X-informatics appears in many domains that are traditionally informatics-irrelevant, such as behavior informatics and health informatics, the role of informatics in data science is multi-fold:

- the representation and modeling of domain-specific information and problems driven by data, i.e., the evolution from data to information through data representation, embedding and encoding;
- the processing of information transformed from data, which may adhere to or extend beyond existing theories and tools in information science;
- the discovery of knowledge and insights in information by creating knowledge discovery and transformation theories and tools; and
- the engineering of data, information and knowledge-based systems.

Given the above interpretation of the role of informatics for data science in domain-specific areas, the following possible functions of informatics for data science are indicated:

- Informatics enables the representation, processing and engineering of domain-specific data-to-information processing problems, such as behaviors and biological systems, to create the corresponding domain-specific information science which is often interdisciplinary, e.g., behavior informatics and bioinformatics. In this way, a domain-specific data problem is converted to a domain-based informatics problem, and general informatics theories can then be applied to the converted domain-specific information for processing and engineering. This is how X-informatics was invented. For example, *behavior informatics* [55] handles behavioral representation, modeling, processing, analysis, engineering and management.
- Informatics provides general theories, tools and systems for directly or indirectly representing, processing and engineering data science problems and systems. In the DIKIW pyramid, as data is next level below information, general theories, methods, tools and systems in informatics are also applicable to data processing and engineering. Their application may be straightforward or built on information converted from data. As an example, mutual information, self information and entropy are fundamental tools in informatics. They are intensively used in data mining and machine learning for building learning metrics and models, and verifying the quality of results.

6.4.3   General Information Technologies

Data science cannot survive without general information and communication technologies (ICT). ICT provides essential theories, methods, and tools for engineering data science problems and systems. While many ICT subjects have the potential to be involved in data science, the following are essential: software engineering, programming, enterprise application integration, networking, communications, and information systems and management.

The above ICT knowledge and tools are required by data science for many purposes, such as software engineering for data system analysis, design, and testing; analytics programming and language support; data and application integration and communications; networks for data computing, transport and services; building infrastructure and architectures to support data analytics and engineering; and information systems, including building data and data warehouse storage, backup and management.

Data science projects and systems need to involve relevant ICT theories, systems and tool sets. Data scientists generally need to understand and grasp some of these techniques in order to undertake data science tasks. In fact, some data roles may be defined that take sole responsibility for the use of these technologies in a data

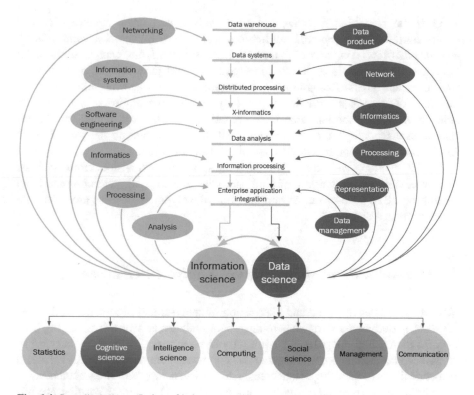

Fig. 6.4 Interdisciplinary fusion of information science and data science

science team to achieve the above goals. More discussion on data science roles can be found in Sect. 10.3.

Figure 6.4 summarizes the main bilateral connections and complementarity between information science and data science in their interdisciplinary fusion and evolution. The interactions address issues and tasks in data science including enterprise application integration, information processing, data analysis, creating various X-informatics, data systems, data warehousing, and supporting distributed processing.

6.5 Intelligence Science and Data Science

Data analysis and *information processing* were first proposed five decades years ago with the aim of achieving an easier and more precise understanding of data and information. Analysis and processing (AP) have played an increasingly important role in handling multiple types of media and objects, addressing an expanding number of domain problems, and achieving a greater number of goals.

Over the last three decades, AP have evolved and have expanded to new areas referred to as *computational intelligence, pattern recognition, data mining, machine learning, computer vision,* and various *data analytics,* such as multimedia analytics. Their aim is to tackle more complex and advanced objects, domain problems, and goals. Recognition, mining, analytics and learning have been the keystone of today's information technology and are the engine of today's data science. Another important area is computational intelligence, which focuses on nature-inspired computational modeling. All these contribute to the formation of *intelligence science,* an advanced stage of information science.

In this section, we summarize the migration and expansion from analysis and processing to recognition, mining, analytics and learning, and computational intelligence. The state-of-the-art and the future development of these areas for data science are also discussed.

6.5.1   Pattern Recognition, Mining, Analytics and Learning

With the rapid development and application of analysis and processing in the past 30 years, pattern recognition, data mining, data analytics and machine learning have become increasingly popular areas not only in science and technology but also in social science and business management. This section summarizes the key roles and responsibilities in these important areas, which form a core part of data science.

Pattern recognition, data mining, data analytics and machine learning are relatively recent developments to add to classic analysis and processing. They are typically concerned with latent variables, relations, structures, frequent phenomena, exceptions and knowledge in data observations. These aim to discover hidden but interesting patterns, trends, clusters, classes, and outliers. In contrast, classic analysis and processing are more focused on observations and explicit findings.

The migration and expansion from observation-based shallow analysis and processing to knowledge and intelligence-oriented deep analytics, mining and learning have been built on to substantially expand the development of AP. As a result, we are seeing more advanced, specialized, and mixed media-based, domain-specific, and goal-oriented discovery.

- *Media-based recognition, mining, analytics and learning*: for example, visual analytics, video analytics, multimedia analytics, text mining, document analysis, multi-source analysis, multi-view learning, cross-media analysis, graph analysis, and spatial-temporal analysis;
- *Domain-specific recognition, mining, analytics and learning*: such as corporate analytics, behavior analytics, marketing analytics, network analysis, social media analysis, bioinformatics, behavior informatics, social analytics, brain informatics,
- *Goal-oriented recognition, mining, analytics and learning*: for example, representation learning, similarity learning, supervised learning, unsupervised learn-

ing, semi-supervised learning, true learning, rule learning, community learning, dimensionality reduction, anomaly detection, predictive analytics, reinforcement learning, deep learning, multi-task learning, sparsity learning, heterogeneity learning, non-IID learning, transfer learning, behavior analytics, intelligence analysis, sentiment analysis, intent analysis, influence analysis, and more advanced computational intelligence analysis and learning.

The latest development and expansion in recognition, mining, analytics and learning directly addresses the key purpose and challenge of data science, that is, to discover data values through handling sophisticated data characteristics and complexities. Related keywords and topics have lately dominated the current interest of communities including data mining (e.g., KDD and ICDM conferences), machine learning (e.g., ICML and NIPS), pattern recognition and computer vision (e.g., ICCV and CVPR), artificial intelligence (e.g., AAAI and IJCAI), and more recently data science and analytics (e.g., DSAA). The state-of-the-art in these areas form the core capabilities and knowledge base of today's data science.

6.5.2   Nature-Inspired Computational Intelligence

Inspired by human and natural intelligence, computational intelligence [249] has evolved into an important area for data science. This is because computational intelligence mimics human intelligence and natural intelligence, and their problem-solving mechanisms, which are constituents of data science intelligence (see more discussion in Sect. 4.3). They address complex problems and complement classic mathematical and modeling methods.

Computational intelligence, which is a subset of artificial intelligence, consists of neural networks, fuzzy logic, and evolutionary computation. Certain types of data science problem-solving can benefit from these technologies rather than others. Computational intelligence may be used for learning data and observations or for representation, reasoning, and optimization of data science problems.

6.5.3   Data Science: Beyond Information and Intelligence Science

In response to the question "what is the relationship between data science and existing information science, and more recently, intelligence science?", we can observe that data science is emerging as the next generation of information science and intelligence science. Data science addresses the challenges and problems at the lowest level of the hierarchical Data, Information, Knowledge, Intelligence and Wisdom (DIKIW) pyramid (see more in Sect. 2.3), which is assumed to be fundamental and original.

The next generation of pattern recognition, data mining, data analytics and machine learning forms the keystone of data science. While there may be disagreement between communities as to what this next generation should look like, we believe that new and effective theories and methods to handle original and fundamental real-life data and problem challenges and complexities that have not been well addressed will be invented. These will involve (1) more sophisticated business problems, (2) more significant data complexities, (3) more significant data environment challenges, and (4) more advanced user requirements.

- *More sophisticated business problems and requirements*: such as intelligent, online, real-time and active application integration, scenario analysis, event and exception detection; early prediction and warning, and active intervention in suspicious scenarios; intelligence information and security. This covers many hybrid problems, and will involve multiple organizations on a large scale, a large number of mixed roles and actors, different domains and areas, and heterogeneous information resources from different systems and media channels. Examples are intelligent global military command and control systems and intelligence analysis and security systems.
- *More significant data complexities*: for example, how large or small is the data required for us to completely represent and describe user relationships in Twitter and Facebook? How can we precisely and completely represent user coupling relationships in Twitter? Answering these questions typically involves the challenge of precise sampling from petascale candidates, learning unclear couplings between users and user behaviors, and estimating, evaluating and verifying the sufficiency and necessity of sample sizes and model effectiveness.
- *More significant data environment challenges*: many of the issues discussed in Sect. 6.5.1 involve a complicated data environment, which may be dynamic, infinite, heterogeneous, hierarchical, and open. This creates significant challenges for existing pattern recognition, data mining, data analytics and machine learning, e.g., how to model and learn the influence of such environments on the underlying data and problems, or how to assure the precision and effectiveness of analytics and learning systems in such and environment.
- *More advanced user requirements*: with the wide and deep involvement of artificial intelligence and intelligent analytics and learning, business operational and entertainment systems, services and applications are becoming increasingly intelligent. Examples are online gaming systems with active learning and user modeling capabilities, and personnel assistants with personalized recommendation and services. However, we are often pushed with irrelevant, duplicated and uninteresting services. How can relevant applications mimic human behaviors and intelligence, and understand human intent and preferences, so that perfect recommendations can be made?

Pattern recognition, data mining, data analytics and machine learning also face more general functional and nonfunctional challenges. These are:

- supporting larger, more, longer, sparser, faster and deeper pattern recognition, data mining, analytics and learning. These aspects will be further discussed in Sect. 6.3.1, as they also challenge today's statistics theories and methods.
- discovering and delivering active, online, real-time, actionable results in data problems embedded with X-complexities and X-intelligences so that the results are directly converted to decision-making actions to improve or change scenarios.
- implementing human-like machine intelligence, so that human-like automated machines can really understand, mimic, cooperate with and complement human beings, and can follow human guidance to take on hazardous jobs in complex environments that are not suitable for humans. More discussion on this topic is available in Sect. 4.6.
- developing more generalized theories and machines that can self-detect, adjust and determine the selection, collection, representation, analysis, mining and learning of the most relevant, sufficient data and problems, and provide theoretical analysis to guarantee the perfect fit between data results (see more about fitting scenarios in Sect. 3.5.3).

Figure 6.5 summarizes the migration and expansion from processing and analysis to informatics and advanced analytics, which essentially covers many areas including pattern recognition, data mining and knowledge discovery, various data analytics, machine learning, text mining, and anomaly detection.

Figure 6.6 summarizes the main bilateral connections and complementarity between intelligence science and data science in their interdisciplinary fusion and evolution. The fusion supports data analytics to various advanced analytics, data-driven discovery, model-based learning, and data+model-based analytics. It further evolves towards deep modeling, deep analytics and deep learning of complex data and problems.

6.6 Computing Meets Data Science

Computing forms a fundamental component of data science, thus the role computing plays in data science and its contributions are discussed in this section. Data science challenges existing computing paradigms, architectures, and systems, and is driving the revolution in computing.

6.6.1 Computing for Data Science

In general, *computing* refers to any calculation-oriented representation and includes design, development, processing and management activities, systems and support, all of which are nowadays often enabled by an algorithm and conducted on a

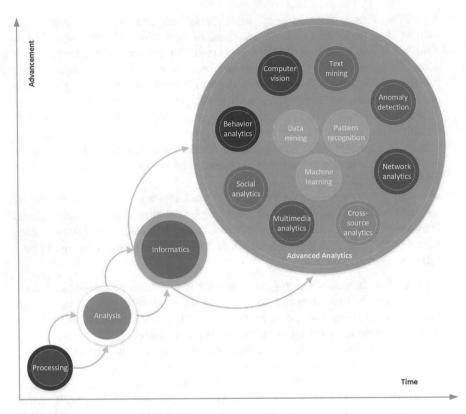

Fig. 6.5 From analysis and processing to advanced analytics and learning towards data science

computer. The definition of computing can be broad or narrow in meaning and scope
[424], and computing is not necessarily computer-based, although that is the general
understanding today.

In data science, computing has a multi-faceted fundamental role in implementing
and achieving the data-to-insight and data-to-decision transformation, including

- providing the support infrastructure, architecture, programming language and
  platform, hardware and software, and enabling systems,
- processing, analyzing and managing data,
- undertaking analytical algorithm design, programming, implementation and
  execution,
- ensuring computing performance for specific and general needs such as scalable,
  efficient and online computing, and
- developing interdisciplinary algorithms for certain domains or involving certain
  disciplines, such as statistical computing and behavior computing.

In today's data science courses, the focus is often on a specific function of
computing, such as programming with R or Python. This is not sufficient for

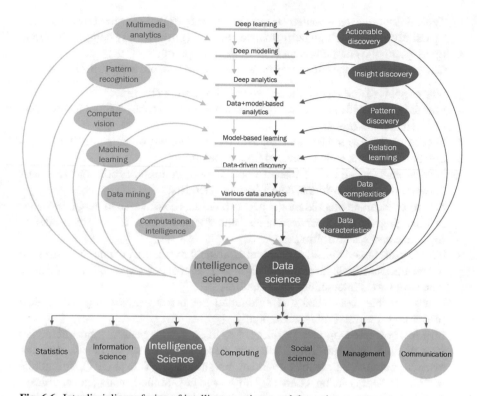

Fig. 6.6 Interdisciplinary fusion of intelligence science and data science

training a data scientist. Relevant computing capabilities include, but are not limited to, thinking, architecture and infrastructure, algorithm design, data structure, computational complexity, programming, data management, enterprise application development and integration, operationalization, and deployment.

Data engineering is another term that is highly relevant to computing. *Data engineering* usually refers to the process and quality assurance of the extraction, preparation, processing and management of data by data engineers for analysis by data scientists.

6.6.2   *Data Science for Computing*

Data science is significantly challenging and is driving the development and evolution of computing. Computing infrastructures, architectures, platforms and tools need to enable faster, larger, more parallel, more distributed, more hierarchical, and more complex processing, analytics and management of big data.

Typical *data science-oriented computing* techniques include distributed computing, parallel computing, high performance computing, in-memory computing, cloud computing, mobile computing, and embedded computing.

Computing challenges driven by data science include, but are not limited to:

- how to analyze and learn petascale data online and in real time;
- how to process and analyze cross-source heterogeneous data without information fusion and transformation;
- how to generalize models and algorithms for different domain-specific (cross-domain) problems with automated domain adaption;
- how to enable online and instant coordination, communication and feedback between models/algorithms, data and data scientists;
- how to overcome potential model generalization, robustness and adaption issues in changing data and environments, and for new data, to ensure modeling resilience and sustainability;
- how to present and visualize data and analytical results that cannot be presented in conventional tools, for example, high-dimensional heterogeneous cross-source data with varied interactions and relations;
- how to enable unattended and automated but privacy-preserving online data processing, analysis and presentation;
- how to form one-stop user profiling, activity/behavior patterns and interaction by real-time consolidation of relevant resources from social media, working systems, entertaining, mobile applications, Internet of Things (IOT), and public services including living, connected home and city council-managed activities;
- how to understand the characteristics, dynamics, uncertainties, principles and feedback/responses in coupled human-natural systems in changing environments, which may involve scientific, socio-economic, cultural, political and environmental aspects and do not follow the usual partial equilibrium models;
- how to enable open data and open access while also supporting analysis, reasoning, decision and recommendation at scale and with privacy preservation;
- how to ensure the transparency and reproducibility of analytical outcomes in sophisticated data and environments;
- how to evaluate and ensure data-driven evidence-based proposals and accurate and reliable discovery, inference and insight when the data, environment and ground truth are unknown;
- how to transform conventional science research and innovation, education, business, production, and decision-making to be data-driven, evidence-based and actionable while also validating domain and expert knowledge.

Figure 6.7 summarizes the main bilateral connections and complementarity between computing and data science in their interdisciplinary fusion and evolution. The interactions provide analytics infrastructure and programming, support high performance analytics, distributed analytics, parallel analytics, scalable analytics, and cloud analytics.

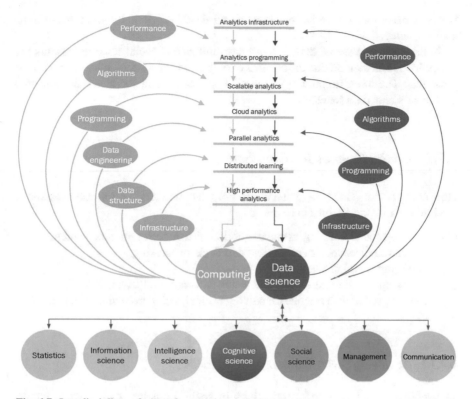

Fig. 6.7 Interdisciplinary fusion of computing and data science

6.7   Social Science Meets Data Science

There are an increasing number of conventional driving forces in business, society, economy, politics and other social areas that promote the interdisciplinary synergy between social science and data science to address the many challenging problems and revolutionize existing social theories, areas and systems. Other drivers are the new economy and technology-driven social revolution, such as the Internet of Things (IoT) connecting and bonding societies, the expanding online social activities in social networks and social media, online shopping, entertaining, dating, and socializing, and the online-based socialization of traditional media and publications.

The synergies between social science and data science are: (1) data science for social science, and (2) social science for data science. The social science subjects include but are not limited to political science, sociology, behavioral science, anthropology, economics, finance, communication arts, human development, marketing, family studies, history, public administration, and culture. While there are many domains that have a social focus, we are here particularly concerned with scientific issues in behavior, economic, social and cultural (BESC) domains when we refer to social science.

In this section, we first discuss the opportunities that social science presents for data science and then outline the benefits of data science for social science and future prospects. The interdisciplinary fusion of data science and social science forms a new area: social data science.

6.7.1   Social Science for Data Science

How could social science benefit and promote data science? This section discusses the issues and opportunities in terms of

(1) involving social thinking, social methodologies, theories, methods and tools for data-driven discovery, management and decision-making;
(2) addressing social problems;
(3) discussing social features in data science and data products; and
(4) inventing social data science for solving critical social problems associated with data.

6.7.1.1   Involving Social Thinking and Methods

In handling data issues and applications in social domains, data science has to involve social thinking, methodologies, theories, methods and tools in data-driven discovery, management, and decision-making.

Social thinking refers to the mental skills, traits, habits and activities involved in thinking about and collaborating with others, sharing workloads and responsibilities in data science teams, developing social norms and governance for managing data discovery projects, and building structures for communication and consensus building.

Social theories and social methods that may be applied in data science consist of experimental design, sensor data selection and sampling, negotiation, planning, team and coalition formation, group decision-making, consensus building, qualitative research and methods, behavior study, organizational behavior, relationship management, leadership, and social evaluation methods and tools.

Typical pitfalls and myths about the involvement of social thinking and methods in data science problem solving include

(1) understanding the limitations of a single science, for example, whether data science findings reflect the true stories in social systems;
(2) adhering to data-driven and evidence-based thinking and exploration, rather than being restricted to the pre-defined and prior knowledge, hypotheses, and expertise obtainable from social science;

(3) maintaining a necessary balance between conflicting thinking and results, especially when it is not possible to evaluate which makes more sense; and
(4) resolving potential inconsistencies and unforeseen events related to prior knowledge, expectations and assumptions; for example, identifying a new understanding of the working mechanisms and constraints in political gaming.

6.7.1.2 Inventing Social Data Science for Social Problem-Solving

Increasing amounts of BESC-related big data are being generated, particularly by BESC-based online, mobile and social applications. They record human BESC activities, interactions and outputs collected from both well-established economic, financial and cultural systems, businesses and platforms, such as stock markets, banking and insurance businesses, and new economy-enabled channels and systems. The collection, transport, and application of this BESC data is achieved through computerized social channels, including distributed sensors, IoT, the Web, mobile devices, and social media websites and applications, as well as connected home, societal and personal devices, such as embedded systems, smart living and city equipment and facilities, and mobile devices.

BESC and other actions and attitudes in society bring about new opportunities for social data science and social problem-solving, for example,

- emerging rich opportunities to invent new learning and analytical theories, techniques, tools and systems on BESC data;
- the creation of data science for particular domains, such as social data science, behavioral data science, economic data science, financial data science, and cultural data science;
- an in-depth understanding of the nature, dynamics, impact and management of social problems; and
- the promotion and communication of data science outputs of social issues to the broader community to achieve greater social impact.

The corresponding social data science research and practice will undoubtedly create new opportunities and subdisciplines in data science, and also promote the evolution of social science by incorporating new ideas, theories, tools, methods and approaches into BESC and other social areas of research.

6.7.1.3 Social Features in Data Science and Products

In sociology, *socialization* plays an important role in ensuring the inheritance and dissemination of social influence, norms, customs, values and ideologies in the process of transforming individuals into societal communities. Individuals follow, conform to and adapt to social systems and are influenced to aim for certain behavioral outcomes. *Society* instead builds consensus, resolves conflicts, and evolves with the summation of individuals' behaviors and the emergence of collective intelligence.

Some transformative social features of the era of data science, compared to more structured and hierarchically concentrated societies and control systems, are openness, flat structures (rather than conventionally hierarchical societies), self-organization, interaction, uncertainty (including undetermined authenticity and soundness), and scale.

Our discussion on these social features in the data world are extended below.

- *Open*: openness is a fundamental feature differentiating the era of data science and data products from conventional sciences and business. It is embodied through key terms that include open source, open data, open government, open access, open science, open innovation, open economy, and open service. The term "open" in this sense means freely available and transparent to the public with limited or no constraints.

- *Flat-structured*: in the data world, structures and categorizations between entities (end users) are flatter than in conventional hierarchical systems such as political systems. "Flat" here means very limited or no hierarchy between entities; every entity in the data world is equal, sharing similar rights and responsibilities. Typical hierarchical categorization in conceptual systems does not have a major role, while equally positioned peer-to-peer relations are often maintained. A typical example is the tagging system used in social media and social networks, in which tags can be anything, and there may be no hierarchy between tags.

- *Self-organizing*: entities act, react and interact within very limited rules (constraints), albeit in a disordered manner, at a local and individual level. It is from the amalgamation of entities that the overall order and collective intelligence emerges. Social media, social networks, crowdsourcing, and open source data products evolve in a similar spontaneous process.

- *Interactive*: interactions are essential elements in data systems. Interactions may take place on multiple levels, e.g., local, medium and global, in different forms, e.g., logical and probabilistic, within and between hierarchical entities, e.g., pixels, regions and views. Interactions are broadly embodied and described as "relationships" or "relations", which may be further represented and formalized in terms of association, correlation, dependency, uncertainty, and broadly coupling relationships.

- *Scalable*: Although the focus of scientific studies is often complexity, scale significantly differentiates the various data products and systems in the era of data science from those previously analyzed in statistics and mathematics, and challenges the methodologies and approaches developed in classic analytical and learning systems. Critical challenges facing conventional statistical, mathematical, and computing-based analytical and learning systems are the need to evolve existing theories, models and tools to large-scale data (especially to terascale and petascale data). Learning and analytics at scale drives the need for, and development of, data science and data products, and data economy emerges when data-driven business moves to large scale data practices, such as high frequency algorithmic trading, crowdsourcing, open source, social media and network, mobile services, and online business.

- *Uncertain*: Data in the open world are uncertain. On one hand, data with free access and data products that enable free collaboration and interaction make data big and appealing to everyone, not only data governors, engineers and specialist users, especially in the public domain of social media and online blogs. This data creates a significant new economy—big data business, represented by enterprises such as Facebook and Twitter. On the other hand, it is increasingly recognized that open data (even open innovation and open science) are probably fundamentally different in nature from classically closed data and data systems (e.g., management information systems in an organization) in terms of authenticity, accountability, transparency, reproducibility, and trustfulness. This becomes a serious issue when data is created, used, manipulated and shared for social and political purposes by special interest groups, e.g., competitors in political campaigns, fraudsters in online business, and hackers in intelligence systems. Consequently, data uncertainty and data product uncertainty become increasingly serious problems when accountability and authenticity are no longer immutable, transparent, or provable. Snke Bartling and Benedikt Fecher have examined the technical aspects of blockchain and discussed its application to research, with a view to strengthening verification processes in science. Their aim is to make research results reproducible, true, and useful while avoiding misinformation, misuse, and over-use.

6.7.2 Data Science for Social Science

Data science will create new opportunities and tools to assist in and strengthen the evolution of social science, making it more evidence-based, quantitative, practical, coherent and unified. Typical opportunities include

- Promoting a paradigm shift and cultural change in social science,
- Creating new social science theories and methods, and
- Addressing emerging social, legal and global issues with data science.

6.7.2.1 Promoting Social Science Paradigm Shift and Cultural Change

In synergizing data science with social science, a number of initiatives and views [144, 400] have emerged from the social science community. They are concerned with the relationships between data science and social science, and the impact, positioning, and prospects of data science on social science.

Data science has the potential to enable a shift in social science from a *hypothesis-driven paradigm* to a *hypothesis-free paradigm*. In general, social science pursues the following approach: a new theory or hypothesis is proposed and corresponding data is gathered to support, explain and test the theory. Data science enables a new way of building social science theories. *Social data science* observes social phenomena, creates exploratory findings in relevant social data, and then builds and evaluates a new social theory (if required) and interprets the observation.

Data science is changing the social culture of the community, and transforming social science research methodologies and epistemological methods. This generates a paradigm shift for social science transformation:

- from holism and system science to reductionism to systematology;
- from qualitative to quantitative research;
- from small world to large world;
- from diversified debate and argument-based research to quantitative evidence-oriented empirical study;
- from discipline-specific research to cross-area, inter-field and cross-disciplinary interactions, collaborations and consensus building within the broad social science family, including sociology, humanity, public relations, marketing, finance, economics, law, and politics; and
- from simulation and story-telling to real data-driven pragmatic approaches and applied empirical work through social analytics.

Data science theories, infrastructure, tools and systems enable challenging social issues to be handled and big social questions to be answered that could not previously be addressed in the social science community by social science theories and tools. This is enabled by

- more powerful computing systems,
- scaled-up social study samples,
- complementary theories and tools for data-driven discovery, and
- enabling technologies for deep analytics and learning.

Typical examples in practice include the development of high frequency algorithm trading strategies and the learning of high-dimensional dependence across markets [418]. The former requires high performance computing to discover trading signals within high frequency data streams, and the latter requires the capture of sophisticated relationships and interactions between heterogeneous market indicators. These requirements are beyond the capability of finance and economics. Other examples include coupled behavior analysis, such as pool manipulation [74] and financial crisis contagion [80].

At the same time, data scientists need to be more cautious about what they can claim from social data research. When large-scale social data is used for evidence-based social discovery, it has good chance of generating statistically meaningful outcomes which appear to be sound and reliable, and the findings may look objective and applicable to other similar themes. An important issue is to check to what extent the data, level, scope and learning objectives of the findings are solid and applicable.

6.7.2.2 Creating New Social Science Theories and Methods

Our society is facing a transformative data deluge, from which new scientific, economic, and social value can be extracted. *Data-driven social discovery* provides quantitative and evidence-based methods and opportunities for problem-solving challenging social issues. This includes the creation of new social science methods uniquely enabled by theories and technologies in the data-driven science, technology, engineering and mathematics (dSTEM) disciplines. Data science can make social science more quantitative, evidence-based, and hypothesis-free; create new ways and thinking to handle social materials and documents, sentiment, organizational behavior, government, and policy; and enhance the qualitative evidence.

Data science can support the move from small world-based social research to large world-oriented social research in studying complex social issues and artificial social issues in a large population community and network. Data science can support experiments and qualitative research in terms of much more sophisticated data volume, variety, timespan, populations, and analysis units, and can understand the respective dynamics, working mechanisms, social influence, and political mobilization.

Data science can handle social systems, structures, problems and questions that cannot be addressed well by existing social theories and methods alone, e.g., algorithmic trading, cross-market manipulation detection, and can understand the dynamics and contagion of financial crises in financial markets. One example is the solution of global social issues by crowdsourcing approaches, in which solutions are developed and refined by groups of people based on particular collaborative modes and tools.

Data science can find better ways to fight severe social problems and challenges, such as fraud, risk management, counter-terrorism, unnecessary hospital readmission, pharmaceutical evaluation, civil unrest, and smart living in urban environments and the home. Data science can also fundamentally change the way political policies are made, evaluated, reviewed and adjusted by providing evidence-based informed policy making, evaluation, and governance.

Data-driven social discovery can assist in experimental social science, provide online facilities for social research, and identify hidden and silent social factors and social system dynamics to disclose the intrinsic nature of social problems. For example, social security data analytics has played a critical role in identifying, recovering, preventing, and predicting social welfare overpayment, fraud, and policy performance in Australia [58], which could translate to billions of dollars' debt per year.

Several initiatives for creating new social science theories, methods, and education have been proposed and are under exploration internationally. Examples are

- the perspective on big data and social science taken by the UK Economic and Social Research Council [153];

- the Social Data Science Lab sponsored by ESRC [354] which focuses on risk and safety, cybersecurity, health and wellbeing, demographics, HPC and scalability, and business and innovation;
- the research on data mining and machine learning methods and algorithms for analyzing social phenomena and social behavior of users on the Web and in online communities at the Computational Social Science (CSS) department [108];
- the Big Data Social Science program at the University of Pittsburgh [27] which is an integrative education and research program in social data analytics; and
- the Data Science for Social Good summer program [142, 184] organized at the University of Chicago, University of Washington, and Georgia Technology University.

6.7.2.3 Addressing Emerging Social, Legal and Global Issues with Data Science

There are some existing and emerging social issues that may be well addressed by data science alone. Other social issues exist in data science, called *social data issues*, which need to be addressed by integrating data science with social science. Here we illustrate some of them.

First, *the imbalance, unfairness and conflict of interest between data contribution and data values ownership*: Imbalanced opportunities and social problems are created by owning and using big data. Since big data is owned and controlled by certain organizations and people, its commercial values, social impact, and commodity are also controlled by those owners. This has formed the data economy, which has been driving the world economy, lifestyle and societal change for the last 10 years. However, it is causing new social problems, such as breaches in the privacy of data contributors, unfairness between the roles of those who contribute data and those who own the data value, and revenue imbalance between the holders of different data roles.

Solving these problems requires new social science, policies and systems; for example, how can the economic and ownership relationships between the various roles in a data system be managed? Data science can acquire, analyze, and present relevant data, evidence, indicators, and recommended solutions. Comparisons between different benefit distribution and interest conflict resolution models can be supported by data science. Data-driven findings can help understand the formation, dynamics and evolution of underlying problems, the interactions between different roles, and various models and modes for the possible management of stakeholder relationships.

Second, *the effects of improper socialization, globalization and governance of data societies*: The term *data societies* refers to virtual data-based social communities. Typical *data societies* are online games-based societies, social networks, social media communities, and mobile messenger-based societies. Global social activities and the lack of proper governance of data societies are challenging social norms and systems, as well as the traditional ethics and management of privacy, accountability, trust, security, law, and ethical practice in physical human societies. Examples are malware, phishing and shilling attacks, online/Web bullying, money laundering, unethical behaviors such as boosting extremism, terrorism, violence, and crimes against children. These issues are becoming more and more serious when new techniques are created to generate new data societies with new governance and operational systems.

To address these issues, social scientists need to study the nature of *data-driven societies*. New *data society governance* theories, systems and mechanisms are required to maintain social justice, ethical practice, and privacy in data societies, and to sustain the balance between data societies and human societies. It is challenging to make the human society norms, governance systems and regulation systems adaptive to the ever-evolving data societies. The differences between data societies and human societies therefore need to be well understood. Active, preventive, and regulated risk management requires customized and effective country, culture, religion, race-independent policies and systems to be built, shared, accepted, and applied by global communities.

To achieve this, social scientists need to work with data scientists or apply data science to collect and analyze relevant data while also taking a social science approach. Data-driven social analytics on relevant data or other resources, especially social information, could identify evidence of breaches or other issues, present the processes, behaviors, and activities of the underlying stakeholders and their interactions, and disclose hidden motivations and intent that may drive the occurrence of the identified issues. This understanding has resulted in the proposal and study of the new concept of X-informatics, such as social network analysis, social media analytics, economic computing, behavior informatics, and cultural computing. These X-informatics have been proposed to address social, behavioral, economic, and cultural issues from the informatics and computational perspectives.

Third, *the abuse, misuse, overuse and underuse of data and data products for criminal, illegal, and malicious purposes*: Accidents and risks of abuse, overuse, misuse, and underuse of data, data findings, data power and their social influence increasingly occur in data societies and data-related business. Examples are the dissemination of misleading and false information online, over-servicing by health-care providers to take financial advantage, the misuse of confidential information by private email services, and the blocking of the access and circulation of certain information online and in social media.

Another typical example can be seen in the prominent role played by digital media in the 2016 US presidential election. The manipulated dissemination of information on social media platforms including Twitter, Wikileaks, and Facebook fundamentally upset traditional social mechanisms from polling to mainstream media, which failed to adequately and effectively capture public sentiment around political events. It also significantly challenged mainstream US political discourse and boosted the populist, extremist, and post-truth politics of groups that were previously on the fringe of political culture.

We are stepping into an era in which news is manipulated and misinformation abounds in social data societies. The misuse of the Internet, mobile and social media technologies and systems carries a high risk bringing disorder not only to data societies but also to human societies. This is because artificial societies and physical societies are so deeply interconnected, co-evolving, and co-influenced. It has never before been possible to spread malicious activities in one society so widely, deeply, and promptly in another society, and to have an unexpected impact on that society while raising awareness and self-reflection in the grassroots society.

The effects of such actions not only compromise the development and evolution of the data society, they also deform human society. If no global action is taken on data society regulation, governance and management, such negative co-influence and co-development could significantly change, damage or destroy existing achievements within human society, including economic and cultural globalization, democracy, fairness, transparency, harmony, and sharing principles. They will also impact the positive effect of appropriate applications of the Internet, social and mobile technologies that enable people to be more globalized, benefit from cross-culture and cross-country trading, and engage in tackling social, economic and living challenges with new technologies, economy and collaborations.

Data science needs to be in the foreground of detecting, analyzing, predicting, attacking, and destroying the criminal, illegal, and malicious activities, behaviors, and objectives. The above activities are often data-driven, hidden, and purpose-driven, thus special data discovery theories and tools and cross-organization, national and global cooperation are necessary to collect relevant data, evidence, and resources. Deep modeling, behavior construction, scenario replay, trace identification, intent modeling, and manipulation detection could be undertaken. This requires collaboration between data scientists, domain experts, and social scientists, as well as representatives of relevant organizations, including data service providers, regulators, legal authorities, and ethicists.

6.7.3   Social Data Science

In the above sections, we have discussed the two-way interactions between the data world and the social world: social science for data science, and data science for social science. The interdisciplinary fusion of social science and data science triggers the emergence of a new area: social data science. *Social data science* is the data-driven science for social principles and social problem-solving theories and methods.

From the data science perspective, businesses and activities related to typical behavior, economy, society and culture (BESC) [28] social domains have attracted significant interest in data science and analytics. This builds on the marriage between social activities and analytics, computing and informatics, and has resulted in new interdisciplinary areas, including behavior computing and behavior informatics [72], economic computing, social computing, social media analysis, social network analysis, and cultural computing, to name a few.

There are several reasons why data science needs social science, and some of the driving factors are as follows:

- To involve social thinking, theories, methods, and tools in data science research and development;
- To establish and maintain social norms, ethics and trust in handling data science problems, and in the data world;
- To access and understand BESC or other social business-generated data and applications;
- To tackle social issues and problems in an open, scalable and deep manner by involving social data; and
- To socialize data science and data products and translate data science into socio-economic benefits for academic, governmental, and non-governmental organizations.

The value of data science for revolutionizing social science is in

- Developing and evaluating data-driven and evidence-based proposals in social and public policy-making;
- Supporting data-driven evidence-based design, analysis and decision-making; and
- Mitigating the limitations and impact of classic social scientific research methodologies and tools, expert knowledge-based hypothesis testing, and qualitative reasoning and decision-making.

This *data-driven evidence-based method* draws on necessary and relevant resources from a range of sources in various forms, extracts knowledge and connections across sources, discloses intrinsic relationships and working mechanisms that cannot be directly disclosed from data observations, supports simulation and trials in controlled social experiments, and incorporates rapid feedback to optimize design, operations and outcomes.

The bilateral interactions and synergy between data science and social science will upgrade and transform resilience and sustainability in social science research and problem-solving, evaluate and optimize the impact, equity, efficiency and efficacy of social policy and social problem management, and ensure the ethical development of data science.

To enable social data science, bilateral learning, fusion and co-evolution of social science and data science are necessary. Data scientists and social scientists need to be more open to each others' ideas and theories. Data science researchers are learning sociology and BESC-related theories and techniques in order to compute and quantify BESC and other challenging data-related social phenomena. Data science courses should include knowledge components and real-problem case studies of social science issues. Social science workers should rethink their existing theories and tools, scale up their approaches, and adopt data-driven thinking and approaches for the next-generation of social science.

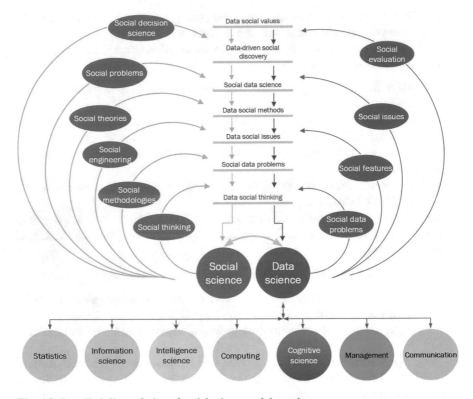

Fig. 6.8 Interdisciplinary fusion of social science and data science

Figure 6.8 summarizes the main bilateral connections and complementarity between social science and data science in their interdisciplinary fusion and evolution, and the formation of social data science. The fusion enables data-driven social thinking, addressing social data problems and data-based social issues, and inventing data-driven social methods and social data science through data-driven social discovery, to achieve the goal of discovering data social values.

6.8 Management Meets Data Science

Management science and organizational management have been widely recognized as closely relevant to data science. Management meets data science for good reasons, including

- There are management roles and responsibilities in data science projects and problem-solving;

- There are management issues in data science that need to be addressed by management theories and methods;
- Data science can provide better or new answers to address the issues and limitations found in classic management theories and systems;
- Data science can promote the evolution of management science to another generation.

Accordingly, in this section, we discuss

(1) How management science and its methods support data science;
(2) How data science benefits and transforms management science; and
(3) A new area: management analytics (or management data science) as the interdisciplinary fusion of management science and data science.

6.8.1 Management for Data Science

Management science refers to the interdisciplinary study of operational and management issues and managerial problems, and the better administration of time, money, resources, operations, and performance. Management science originated in operations research; it has unifying effects on, and strong connections with, a broad range of areas and subjects including communications, accounting, micro/macro-economics, organizational processes, psychology, probability, statistics, and optimization, as well as engineering, computing, business, marketing, finance, and other sciences.

Typical management theories and methods consist of operations research, mathematical modeling, applied probability and statistics, econometrics, business analytics, optimization, and dynamical systems theory.

The role of management in enabling data science is multi-faceted. Management science

- directly addresses management issues in data science problems and systems; and
- consists of valuable management thinking, theories and methods which are necessary for building management-related disciplinary components in data science.

We discuss these aspects below.

6.8.1.1 Addressing Management Issues in Data Science

Data science problems and systems naturally involve management issues, forming *data science management*, which are related to the management of data science input (data, resources, models and methods), data science projects and processes, data science output (the resultant models and methods; identified knowledge, insights and intelligence; and deliverable data products).

Data science input management involves data management, resource management, requirement management, and objective management.

- *Data management* and *resource management* is to collect and manage relevant data, information, resources, devices, equipment, infrastructure, and computing facilities that may be required to conduct a data science project.
- *Requirement management* is to acquire and manage functional and nonfunctional requirements related to data science problem-solving and projects.
- *Objective management* is concerned with objectives, goals and milestones to achieve the objectives in a data science project.

Data science projects and processes involve project management, data team management, model management, process management, communication management, risk management, and quality management.

- *Project management* is concerned with the scope of the project, goal definition, milestone definition and check, workload assignment, costs, budgets, and related issues, as well as project planning, progress reviewing, timetabling, and so on.
- *Team management* involves the definition, selection, and management of data science roles and responsibilities in a data science project, collaboration workload assignment, work patterns, team management, hierarchical leadership, reporting, scheduling, performance, and reflection.
- *Model management* is concerned with the models, methods, algorithms, systems, and programming that directly manipulate data to achieve the defined business objectives.
- *Process management* deals with the definition of processes, goals and activities for each step, the personnel required to undertake each activity, connection and transfer between procedures, and the data, information, documentation, and models associated with each step.
- *Communication management* handles the documentation of the project, process, roles, input and output, and drives the communication between roles, activities, and stakeholders.
- *Risk management* handles the possibility and severity of risk, including the effects associated with each step of the process and the impact on personnel, the data selected, the modeling method, the communication design, and the evaluation mechanisms. Risk may be evaluated in terms of a specific aspect of the project or the overall risk from the perspective of resources, technical or economic component, timing, market value, and more.
- *Quality management* defines existing or potential quality issues and develops measures for quantifying quality issues.

Data science output management consists of knowledge management, product management, testing management, and possible deployment management.

- *Knowledge management* administers the resultant intellectual property of the delivered models, algorithms, codes and systems; the identified findings including patterns, rules, exceptions, and other analytical results; the lessons and insights gained in a data project concerning optimal operation, management, and decision-making; the evaluation and reflection on the scope and objectives, team, process, communication, risk control, and quality assurance of a project.

- *Product management* refers to the management of data products or broad data-related deliverables. This may involve product definition and specification, as well as issues related to lifecycle, quality, end users, usability, market value, and market segment.
- *Testing management* of data science output oversees the various testing methods on different granularities, test specifications, test result analysis, testing-based adjustment, refinement and optimization.
- *Deployment management* is to manage things related to data output deployment. This may involve the definition and specification of the output or data product, scheduling, stakeholder relationship, modes of execution, and problem reporting and management on deployment.

6.8.1.2 Building Management Components in Data Science

The discussion in Sect. 6.8.1.1 on management issues related to data science projects and problems illustrates the necessity of building management-related disciplinary components within data science from disciplinary and knowledge perspectives. Data science as a new discipline requires such components to ensure its completeness and self-containment.

On one hand, respective subjects and disciplinary components in classic management science will still play an important role in data science, which comprises the different types and functions of management discussed in Sect. 6.8.1.1. Typically, data management, project management, team management, product management, quality management, deployment management, and risk management are also essential features of data science projects and problems, creating a large area of data science management.

On the other hand, data science projects and problems have specific characteristics that differentiate them from ordinal IT projects and even business intelligence projects. Comparative characteristics are that they usually feature: (1) much wider involvement of both directly and indirectly related resources, as well as private and public resources, (2) many more diversified roles, from data to business-oriented roles, and from technical to management-related roles, (3) the involvement of people with higher qualifications and higher levels of expertise who can cope with deeper, more challenging tasks, and (4) data science teams that are encouraged to create a more personalized, dynamic, non-standardized workplace culture and work patterns.

Specific or customized management components need to be invented to cater for both general and specific management needs in data science. As a result, *data science management* may build specific management theories and methods for areas such as data management, data role management, analytics management, data process management, data issue management, data product management, and management analytics.

- *Data management* should have greater technological scope than database management. In data science, data management involves multiple types, channels, modals and domains of data and resources. It also has to manage heterogeneous data sources and data generated during data science projects, as well as address critical data challenges such as volume, variety, velocity, and veracity, and the distributed management of data, resources, facilities associated with a distributed data project.
- *Data role management* needs to manage the responsibilities, status, and performance of the data roles required to conduct a data science project (see more discussion about data roles in Sect. 10.3).
- *Analytics management* is concerned with the management of analytical goals, roles, processes, models (including algorithms), results and evaluation in an analytics task.
- *Data process management* handles aspects and issues related to the process of data manipulation and analytics, which may be a trial and error process, a training to testing and validation process, or an iterative and generative process.
- *Data issue management* tackles issues related to data quality, data ethics, and social issues in a data science project.
- *Data product management* handles issues related to products and addresses the difference between data products and traditional products. It involves different product modes, user communities, sale channels and modes, testing methods, and marketing strategies.
- *Management analytics* is a new area that analyzes management-related data to detect and predict management issues in a data science project.

6.8.2   Data Science for Management

In management communities, incorporating and integrating data analytics, data science and relevant technologies into management science and engineering has become a new trend, driving the emergence of *evidence-based management*. Business analytics and management statistics are the usual vehicle for furthering this vision and mission.

Data science can significantly upgrade management research and handle challenging management issues that may not be effectively addressed by conventional management theories and methods. This can involve many business domains and areas [180]. The ways in which data science could contribute to management research enhancement and revolution are multi-faceted and could

- provide better answers and methods to address management questions than was previously possible;
- pose new questions to upgrade and reinvent existing management theories; and

- address questions from new perspectives for next-generation management science.

These aspects are briefly itemized and discussed below.

Providing better answers and methods to address existing management questions: Data-driven research on management data, behaviors and business can provide more precise, quantitative, and extensive sample-based evidence for improving management operation, organization, strategy, behavior, risk, and resources, as well as contributing to decision design, review, adjustment and performance.

It can also provide clearer causal mechanisms and more informed organizational and management responses to risk, emergencies and disasters; more quantitative modeling, estimation, prevention and intervention in failures, accidents, at-risk behaviors, and systemic resilience in such areas as transport, insurance, and health, as well as early identification and warning of management problems and risk, and strategy recommendations for retaining and caring about important employees, better staff leave management, and better human resource management. This requires more advanced analytics to enhance and revolutionize existing business analytics and management statistics.

Posing new questions to upgrade and reinvent existing management theories: Existing management theories and methods may fail to address some complex management issues, or may generate misleading answers. A typical example is the unreliable polling in the 2016 US presidential election, in which the election result ran counter to poll predictions. This demonstrates the limitation of existing methods in the data era. The application of advanced data theories and technologies can disclose hidden phenomena behind observable polling and survey results, capturing comprehensive community-wide explicit and implicit movement and sentiment by involving polling/survey-independent data. Data-driven approaches can inform more personalized and targeted advertising, marketing and recommendation which are usually more public and general and often lead to irrelevant, duplicated, wrongly targeted results. Evidence, indications and informed arguments can be extracted by data science from relevant data and business, and the roles, processes, and organizations and media channels can enable more informed decision processes and results about operations and strategies, and can gain a more accurate understanding of corporate sentiment and relationship management. These examples show the opportunities available as a result of strengthening and reinventing management theories and methods.

Addressing questions from new perspectives for the next-generation management science: The significant changes driven by data economy and data societies, and the integration of the data world and natural human/business worlds pose new questions that fundamentally challenge existing theories, methods and systems in management. New questions emerge that have to be addressed by new theories. Examples are the active, systematic and comprehensive reporting of corporate productivity, working patterns, and effects of team collaboration in a multinational enterprise. Achieving this requires thinking beyond traditional business analytics and management statistics, and involving large scale operational, management, production, marketing, performance, communication, and public data. Other examples include quantifying the efficiency and optimal supply/demand planning of public utilities such as water, energy and transport; the proactive and tailored strategies and systems for the emergency management of very large gatherings; highly effective detection, reasoning, evaluation and mitigation strategies for regional and global environmental, pollution and climate change management.

Large scale management data science needs to be invented. In management data science, enterprise business analytics theories and methods have to integrate multiple disciplinary knowledge, skills and tools. Default experimental settings should include multiple sources of business data, social and sentiment data, and public data to build a comprehensive and deep picture of management-related issues, processes, behaviors, causes, effects, together with detection, prediction, intervention processes, systems, strategies, and policies.

6.8.3   Management Analytics and Data Science

Management analytics (or management data science) is a very important new area worthy of significant effort, since management data is comprehensive and involves many different entities, aspects, steps, roles, and performance indicators, as discussed above. *Management analytics* is the data-driven interdisciplinary study of operational and management issues and managerial problems through the unification of theories, methods and tools in management science, analytics science and data science.

The analytics of management data in a data science project may serve many purposes, for example,

- detecting and predicting data quality issues and associated factors and indicators, as well as devising strategies for improving data quality,
- identifying, analyzing, quantifying and predicting team performance and collaboration productivity, discovering low-performing and less collaborative roles and areas, suggesting strategies to convert a low-performing team to high-performing team,
- detecting, analyzing, and predicting issues, blocks and poor performance in data processes, identifying the reasons and determining possible actions to adjust and optimize these processes,
- predicting, discovering, and quantifying risk factors, risk areas, and risk severity and likelihood, and recommending strategies and actions to mitigate risk,
- analyzing the reasons for poor decision-making outcomes from the process, information and evidence provided, examining decision quality, and providing evidence and supporting information for coherent decision-making.

Figure 6.9 summarizes the main bilateral connections and complementarity between management science and data science in their interdisciplinary fusion and evolution. The fusion introduces management thinking, theories, and methods,

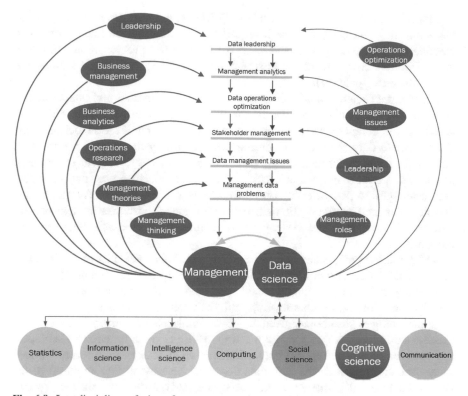

Fig. 6.9 Interdisciplinary fusion of management science and data science

including operations research, business analytics, business management techniques, and leadership, into data science. It enables management data problems, data management issues, and data managerial issues to be addressed, and supports the optimization of data operations. It also enables management analytics to be addressed, and the building of leadership in data science teams and research.

6.9   Communication Studies Meets Data Science

The term *communication studies* refers here to *human communication, business communication* and the role of communication, presentation, visualization, and interaction with executive stakeholders. The other functions related to information-based communications, including data communication, telecommunication, and information theory-based communication, are included in the informatics and computing disciplines of the trans-disciplinary data science family.

While communication studies play a critical role in any scientific activity, they are particularly important for data science for several key reasons.

First, data is at the lowest level of the DIKIW framework (see Fig. 2.1). It has weak meaning and purpose, thus it requires good communication capabilities to elaborate on its meaning, use, and value for higher level processing and management. Different communication skills may result in alternative interpretations and uses of the data's potential.

Second, data science particularly aims to achieve a broader and deeper understanding than information science of the implicit and invisible nature and characteristics of complex data, which requires good communication between the cross-function data science team during the exploratory data science journey. A critical challenge in communicating data science results is to present the implicit and invisible nature and characteristics of complex data in a way that is visible and meaningful to data science non-professionals, especially business leaders and decision-makers. Communication attributes thus include visualization and presentation skills, and tools and systems for disseminating results and demonstrating outcomes.

Third, talking professionally and confidently with data scientist peers, business people, social science workers, customers, leadership and management requires compelling and effective interpersonal, oral and written communication skills. The language of the communication and the tools, presentations, and manner of approaching a stakeholder need to be customized and personalized for that specific stakeholder at that specific meeting. Multiple cross-function stakeholders are involved in a meeting, and data science communicators and presenters need to balance and prioritize the materials they provide, paying due attention to the various interests and concerns, agenda, schedule, and question and answer process. A critical challenge in communicating with cross-function groups is to balance the points of interest, knowledge level, and gaps in understanding.

In addition, data science needs to be actionable [54, 59, 77], i.e., *actionable data science* is capable of providing evidence and indicators for social scientists to interpret the findings and values, and for business leaders make decisions. Business people, and business owners and decision-makers in particular, are therefore core members of a data science team. They often own or sponsor data science projects, determine how data science results can be best operationalized, and "sell" the values of data science to the wider community. Accordingly, people from many different backgrounds and business structures collaborate with one another in a data science team (see more discussion in Sect. 10.6). Ensuring and enforcing good communication is thus critical, since the team of members may speak many different languages and have different interests and objectives.

Lastly, to make data science actionable, it is critical to effectively communicate the findings and values of data science in a way that can be easily grasped and adopted by business owners, leaders, and decision makers. To achieve this, capable communication with leadership is a must-have ability of a good data scientist.

More discussion about the skills required for data science communication can be found in Sect. 10.4.2.

Figure 6.10 summarizes the major bilateral connections and complementarity between communication studies and data science in their interdisciplinary fusion

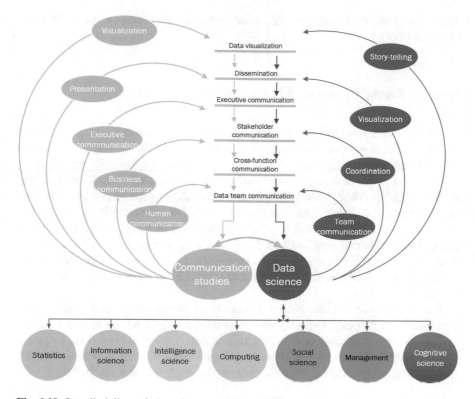

Fig. 6.10 Interdisciplinary fusion of communication studies and data science

and evolution. The fusion enables effective data team, cross-function, stakeholder, and executive communication, and the dissemination and visualization of results and data products.

6.10 Other Fundamentals and Electives

The disciplines discussed above cover a very wide range of theories, knowledge, skills and tools that are directly or indirectly relevant to most data science problems and systems. However, since data science builds on a fusion of existing sciences, other theories, methods and tools may be relevant or necessary for any given data science problem or system investigated.

This section discusses other such areas and disciplinary knowledge, domain knowledge and expert knowledge, together with the innovation and practice that may be needed in data science.

As data science builds on a fusion of existing sciences, in addition to the above discussed disciplines and areas, others may also be involved in demand of specific data science problems and systems. Below, several such areas and disciplines are briefly discussed.

6.10.1 Broad Business, Management and Social Areas

In Sect. 6.7, we discuss the relationships and complementarity between social science and data science. The relationships discussed in Sect. 6.8 are extended to data science with management. Here, we expand the discussion to several areas related to business, management and social science which may be closely relevant to data science. They include sociology, psychology, management, leadership, economics, finance, marketing, communication studies, and behavior science.

These fields contribute qualitative and quantitative thinking, theories, methods, designs, and tools that may be needed in data science for general purposes such as communication, management and leadership in data science teams, and for domain-specific data problems such as economic, financial, marketing and behavioral data analysis.

It is necessary for the above relevant areas to be incorporated into the disciplinary framework, theoretical systems, problem-solving research methods, and tool sets of data science, according to demand. This is especially necessary if data science is to effectively tackle business and social problems, capably handle social and management issues, and cultivate data values for business and social benefit.

6.10.2 Domain and Expert Knowledge

We would like to highlight the importance of domain knowledge and expert knowledge in this section, as these frequently very important for effectively undertaking a data science project, and making data science findings actionable [54, 59]. It is also significant because data science problems and systems naturally involve domain and human complexities (see discussion in Sects. 1.5.1 and 4.2 on domain complexity and human complexity) and domain and human intelligence (see discussion in Sect. 4.3 on domain intelligence and human intelligence).

Domain knowledge refers to both domain-independent and domain-specific understanding of a domain. *Expert knowledge* refers to domain understanding that is built on certain expertise and special awareness about a domain. As data science problems are usually embedded with X-complexities and X-intelligence, involving both domain knowledge and expert knowledge is essential in data science problem-solving.

Before undertaking a domain-specific data science problem, and during its execution, we need to acquire, learn, and reflect on the available meta-domain knowledge, facts, information, conceptual systems, specifications, principles, theories, methods, systems, tools, and lessons. When a data problem is complex, it is critical to seek advice from domain experts about the best way to address the data problem. Domain experts may be involved in data science teams as business analysts, model deployment managers, or decision strategists. In this way, expert knowledge is directly embedded in the data science team and problem-solving process, which assists with defining, executing and managing responsibilities related to business understanding, goal definition, consultation on solutions, experimental design, evaluation methods, and so on.

6.10.3 Invention, Innovation and Practice

Data science as a new scientific field requires the expenditure of considerable effort to build an original research culture that has profound creativity, significant innovation, and reproducible practice.

Data science creates enormous opportunities for inventing breakthrough theories and methods, and for developing cutting-edge technologies that can crack the various challenges and complexities. Successfully engaging in these opportunities requires systematic, professional and long-lasting training at multiple levels. The aim of data science training is to build the mindset, cognitive capabilities, and thinking processes for posing, designing, implementing, evaluating and refining novel ideas, unique observations of market needs, original proposals, creative designs, and innovative implementation.

Special effort should be made to cultivate data science thinking, especially creative and critical data science thinking (as discussed in Chap. 3). People in data science roles should be empowered with personalized, profound, and promising observations and insights about new and better ways to explore challenges, complexities and prospects in data input, exploration, and output.

Data science practices are built on and implement data invention and innovation. Skilled practice requires the formation of data science competence and well-trained competent personnel in the various data science roles. Data scientists as professionals need to receive systematic and practical data science training and education to ensure they have a comprehensive understanding of data science. They also need a substantial grasp of foundational interdisciplinary knowledge and advanced analytics technologies[1] related to the various roles and responsibilities, and they must accumulate experience in relevant domains and projects. Learning from well-grounded and best practice data science case studies is another important task in the training of high profile data scientists.

Lastly, there are many confusing and conflicting arguments and viewpoints surrounding data science and relevant communities, which have led to a number of myths and pitfalls. This is understandable since data science is in such an early

---

[1]See more about the overview of analytics and learning techniques in Chap. 7 and detailed discussion about analytics and learning technologies book [67].

stage of development; nevertheless, inefficiency, risk, high cost or even destructive practices and errors in data science projects could result if improper, misleading, and imperfect strategies, designs and practices are mistakenly implemented. Data scientists thus need to be aware of the various pitfalls and make reasonable judgments about what is appropriate or inappropriate, good or bad, applicable or inapplicable. Achieving this is often difficult, especially when data science tasks are substantially challenging, and this is addressed more fully in Sect. 2.9. Seminars and workshops may be organized for data science teams and especially data science executives to raise awareness of the challenges and encourage strategies to mitigate or avoid possible bias and negative effects.

6.11   Summary

What lays the foundation for data science? This has been a widely debated question in many communities. The nature of data science as a trans-disciplinary, cross-domain, and often problem-dependent new science determines that many disciplines and bodies of knowledge should or could be more or less relevant to this new discipline.

This chapter has selected several disciplines as the hard foundation of data science, and has considered a number of others as the soft foundation. The hard foundation comprises cognitive science, statistics, information science, intelligence science (including computational intelligence), data mining, data analytics and machine learning, and computing. The soft foundations are composed of social science, business and management, communication studies, domain and expert knowledge, and invention, innovation and practice.

We understand it is not possible to include every relevant discipline and area in this framework, and what should be included or excluded is always debatable. The purpose of this chapter has been to paint a reasonable picture of the key components for building data science as a new scientific discipline, and their roles and connections within data science. Detailed discussion of each discipline is excluded, and readers can explore the various disciplines as they wish.

Specific technologies and methods for analytics and learning that align with the foundational structure built in this chapter are outlined in Chap. 7 about an overview of classic and advanced analytics and learning techniques and applications. More discussion on technologies and applications of data science is available in book [67]. The problems and techniques introduced there substantially support and complement the material in this chapter, thus these three chapters will hopefully together form a comprehensive, relevant and hierarchical knowledge map for data science foundations, technologies, and applications.

Chapter 7
Data Science Techniques

7.1 Introduction

In the age of analytics, data analytics and learning form a comprehensive spectrum and evolutionary map that cover

- the whole life cycle of the data from the past to the present, and into the future,
- the analytics and learning from the perspective of known and reactive understanding to unknown and proactive early prediction and intervention, and
- the journey from data exploration to the delivery of actionable [77] insights through descriptive-to-predictive-prescriptive analytics.

In this new era of data science, critical questions to be answered are

- what are the major data characteristics and complexities?
- what is to be analyzed?
- what constitutes the analytics spectrum for understanding data? and
- what form does the paradigm shift of analytics take?

The focus in this chapter, in addition to addressing these questions, is on providing a review and overview of the key tasks, approaches, and lessons associated with various stages, forms and methods of analytics and learning. Highlighted are the paradigm shift and transformation

- from data to insight, and decision-making;
- from explicit to implicit analytics;
- from descriptive to predictive and prescriptive analytics; and
- from shallow learning to deep learning and deep analytics.

Before introducing the above, a high level overview of data science techniques is given. Lastly, general discussion on the marriage of analytics with specific domains, that is, forming X-analytics, is also provided. The chapter concludes with a discussion of the relevant analytics and learning techniques, approaches and tasks.

© Springer International Publishing AG, part of Springer Nature 2018
L. Cao, *Data Science Thinking*, Data Analytics,
https://doi.org/10.1007/978-3-319-95092-1_7

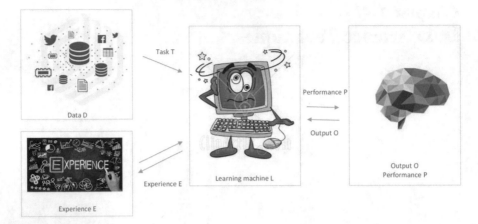

Fig. 7.1 The definition of general machine learning tasks. Note: the diagrams embedded in this figure are from Google search

7.2   The Problem of Analytics and Learning

Data analytics and machine learning are subfields of computer science which play a fundamental role in the innovation and development of modern artificial intelligence and machine intelligence systems.

As shown in Fig. 7.1, a machine learning method is often expressed as a computer program to "learn from experience E with respect to some class of tasks T and performance measure P if its performance at tasks in T, as measured by P, improves with experience E" [295].

A machine learning activity generates a suitable *learning machine* L which obtains the best performance P in undertaking task T on data D, based on experience E. An objective function F is defined to achieve the optimum performance P of solving the learning task T.

7.3   The Conceptual Map of Data Science Techniques

Since data analysis and machine learning were proposed several decades ago, analytics and learning have experienced a significant transformation in their development in terms of target problems, tasks, learning paradigms, and tools. It is quite challenging but important to categorize the analytics and learning 'family'. This section attempts to draw a conceptual map of the analytics and learning-centered data science discipline.

There are different types of data inputs, problems, analytical/learning tasks, experiences, evaluation methods, and outputs (e.g., learning a signal or feedback from data or experience) in data science. Analytics and learning approaches can thus be categorized in terms of their foundations, learning tasks, methods, and business problems.

Figure 7.2 summarizes the main methods, tasks and objectives in machine learning, knowledge discovery, and general data analytics, and the foundations of analytics and learning, and enabling techniques for data science. They can be categorized into the following major groups of techniques:

- Foundations of data science in particular analytics and learning: these include theoretical foundations and tools for analytics and learning in such areas as algebra, numerical computation, set theory, geometry, statistical theory, probability theory, graph theory, and information theory.
- Classic research on analytics and learning: which consists of such areas as feature engineering, dimensionality reduction, rule learning, classic neural networks, statistical learning, evolutionary learning, unsupervised learning, supervised learning, semi-supervised learning, and ensemble learning.
- Advanced research on analytics and learning: which includes such areas as representation learning, Bayesian networks, kernel machines, graphical modeling, reinforcement learning, deep learning (deep neural networks and deep modeling), transfer learning, non-IID learning, X-analytics, advanced techniques for optimization, inference and regularization, and actionable knowledge discovery.
- Enabling techniques for data science: these include artificial intelligence techniques, intelligent systems, intelligent manufacturing techniques, big data and cloud computing techniques, data engineering techniques, Internet of Things techniques, and security techniques.

In the following sections, the main focus is on categorizing and summarizing relevant techniques in the main categories of the data science discipline. More details about classic and advanced techniques for analytics and learning are available in book [67].

7.3.1   Foundations of Data Science

In addition to the broad discussion on multiple disciplinary techniques for data science in Chap. 6, Fig. 7.3 further highlights the fundamentals for analytics and learning. The following aspects of fundamentals are listed, which are commonly required in many analytics and learning tasks and approaches.

- *Algebra*: involves foundations in linear algebra, abstract algebra, group theory, field theory, measure theory, and logic. These include mathematical tools for processing matrices, tensors, norms, eigendecomposition, algebraic structures (such as fields and vector spaces), interaction patterns between objects and their environment, and propositional logic.
- *Numerical computation*: consists of foundations and tools to support the processing of value functions, interpolation, extrapolation and regression, equations,

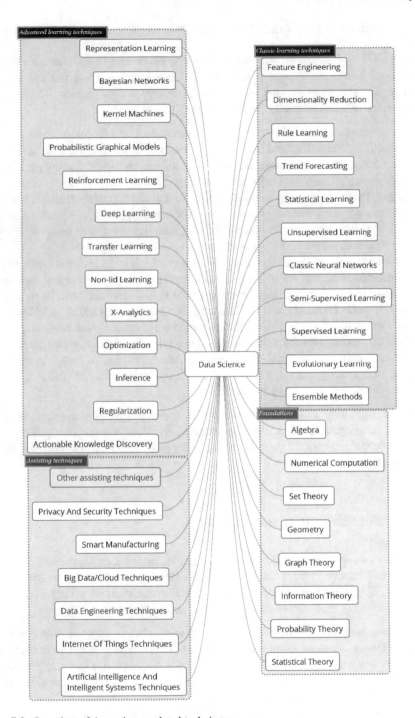

Fig. 7.2 Overview of data science-related techniques

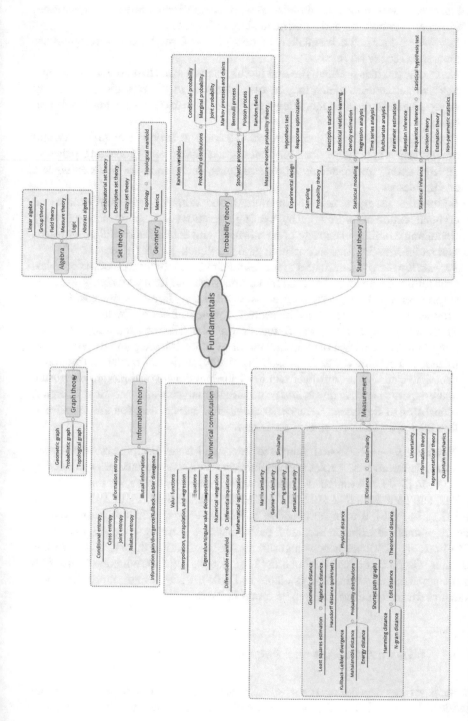

Fig. 7.3 The foundations of analytics and learning

eigenvalue decomposition, singular value decomposition, numerical integration, differential equations, and mathematical optimization.

- *Set theory*: forms the foundation of combinatorial set theory, descriptive set theory, and fuzzy set theory.
- *Geometry*: enables the analytics and learning approaches that are built on topology and geometrics, which may involve algebraic topology such as homotopy groups, homology and cohomology, and geometric topology for manifolds and maps.
- *Graph theory*: consists of geometric graph theory, probabilistic graph theory, and topological graph theory for representing, analyzing, and processing pairwise directed and/or undirected relations, structures, and hierarchies between objects and problems.
- *Information theory*: consists of theoretic foundations for the quantification and communication of information by creating measures such as information entropy (including conditional entropy, cross entropy, joint entropy and relative entropy), mutual information, and information gain (Kullback–Leibler divergence).
- *Probability theory*: contributes to the foundations for handling random variables, probability distributions for conditional, marginal and joint probability, stochastic processes including Markov processes and chains, Bernoulli process, Poisson process and random fields, and measure-theoretic probability theory.
- *Statistical theory*: is the core foundation of analytics and learning. Statistics contribute to the design of experiments for hypothesis testing and response optimization, sampling methods, probability theory, statistical modeling (including descriptive statistics, statistical relation learning, density estimation, regression analysis, time series analysis, and multivariate analysis), and statistical inference (consisting of Bayesian inference, frequentist inference, decision theory, estimation theory, and non-parametric statistics).

More extensive discussion about the inter- and trans-disciplinary foundations of data science and their roles and interactions with data science are available in Chap. 6. In [361], relevant mathematical tools, including algebra, set theory, partial orders, and combinatorics, are introduced. Multivariate data analysis can be found in [232]. In [188], a brief introduction to linear algebra, probability, information theory, numerical computation, and basic machine learning concepts are given. The book by [280] introduces algorithms for information theory, inference and learning. Graph theory can be found in [124]. An extended introduction to statistics and computational statistics is available in [170, 178]. Relevant knowledge about statistics for machine learning can be found in [21].

7.3.2   Classic Analytics and Learning Techniques

In the past three decades or so, many techniques have been developed for analytics and learning, which we called classic techniques. They cover feature engineering, dimensionality reduction, rule learning, classic neural networks, statistical learning, evolutionary learning, unsupervised learning, supervised learning, semi-supervised learning, and ensemble learning. We briefly summarize these techniques below.

Feature engineering refers to the process and techniques for extracting, constructing, mining, selecting and enhancing features that make data analytics and learning effective and efficient. A *feature* is a term that is often interchangeable with such terms as *attribute* and *variable*, which are the preferred terminologies in some disciplines. Each attribute (variable) captures a characteristic of the underlying data and/or problem. Typical issues in feature engineering include the analysis and processing of feature relevance, discrimination, noise, redundancy, bias, and explosion. These may involve the additional challenges of handling the hierarchy (subspace), structure, dependence, and dimensionality.

Dimensionality reduction is the process of reducing the number of input features by mapping them into a lower-dimensional feature space that is more effective and efficient for handling the curse of dimensionality problem. *High dimensionality* may become a serious issue in almost all learning problems and tasks, particularly in feature engineering, unsupervised learning, supervised learning, and optimization. Typical dimensionality reduction techniques include feature selection and extraction, data mapping and scaling, discriminant analysis, value decomposition, regression analysis, and relation analysis.

Rule learning refers to a general analytics and learning method that identifies, extracts, learns or evolves rules from a set of observations. A *rule* may be represented in terms of a structure like 'antecedent' → 'consequence' where → implies or co-occurs (or "IF 'condition' THEN 'result'"), to represent, present or apply knowledge from data. In particular, *rule induction* extracts formal rules from data which represent full or partial patterns in the data. Typical tasks and methods for rule learning include association discovery (including association rule mining— or frequent itemset mining—and frequent sequence analysis), learning classifier systems, and artificial immune systems, as well as a range of paradigms such as horn clause induction, rough set rules, inductive logic programming, and version spaces for inducing rules.

Statistical learning refers to a collection of theories and techniques for modeling and understanding complex data, in particular, finding an appropriate predictive function, based on statistical theories and functional analysis [207, 410]. Statistical learning theories form the core foundation of data analytics and machine learning. General modeling and predictive tasks in statistical learning include resampling methods, linear regression, shrinkage approaches, clustering, classification, tree-based methods, and support vector machines.

Unsupervised learning refers to learning processes and approaches for which no labels are given, and for which a learning system estimates the categorization and structures in its input. Unsupervised learning can discover hidden patterns, detect outliers in data, or conduct feature learning. Typical unsupervised learning approaches include clustering, latent variable models, data exploration-based methods, and sparse coding.

Supervised learning refers to the process and techniques for assigning a label to each object by inferring a function from the data with supervision from a "teacher" (i.e., assigned labels on training samples). Typical supervised learning tasks and approaches consist of regression, classification, and ensemble methods.

Forecasting refers to the process and techniques for predicting or estimating the future based on past and present data. Forecasting is important for business management and decision-making, and forms the basis for estimating and planning research and development, capacity, manufacturing, inventory, logistics, manpower, sales and market share, finance and budgeting, and management strategies.

Trend forecasting refers to the prediction or estimation of future trends based on the analysis of past and present data (typically time series data, longitudinal data, and cross-sectional data), which is the main task in forecasting. *Trend* refers to *patternable trends*, such as prevailing tendencies in style or direction, popularity, or seasonal and cyclic behaviors, or *exceptional trends* such as drift or change. Typical trend forecasting tasks and techniques can be categorized into qualitative forecasting, quantitative forecasting, and modern predictive modeling.

Evolutionary learning refers to analytical and learning techniques and processes that are built on the mechanisms inspired by genetic, biological, and natural systems. Typical genetic, biological, and natural systems involve working mechanisms (also called operators) such as selection, cross-over, mutation, recombination, and reproduction. An evolutionary algorithm takes a search heuristic that mimics the process and working mechanisms of natural systems to generate new individuals that better fit and better approximate the given problem-solving solution.

More detailed discussion on these classic techniques is available in the book—Data Science: Techniques and Applications [67], also by the author of this book.

7.3.3   Advanced Analytics and Learning Techniques

There are many recently developed techniques for analytics and learning which we refer to as advanced techniques. Here, we briefly introduce the following: representation learning, kernel methods, Bayesian methods, probabilistic modeling methods, deep learning, reinforcement learning, non-IID learning, transfer learning, actionable knowledge discovery, and optimization techniques.

Representation is a critical process for achieving desirable learning objectives, tasks, results and systems. Representation learning has received significant and increasing attention from many relevant communities including computing, statistics, and learning. The objective of representation learning is to learn latent data, information and knowledge representations. It can be categorized into shallow representation and deep representation, depending on the depth of the representations.

Kernel methods [213, 262] rely on the so-called "kernel trick", in which original features are replaced by a kernel function. Here, *kernel* mathematically denotes a weighting function for a weighted sum or integral of another function. A kernel method involves a user-specified kernel function to convert the raw data representation to a kernel representation, i.e., in the form of a "feature map" (also called a mapping function) from the raw feature space to the kernel space. A user-specified kernel function can be interpreted as a similarity function that measures the similarity over pairs of data points in the raw data.

Probabilistic graphical models combine probability theory and graph theory in a flexible framework to model a large collection of random variables that are embedded with complex interactions. Probabilistic graphical models such as Bayesian networks and random fields are popularly used, for good reasons, to solve structured prediction problems in a wide variety of application domains including machine learning, bioinformatics, natural language processing, speech recognition, and computer vision. Typical probabilistic graphical models include Bayesian networks and random fields.

Bayesian networks, also called belief networks or directed acyclic graphical models, are probabilistic graphical models. A Bayesian belief network [260, 304] is a probabilistic dependency model that uses Bayesian probabilities to model the dependencies within the knowledge domain; it infers hypotheses by transmitting probabilistic information from data to various hypotheses. A Bayesian belief network is represented as a directed acyclic graph, in which a *node* represents a stochastic variable and an *arc* connecting two nodes represents the Bayesian probabilistic (causal) relationships between the two variables. Each node (variable) may take one of a number of possible states (or values). Here, *belief* measures the certainty of each of these states and refers to the posterior probability of each possible state of a variable, namely the state probabilities, by considering all the available evidence (observations). When the belief in each state of any directly connected node changes, the belief in each state of the node is updated accordingly.

Over the last decade, deep learning has been increasingly recognized for its superior performance when sufficient data and computational power are available. *Deep learning* refers to "a particular kind of machine learning that achieves great power and flexibility by learning to represent the world as a nested hierarchy of concepts, with each concept defined in relation to simpler concepts, and more abstract representations computed in terms of less abstract ones." [188] As a result, deep learning "allows computer systems to improve with experience and data". Deep learning research has evolved and been rebranded many times since its first recognition in the artificial intelligence and machine learning communities. Today's deep learning research essentially consists of deep neural networks and deep learning research [188]. Key principles and insights that enable deep neural networks include representation power, hidden power, depth power, backpropagation power, and training + finetuning mechanisms. Deep learning has made significant progress in many areas and applications, including speech recognition, audio and music processing, image and video recognition, multimodal learning, language modeling, natural language processing, information retrieval, and sequence modeling. Representative deep network architectures can be categorized into (1) supervised learning architectures, such as convolutional neural network (CNN) and recurrent neural network; (2) unsupervised learning architectures, such as autoencoder and generative adversarial networks; and (3) hybrid learning architectures, such as the pre-training + finetuning model DBN+DNN. More discussion on deep learning is available in [188] and in [67].

Reinforcement learning is a general-purpose framework for intelligent problem-solving and decision-making. It had its renaissance in the last decade, migrating from classic reinforcement learning research to advanced reinforcement learning research. In a broad sense and from the understanding and management of learning and decision-making perspectives, *reinforcement learning* refers to computational approaches that enable automated goal-directed decision-making by an individual learning from interactions with its environment. Reinforcement learning concerns making decisions about what to do, i.e., how to map situations to ideal actions, in order to maximize a reward signal to achieve an individual's global and long-term goals. It is about characterizing, solving and optimizing a learning problem, rather than characterizing learning methods and investigating how learning takes place. A learner tries and then discovers which actions ideally impact or produce the expected reward, which may affect subsequent rewards. Typical reinforcement learning methods include dynamic programming, simple Monte Carlo methods, unification of Monte Carlo, temporal-difference learning, and function approximation methods such as artificial neural networks. Advanced reinforcement learning is absorbing significant ideas and supporting tools from other disciplines such as neuroscience, statistics, physics, classic machine learning, and optimization, and emerging areas such as deep learning, advanced analytics and learning, big data technology, intelligence science, and complex systems.

It is often assumed in statistical analysis that data is independent and identically distributed (referred to as IID, or i.i.d.. This IID assumption has been dominantly adopted in analytics and learning, and existing analytical and learning systems have been built on the IID assumption of data and problems. *IID learning* relies on the assumption that all objects are IID, and applies this assumption to objects, object attributes, attribute values, learning objective function determination, and evaluation criteria. However, it has been widely accepted that real-life data and problems are not IID (and are thus called non-IID), i.e., data is non-independent and/or non-identically distributed. In a non-IID data problem, *non-IIDness* refers to any *coupling* and *heterogeneity*. Non-IID learning is the learning system that addresses the non-IIDness in complex data and problems. Non-IID learning applies to almost all analytical and learning tasks and other AI tasks, including memory emulation tasks, analytical, statistical, and learning-based tasks, interaction and recognition tasks, and simulation and optimization.

Transfer learning [321, 420] has been proposed as a general learning methodology for addressing business problems and research issues that involve multiple sources of data (channels, views, databases, etc.) to enhance the analytics and learning in one source by taking advantage of the value in other related data sources. *Transfer learning* is a machine learning methodology that either utilizes the richer and/or labeled data, or transfers the knowledge learned in some domains (called *source domains*) to assist with appropriate learning in other relevant domains (called *target domains*) with significantly limited or less informed (labeled) data. The condition for applying transfer learning is the invariance that exists across domains and/or tasks. This invariance can be used to associate the source domain/task with the target domain/task, thus the knowledge learned in the source domain/task can be transferred to the target domain/task. The relevant learning methods related to or involved in transfer learning include: learning to learn [383], lifelong machine learning [92], domain adaptation [109, 120], self-taught learning, knowledge transfer, knowledge consolidation, inductive transfer, active learning, context-aware learning (also called context-sensitive learning), multitask learning, metalearning, increment/cumulative learning, one-shot learning, and zero-shot learning. Not all of these are part of the transfer learning family, but there are similarities and differences between each of them and transfer learning.

Actionable knowledge discovery (AKD) addresses the problems of existing analytics and learning and various data characteristics and complexities (as discussed in Sect. 5.5.2), with the goal of achieving data-to-insight-to-decision transformation by data-driven discovery (more on this in Sect. 7.4). *Actionable knowledge* [78, 111] refers to knowledge that informs or enables decision-making actions. The term *actionable knowledge discovery* (AKD) refers to the methodologies, processes, and tools that discover and deliver actionable knowledge from data [57]. AKD undertakes *domain driven data mining* [54, 77] to address the various gaps in classic data mining methodologies and algorithms, in particular, the limited actionability of discovery results. Here, *actionability* [78] refers to the power to work, which is an optimal outcome and objective of AKD through the best integration of six core dimensions: problem, data, environment, model, decision and optimization.

Analytics and learning results often face problems of underfitting and overfitting, which lead to bias and variance in learning capacity. This may be caused by the fact that no models can exactly capture the X-complexities and X-intelligence of data and business problems, thus error minimization and the generalization of models is a critical issue. Appropriate techniques are therefore required to ensure this generalization, which includes regularization, optimization, and sometimes also inference as optimization. While optimization, regularization and inference focus on their respective goals, in practice, there are often connections and complementarity between optimization, regularization and inference. They all aim for a certain common performance (e.g., accuracy, robustness, convergence and efficiency) of algorithms to achieve the desired solutions. The respective methods fit their corresponding problem structures, underlying mechanisms, and performance characteristics. Optimization is often essential for finding the best solutions in analytics and learning. Many techniques and algorithms are available to address the respective optimization problems and conditions. They involve multiple areas of research, including mathematical optimization (in particular numerical analysis and optimization), statistical optimization (in particular risk minimization, approximate optimization), and computational methods (including evolutionary computing, and tricks and mechanisms applicable in learning). Depending on whether constraints are applied on the input values, optimization can be categorized as unconstrained optimization or constrained optimization. In accordance with the domain of inputs, optimization can be categorized as discrete optimization or continuous optimization. The type of objective function and constrained conditions also determines the optimization method, which can be categorized as convex optimization (including linear optimization, quadratic optimization and cone programming) and nonconvex optimization. In addition, the model developed for optimization may involve different volumes or proportion of samples or have an assumption on the certainty of the feasible domain, leading to deterministic optimization, batch optimization, mini-batch optimization and stochastic optimization. Lastly, many real-life optimization problems are *NP-hard*[1] and cannot be solved with exact solutions in polynomial time. Approximation methods tend to approximate optimal solutions to such problems in polynomial time with provable guarantees on the approximate solutions.

More details about the above techniques are available in [67].

7.3.4   Assisting Techniques

The proper functioning of data science requires many other assisting techniques which are necessary but do not directly handle analytics and learning. Here we briefly introduce the following techniques: artificial intelligence techniques, intelligent systems, intelligent manufacturing techniques, big data and cloud computing techniques, data engineering techniques, Internet of Things techniques, and security techniques.

7.3.4.1   Artificial Intelligence and Intelligent Systems

Artificial intelligence (AI) [344] refers to the intelligence demonstrated by machines rather than humans or other living things. AI research studies intelligent systems that can perceive environment, take actions, and maximize the achievement of goals. Since the first proposal of AI in the 1950s, AI research has experienced several resurgent waves, and has evolved into more and more sub-fields, becoming a major area in computer science and the general IT field. Traditional AI research has focused on developing techniques for perception, reasoning, knowledge representation planning, natural language processing, computational intelligence, neural networks, and search and mathematical optimization.

Although these AI techniques are still used and continue to be developed, today's AI research is mainly grounded on high-performance computing infrastructure, large amounts of data, interdisciplinary theoretical advancement, and advanced analytical and learning systems and algorithms. This can largely be seen in the extensive research and development in big data analytics, large-scale machine learning, computer vision of large visual analysis tasks, deep learning of large image and textual data, and biomedical analytics of large biological and medical data.

---

[1]NP-hard problems refer to Non-deterministic Polynomial acceptable problems, which indicates that they are at least as hard as the hardest problems in NP.

The current resurgence of AI and intelligent systems is predominantly data science-enabled, and learning and optimization-driven. AI research and intelligent systems have migrated to *data-driven AI*, which discovers and utilizes data intelligence to infer and optimize complex problem understanding and resolution; *human-machine-cooperated AI*, which hybridizes human intelligence and machine intelligence for joint problem-solving of complex problems; and *metasynthetic AI*, which builds on the metasynthesis of various types of intelligence.

Both classic and advanced AI research and techniques can be categorized as symbolic AI, connectionist AI, situated AI, nature-inspired AI, social AI, or metasynthetic AI [62] at all stages of AI research.

- *symbolic AI* which represents *symbolic intelligence* in terms of symbols and logic reasoning.
- *connectionist AI* which represents *connectionist intelligence* in terms of connectionism and networking, especially artificial neural networks.
- *situated AI* which represents *situated intelligence* in terms of multi-agent systems and the interactions within a system and between agents and environment.
- *nature-inspired AI* which represents *natural intelligence* in terms of mimicking the working mechanisms in natural, biological and evolutionary systems.
- *social AI* which represents *social intelligence* in terms of social interactions and collective intelligence in problem-solving.
- *Metasynthetic AI* which represents metasynthetic intelligence by synthesizing human intelligence with other ubiquitous intelligence, including data intelligence, behavior intelligence, social intelligence, organizational intelligence, network intelligence, and natural intelligence according to the theory of metasynthetic engineering [62, 333].

Today's AI has been widely and deeply employed in many existing and emergent business domains and applications, and has demonstrated its significant value for the strategic transformation of existing industries and businesses, and for the generation of new AI-driven businesses. Typical applications are healthcare data and medical imaging data analysis-based applications, driverless car design, cashless e-payment services, sharable businesses (e.g., sharable bikes and cars), smart cities and smart homes, and new-generation financial technology driven by data science and intelligent technologies. In principle, AI and intelligent systems can be applied to almost any domain or purpose, using either more or less data. This application to the economy relies on the integration of intelligent systems, smart manufacturing, big data analytics, e-payment systems, data-driven demand-supply management, and data-driven customer relationship and marketing management.

7.3.4.2 Smart Manufacturing

Smart manufacturing is both the application of data science in manufacturing businesses and an enabling technology for conducting data science in many applications.

Smart manufacturing (or *intelligent manufacturing*) [457] refers to advanced forms of manufacturing that apply sophisticated information technologies to the whole manufacturing process and product life cycle to optimize and advance traditional manufacturing technologies and processes. Here, "smart" indicates that manufacturing enterprises effectively apply advanced intelligent systems to the whole manufacturing ecosystem, rapidly producing new products, dynamically responding to global market supply and demand change, and effecting real-time optimization of manufacturing production and supply-chain systems. Intelligent technologies driving smart manufacturing include digital manufacturing, advanced robotics, industrial Internet of Things and sensor techniques, enterprise application integration, big data processing techniques, industrial connectivity devices and services, rapid prototyping, collaborative virtual factory, advanced human-machine interaction, virtual reality devices, intelligent supply-chain management, and cyber-physical system communication. Data science and artificial intelligence are the main drivers for achieving the so-called Industry 4.0 plan [449].

7.3.4.3 Big Data and Cloud Computing Techniques

Big data [288] is the form of data that is so large and complex that it cannot be processed and analyzed by traditional data processing and analysis theories, infrastructure, algorithms and tools. *Big data* generally refers to voluminous and complex data, and the relevant technologies and systems to acquire, store, transfer, query, share, manage, process, analyze, present, and apply such data. This understanding of *big data technologies* exceeds the multiple V-based definition of big data, in which big data is understood to refer to data described by volume, variety, velocity, veracity, and value. The value of data has to be disclosed by advanced analytics and machine learning, through various X-analytics, building on domain-specific data and problems. Accordingly, big data technologies consist of new-generation technologies and tools for undertaking the above operations on large volumes of complex data.

As distributed and parallel computing has become essential in managing, processing and analyzing big data, *cloud computing* has emerged as a new generation of infrastructure, platforms, services and applications to store, access, share, compute, and analyze big data over the Internet or local area networks. Core technologies to enable cloud computing consist of computing architectures, service models (involving infrastructure, platform, software, and functionalities as services), deployment models (private, public or hybrid cloud), and security and privacy. Cloud computing technologies transform the traditional computing paradigms from enterprise-owned private data and computing to service-based centralized and dedicated data and computing. Cloud computing thus enables more cost-effective operations, more professional management of data and computing, device and location independent access and usage, and more scalable and efficient computing. Typical cloud computing platforms and packages consist of the Apache open source toolset, Amazon Web Services Cloud, Google Cloud platform, Microsoft Azure, and many other cloud services operated by specific vendors. In addition to tools for implementing general data engineering and management functionalities such as service, storage, access, sharing, query, and management, most cloud platforms involve more or less analytics and machine learning modules and form cloud analytics capabilities.

A typical big data and cloud computing platform is the open source Apache framework and ecosystem based on Hadoop. Hadoop generally refers to a collection of software packages built on the Hadoop framework, which consists of a storage system—Hadoop Distributed File Systems (HDFS), and a processing system—MapReduce programming model. Large data (files) are split (mapped) into blocks and distributed across nodes in a large computing cluster for processing in parallel. The processed results from distributed nodes are summarized (reduced) to form the answers to the original processing task. The Hadoop-based cloud computing ecosystem consists of many software packages to handle different and specific functionalities. Examples in the Apache Software Foundation [10] consist of

- Apache Spark, a cluster-computing architecture for cluster-based distributed storage, programming, cluster management (including task dispatching and scheduling), application programming interface, and fault tolerance.
- Apache Ambari, a Hadoop management web interface for provisioning, managing and monitoring Hadoop clusters.
- Apache HBase, a non-relational distributed database for column-based key-value-oriented storage, compression, and in-memory operation in the Cloud Bigtable form.
- Apache Hive, a Hadoop-based data warehouse for undertaking SQL-like data summarization, query and analysis.
- Apache Pig, a software for programming and executing MapReduce jobs on very large data on the Apache Hadoop framework.
- Apache Zookeeper, a centralized service platform for distributed service configuration, synchronization, and naming registry for large distributed systems
- Apache Sqoop, for data connection and transfer between relational databases and Hadoop.
- Apache Oozie, a workflow management system for handling workflow, scheduling and Hadoop jobs.
- Apache Storm, a distributed stream processing framework for stream processing.

7.3.4.4   Data Engineering Techniques

Data engineering techniques prepare data for analysis and learning. They design, build, manage, and evaluate data-intensive systems data science workload from the engineering aspect. Today's data engineering technologies are seamlessly integrated with big data technologies and cloud computing.

Specifically, *data engineering* provides technologies and solutions for database, data warehouse, hardware and memory storage system architecture, construction, management, privacy and security; data, metadata and application integration and interoperability; distributed, parallel, high-performance, and peer-to-peer data management; cloud and service computing; data and information extraction, cleaning, retrieval, provenance, workflow, query, indexing, processing, and optimization. These processes may involve different types of data, e.g., strings, text, keywords, streams, temporal data, spatial data, mobile data, multimedia data, graph data, web data, and social networks and social media data.

7.3.4.5   Internet of Things

The Internet of Things (IoT) [450] refers to the ecosystem that connects physical devices, home appliances, vehicles, wearable equipment, and other equipment to create a network via the Internet, wireless networks, or other networking infrastructure for exchanging data, remotely sensing or controlling objects in the network. IoT techniques connect physical systems with virtual cyberspace to form cyber-physical systems, which can essentially connect with and integrate the Internet (including mobile networks and social networks) with everything in the physical world, e.g., homes, cities, factories, transport systems, mobile vehicles, humans, and other living entities or objects.

Enabling technologies for IoT consist of IoT network infrastructure; data, information, application and device integration and fusion; wireless networks; sensors and sensor networks (including RFID technology); data storage, access sharing and connection; object remote access, connection, and management; remote locating, addressing, and accessing; tele-operations and remote control; and data and object security, surveillance and privacy.

IoT techniques involve a growing number of applications, e.g., smart homes, smart cities, smart grids, intelligent transportation, and industrial domains (referred to as industrial IoT). Data science plays a critical role in IoT for intelligent analysis, alerting, prediction, optimization, and the control and intervention of devices on the IoT. IoT also expands data science to traditional non-digitized things and areas and enables their data-driven transformation.

7.3.4.6   Security and Privacy Protection Techniques

As discussed in Sect. 4.8.1, data privacy and security are important assurance aspects of data science, technology and economy. They may incur severe ethical, social, economic, legal or political challenges if they are not effectively protected.

Privacy and security protection technologies consist of software, hardware, management and governance, and regulation for preserving and enhancing the privacy and security of devices, systems, networks, and to protect networking, sharing, messaging, communications, and payments. Effective and efficient techniques and tools are required to predict, detect, prevent and intervene in risky behaviors and events. Other mechanisms and actions for preserving and enhancing privacy and security include creating laws and policies for data protection and authentication, as well as active and automated regulation and compliance management.

Sufficient privacy and security protection will ensure the healthy development of data science, technology and economy. This becomes increasingly important at a time when anti-security and anti-privacy are becoming professions, and as professional bodies and individuals utilize and develop increasingly advanced anti-security and anti-privacy technologies and tools.

7.3.4.7 Other Assisting Techniques

Many other technologies and tools may be required to implement data science, especially when data science is conducted within a specific domain. Some such assisting techniques and tools may already exist; others will need to be developed to support domain-specific data science tasks. For example, to undertake data science innovation in traditional manufacturing businesses, existing production devices, production lines, factories, workflows, processes, scheduling, management, governance, and communications may have to be digitized, networked, quantified, and automated. This revolution in manufacturing will require the development of corresponding sensors; virtual and networkable devices, lines and factories; intelligent and networked enterprise resource planning systems, management information systems, messaging and communication systems, and risk and case management systems. Other examples include digitized and networked payment systems, supply chain management, demand-supply management systems, dispatching and scheduling systems, identification, authorization and certification systems, and risk and security management systems.

7.4 Data-to-Insight-to-Decision Analytics and Learning

As shown in Fig. 7.4, *data-to-insight-to-decision transfer* has taken place at different times and in different analytic stages along the *whole-of-life span of analytics*. This can also be represented in terms of a range of analytics goals (G) and approaches (A) designed to achieve the *data-to-decision* goal in different data periods—past, present, and future—and for the purpose of generating actionable decisions along the timeline of conducting analytics and during the life of analytics.

- Data-to-Insight: *Data-to-insight* transfer seeks to form a valuable and deep understanding of the true nature, inner character or underlying truths about data and business, which usually involve hidden data complexities and/or business dynamics and problems and their nature and solutions, but are interesting to be deeply, richly and precisely explored.
- Data-to-Decision: The aim of *data-to-decision* transfer is to discover, define and recommend decisions that are informed, supported and verified by data despite

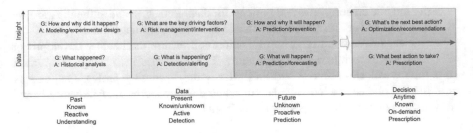

Fig. 7.4 Data-to-insight-to-decision whole-of-life analytics

being invisible. Such decisions are valuable for taking action to achieve better business performance and benefits.

7.4.1 Past Data Analytics and Learning

First, for *past data*, we aim to understand stories that are known to have taken place in the historical data and business; the analytical findings are used for business in a reactive way (here *reactive analytics* means the findings and insights are identified after the event to understand and tackle the problems that have taken place).

The goals and approaches to past data are as follows.

- Goal: The main purpose in past data analytics is to explore "what happened" in the data and business;
- Approach: Historical data analytics methods are used to explore what happened.

Accordingly, the insight extraction from past data is as follows.

- Goal: The purpose is to gain insights into "how and why it happened";
- Approach: Data modeling, experimental design and hypothesis testing are typically applied to utilize the insights gained by this approach.

By understanding past data, this stage is able to focus on "we know what we know" to gain a reactive understanding of what took place and to take appropriate action.

7.4.2 Present Data Analytics and Learning

Second, for *present data*, we are interested in both known and unknown stories in the data and business. Active understanding of hidden data indication is conducted, to detect what is happening in business.

The goals and approaches for present data are:

- Goal: The aim at this stage is to explore "what is happening";
- Approach: Typical methods such as real-time and current detection and alerting are used to understand "what is happening".

The insight discovery through present data analytics is as follows.

- Goal: The purpose is to generate insights about "how and why it happens";
- Approach: Applying the insights is typically dependent on risk management and intervention.

This stage typically addresses "we know what we do not know", with alerts generated about suspicious events, or interesting groups or patterns presented in the data and business. The insights are extracted for decision-making purposes, such as real-time risk management and intervention, to address the question "what are the key driving factors?"

Both past and present data are usually involved in real-time detection and analysis.

7.4.3   Future Data Analytics and Learning

Third, for *future data*, we focus on mainly unknown stories in data and business. This approach relies on proactive and predictive analytics and data modeling.

On present and future data, we are concerned with the following.

- Goal: To investigate "what will happen" in the future;
- Approach: Typical methods for predictive modeling and forecasting are focused to understand "what will happen".

The insight gained from future data analytics has the following objectives.

- Goal: To gain a deep understanding of "how and why it will happen";
- Approach: The extracted insights are typically used for prediction and prevention of future unexpectedness.

This stage gains a deep understanding of the problem that "we do not know what we do not know" and seeks to find solutions to by estimating the occurrence of future events, grouping and patterns, and undertaking and achieving proactive understanding, forecasting and prediction, and early prevention.

When necessary, data from the past, present and future may all be involved.

7.4.4   Actionable Decision Discovery and Delivery

Lastly, our focus is on consolidating anytime data to recommend and take actions for *actionable decision-making*. At this stage, we gain a fairly solid understanding of data and business problems, and on-demand mixed analytics are undertaken to provide prescriptive actions for business.

At this stage of data understanding and analytics, we are concerned with the following.

- Goal: To investigate "what is the best action to take" in business;
- Approach: Typical methods may include simulation, experimental design, behavior informatics, and prescriptive modeling to identify the next most appropriate actions (often called *next-best* actions).

Correspondingly, we look for insights that aim for the following.

- Goal: To understand and detect "what is the next best action";
- Approach: The next most appropriate actions to take are applied to optimization and recommendation to improve business.

Prescriptive analytics and actionable knowledge delivery thus interpret findings from past, present and future data, and enable the corresponding optimal actions and recommendations to be put in place based on the findings. This addresses the problem of "how to actively and optimally manage the problems identified" by making optimal recommendations and carrying out actionable interventions.

In practice, as shown in Fig. 7.4, real-life analytics involves past data analytics, present data analytics, and future data analytics. Enterprise analytics requires whole-of-life analytics to implement and achieve the data-to-insight-decision transformation.

7.5   Descriptive-to-Predictive-to-Prescriptive Analytics

The paradigm shift from data analysis to data science constitutes the so-called "new paradigm" [209, 305], i.e., data-driven discovery. The history of analytics from an evolutionary perspective spans two main eras—the era of explicit analytics, and the era of implicit analytics [67]. Analytics practices have seen a significant paradigm shift in three major stages:

(1) Stage 1: descriptive analytics and reporting;
(2) Stage 2: predictive analytics and business analytics; and
(3) Stage 3: prescriptive analytics and decision making.

Two other analytical terms: *diagnostic analytics* and *cognitive analytics*, are sometimes used in business, in addition to the above three analytics paradigms:

- Diagnostic analytics: the diagnosis of causes, or "why it happened". This function is often included in descriptive analytics, however.
- Cognitive analytics: the act of causing something to happen. This function is often embedded in prescriptive analytics.

Figure 7.5 shows the analytics paradigm shift.

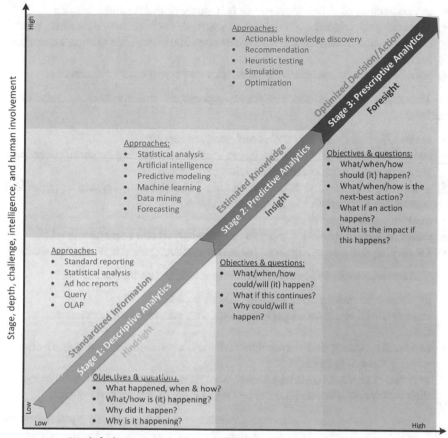

Fig. 7.5 Three stages of analytics: descriptive-to-predictive-to-prescriptive analytics

We briefly discuss the objectives, main approaches and benefits of the above three stages below.

7.5.1 Stage 1: Descriptive Analytics and Business Reporting

Descriptive analytics is the preliminary stage of advanced analytics. *Descriptive analytics* summarizes what has happened or is happening, and characterizes the main features in business.

The main objectives and goals at this stage are:

- answering the questions "what happened?" and "why did it happen?"
- summarizing the main features and trends in business;

- understanding what has happened or is happening in data and business;
- generating descriptions of business operations, performance, dynamics, trends, and exceptional scenarios, with implications about possible driving factors and reasons; and
- presenting regular and periodic summaries and statistics of business concerns.

The main reporting and analytical approaches and methods at this stage include:

- major effort on explicit analytics and standard reporting;
- statistical analysis of data indicators related to or reflecting business operations, performance, dynamics, trends and exceptions;
- generating standard, ad hoc and/or cubic (OLAP-based) reports as periodical summaries and statistics in relation to business;
- querying, drilling up/down;
- factor analysis, correlation analysis, trends analysis and regression; and
- identifying and generating alerts for business management and decision-making.

The consequences and benefits of descriptive analytics for business are:

- to summarize and describe hindsight about business;
- to understand routine and periodical business characteristics, trends and evolution;
- to identify factors and correlations driving the routine and periodical normal development of business;
- to detect exceptional and risky events, occasions, areas and scenarios and identify their implications beyond routine and regular development and trends.

Business reports (often standard analytical reports) generated by dashboards and automated processes are the means for carrying findings from analytics to management.

7.5.2   Stage 2: Predictive Analytics/Learning and Business Analytics

Predictive analytics are a key element of advanced analytics which make predictions about unknown future events, occasions, or outcomes.

The main objectives and goals at this stage are:

- answering the questions "what could/is likely to happen?" and "why could/will it happen?"
- generating a reasonable estimation of the trends and future directions of business;
- predicting what will happen and why it will happen in business;

- estimating the future development of business operations, performance, dynamics, and exceptional scenarios, with implications about possible driving factors and reasons; and
- identifying probable future risk and exceptions.

The main analytical and learning approaches and methods at this stage include:

- forecasting, in particular focusing on regression analysis techniques (including linear/logistic regression, time series analysis, regression trees, and multivariate regression);
- statistical prediction and forecasting of future business operations, performance, dynamics, trends and exceptions; and
- machine learning, artificial intelligence and data/text mining techniques for predictive modeling (classification methods such as neural networks, support vector machines, Naive Bayes and nearest neighbour-based methods; pattern-based methods; computational intelligence methods such as fuzzy set and evolutionary computing; trend prediction methods; and geospatial predictive modeling).

Predictive analytics may bring benefits and impact to business such as:

- identifying insights about the future;
- understanding the likelihood of future business trends and evolution;
- identifying probable factors that will drive the future development of the business;
- predicting exceptional and risky events, occasions, and scenarios, and determining possible driving factors.

Predictors, patterns, scoring and other findings are created for and presented through dashboards and analytical reports to business managers and decision-makers to understand projected future trends and directions, and the reasons behind them. Analytical reports are delivered to predict trends and exceptions, and to explain the underlying factors.

7.5.3 Stage 3: Prescriptive Analytics and Decision Making

Prescriptive analytics is the most advanced stage of advanced analytics. It suggests the optimal actions to take in decision-making.

The main objectives and goals at this stage are:

- to answer the questions "what should happen?" and "why should it happen?";
- to estimate possible impact, i.e. "what would be the impact if an action is taken?"; and
- to recommend the possible next-best action if a particular action has already been taken, or something untoward happens.

To achieve its objectives, this stage may take the following approaches and methods:

- optimization to generate optimal recommendations;
- simulations to obtain the best possible scenarios and understand the impact of taking action;
- heuristic testing of different options and possible effects; and
- actionable knowledge discovery [57, 59, 77] to ensure that the recommended options are actionable.

Prescriptive analytics contributes directly to business decision-making, and is thus more beneficial for problem-solving and business impact. It achieves this by, for example:

- enabling the immediate application of analytical results by recommending optimal actions to be taken on business change;
- estimating the effect of one option over another, and recommending the better option for prioritized actions;
- promoting outcome-driven analytics and transforming actionable analytics and learning for effective problem-solving and better business outcomes.

Prescriptive decision-taking strategies, business rules, proposed actions and recommendations are subsequently disseminated to decision-makers for the purpose of taking corresponding action.

7.5.4 Focus Shifting Between Analytics/Learning Stages

Figure 7.6 illustrates the shift in focus between three stages of analytics: descriptive analytics to predictive analytics and prescriptive analytics.

At the descriptive analytics and business reporting stage, limited implicit analytics effort is made for hidden knowledge discovery, and even less effort is made in actionable knowledge discovery. Descriptive analytics is mainly (80% of effort) achieved by using off-the-shelf tools and built-in algorithms for explicit analytics.

At the predictive analytics stage, significantly more effort (less than but close to 50%) is made in implicit analytics, which focuses on predictive modeling. Descriptive analytics still plays an important role here, particularly for business analytics. *Business analytics* elevates the process to another level, which aims to gain an in-depth understanding of business through deep analytics. By contrast, classic business analytics, which is widely adopted in business and management, mainly focuses on descriptive analytics. Actionable knowledge discovery receives greater attention, with more effort being made to apply forecasting, data mining and machine learning tools for deep business understanding and prediction.

At the stage of prescriptive analytics, the major (80% plus) effort is focused on the suggestions and delivery of recommended optimal (next best) actions for domain-specific business decisions. This is accompanied by discovering invisible and actionable knowledge and insights from complex data, behavior and environment and by implicit analytics in a specific domain by considering its specific X-complexities. About 50% of the effort is made on implicit analytics and actionable knowledge discovery. As a result, innovative and effective customized algorithms and tools are invented to deeply and genuinely understand domain-specific data and business, and significant effort is expended to discover and deliver actionable knowledge and insights. This forms the scenario of personalized analytics and learning, tailored for specific data, domain, behavior and decision-making. In contrast, relatively limited (less than 20%) effort is expended on explicit analytics, since this is conducted through automated processes and systems.

During the paradigm shift (as shown in Fig. 7.6), a significant decrease is seen in the effort expended on routine explicit analytics, which is increasingly undertaken by automated analytics services. By contrast, a significant increase in effort is seen in implicit analytics and actionable knowledge delivery [77]. The shift from a lower stage to a higher stage accommodates an increasingly higher degree of knowledge, intelligence and value to an organization, but it also means there are more challenges to face.

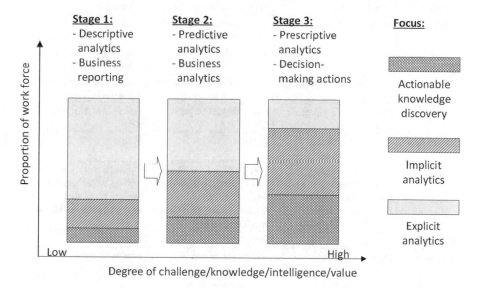

Fig. 7.6 Focus shift from descriptive to predictive and prescriptive analytics

7.5.5   Synergizing Descriptive, Predictive and Prescriptive Analytics

Table 7.1 summarizes the key aspects and characteristics of descriptive, predictive and prescriptive analytics.

The summarization and comparison of descriptive analytics, predictive analytics and prescriptive analytics shown in Fig. 7.5 and Table 7.1 clearly show the complementarities between them. The complementary capabilities are embodied in terms of the objectives, challenges and questions to be addressed, data timeliness, approaches and foci, and levels of intelligence and automation in respective stages.

Table 7.1 Descriptive, predictive and prescriptive analytics

Categories	Descriptive analytics	Predictive analytics	Prescriptive analytics
Definition	Summarize what happened or is happening, and characterize the main features	Make predictions about unknown future (events, occasions, or outcomes)	Optimize indications and recommend best actions for smart decision-making
Objectives	Understand what happened or is happening in data and business	Predict what will happen and why it will happen in business	Suggest the next-best actions and estimate possible impact
Challenge	Low to standard	Medium to advanced	Advanced
Questions	What happened, when and how? or	What could/will happen, when and how?	What is the next-best action to take, when and how?
	What is happening, when and how?	What if this continues?	What happens if an action is taken?
	Why did it happen or is it happening?	Why will it happen next?	What is the impact if it happens?
Data	Past data and/or present data	Past data and/or present data	Past data and/or present data
Timeliness	Historical and/or present/real-time	Future	Future
Approaches	Statistical analysis, standard reports ad hoc reports, dashboards/ scorecards query, OLAP, visualization	Predictive modeling, data/ text mining, forecasting, statistical analysis, artificial intelligence, machine learning	Optimization, simulation, testing, experimental design, heuristics, actionable knowledge discovery
Intelligence	Hindsight	Insight	Foresight
Automation	High	High to standard	Low with human involvement
Advantage	Low to standard	Medium to advanced	Advanced
Output	Reporting, alerting, trends, exceptions, factors and implications	Likelihood of future, unknown events or unknown outcomes	Optimal actions, decisions or interventions, better effect and impact
Decision power	Low	Medium	High

In the real world, enterprise analytics often requires all three-stages of analytics. Many analytics case studies follow the descriptive-to-predictive-to-prescriptive analytics transition, as shown in Fig. 7.5, conducted to address different objectives by different approaches. How can the three-stage analytics be synergized in enterprise analytics practices? The following observations offer some direction in undertaking complex data analytics projects. For simple and specific projects, and scenarios that are well understood, such as the predictive modeling of credit card fraud, direct and targeted analytics may be executed.

At the very beginning of a data analytics project, descriptive analytics may be explored on selected samples or sub-topics of data to understand data characteristics and complexities, as well as the main challenges and the level of those challenges. Predictive analytics are then arranged. In general, prescriptive analytics are arranged as the most advanced and late stage of data science project.

For each stage, a comprehensive understanding of analytics plan, applicable approaches, data manipulation, feature engineering, and evaluation methods need to be developed. It is necessary to understand the differences between stages in terms of analytical functions, platform support, data processing, feature preparation, programming, evaluation, and user interactions.

On one hand, the synergy of conducting three stages of analytics when developing an enterprise analytical application is required to incorporate system-level and unified requirement analysis, analytics infrastructure and architecture selection, analytics programming and project management tool selection and configuration, enterprise and external resource acquisition, sharing, matching, connection and integration, feature engineering (including selection, fusion and construction), data governance, and deployment and case management arrangements.

On the other, as each analytics stage involves diversified tasks and purposes, and addresses different challenges and objectives, the support for individual modeling, infrastructure and programming platform configuration, data preparation, feature engineering, and project management, evaluation, deployment and decision-support arrangements can be quite divided and must be specified and customized. This is in addition to the overall connections and collaborations within an enterprise solution.

Skill and capability sets are complementary to all of the above, thus data analysts with appropriate backgrounds, qualifications, knowledge, skill and experience are required to take on the respective roles and responsibilities. This requires the formation of a collaborative data science team, with suitable team management and project management methodologies, tools and performance evaluation systems.

From the perspective of the management of analytical complexities and the feasibility of application and deployment of respective analytics, a systematic plan is required to determine the business scope and area, analytical depth and width, timeliness and data coverage, as well as the analytical milestones and risk management for each type of analytics. This involves the recruitment of data science teams to fulfil the respective plans, cross-team collaboration, communication, and consolidation.

Lastly, as each type of analytics specializes in addressing different business problems and objectives, the collaborations and communications between respective analytical stages (or analytics team) and business and management teams will differ; for example, involving different areas and levels of operations and management. Synergizing the three stages of analytics approaches thus requires careful planning and the implementation of plans for business communications and decision-making.

7.6 X-Analytics

The data-oriented and data-driven factors, complexities, intelligences and opportunities in specific domains compose the nature and characteristics of data analytics, and drive the application, evolution and dynamics of data analytics.

7.6.1 X-Analytics Spectrum

The application association of data analytics in specific domains has created the phenomenon of domain-specific analytics. *X-analytics* is the general term for analytics in domain-specific data and domain problems, where X refers to a specific domain or area.

As shown in Fig. 7.7, typical *domain-specific X-analytics* consists of

- data type and media-based analytics, such as audio and speech data analysis, transactional data analysis, video data analytics, visual analytics, multimedia analytics, text analysis and document analysis, and web analytics;
- intensively explored core business-oriented analytical areas, such as business analytics (general business and management), risk analytics, financial analytics (including accounting, auditing, banking, insurance, and capital markets), government analytics (including social security, taxation, defense, payment, financial service, border protection, statistics and intellectual properties-based), military analytics, marketing analytics, sales analytics, portfolio analytics, operations analytics, customer analytics, health analytics, medical analytics, biomedical analytics, and software analytics;
- popularly explored new business-oriented analytical areas in recent years, such as social analytics (including social network analysis and social media analysis), e-commerce analytics, behavior analytics, mobile analytics, security analytics and intelligence analytics, location-based analytics, and service analytics;
- in recent years, previously rarely explored business-based analytical areas such as manufacturing analytics, logistics analytics, brain analytics, city utility analytics (including energy, water, and transport), agricultural analytics, security analytics and intelligence analytics, learning analytics and education analytics;
- emergent analytical areas such as behavior analytics, people analytics, work analytics, leave analytics, performance analytics, living analytics, urban analytics, and cultural analytics.

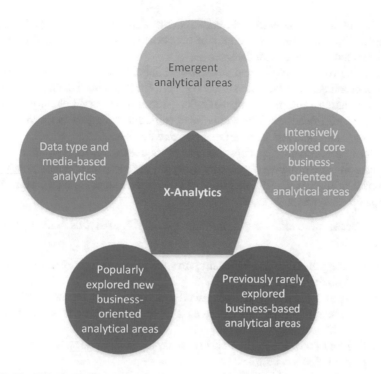

Fig. 7.7 The X-analytics family

These analytical areas and applications largely drive the development, evolution and popularity of data analytics, and the transformation from data analysis to advanced analytics, and from statistics to data science.

7.6.2   X-Analytics Working Mechanism

Typically, a domain-specific X-analytics is formed to address the following aspects:

- understanding and identifying domain-specific complexities and intelligence; e.g., identifying and modeling properties related to behavior in behavior analytics.
- specifying and formalizing domain-specific problems and challenges, and converting them to research issues; e.g., identifying behavior-related problems such as abnormal group behaviors or the impact of negative behaviors, and describing the corresponding behavior-based research issues, such as detecting abnormal group behaviors and quantifying the impact of negative behaviors.

- specifying and constructing domain-specific data structures; e.g., building a behavior model to represent behavior properties and processes, and converting business data to behavioral data.
- identifying research topics and developing analytical theories, models and tools on top of the constructed data structures; e.g., for behavioral data, such topics as behavior pattern mining, group behavior pattern discovery, behavior modeling, and behavior impact modeling may be specified and studied to establish corresponding analytical and learning theories and models.
- developing evaluation systems including measures, processes, and tools to evaluate the performance of relevant research theories, models and results; e.g., behavior utility to quantify the utility of behavior sequences and the significance of specific behaviors.

The process of conducting a domain-specific X-analytics may accordingly include:

- understanding the domain by focusing on specific domain problems and learning relevant domain knowledge;
- identifying and quantifying the specific domain complexities and intelligence;
- converting specific domain complexities to valuable research issues and questions;
- modeling and formalizing domain-oriented problems by proposing corresponding representation structures;
- preparing the data and establishing the corresponding data structures;
- constructing and selecting features and conducting feature analysis and mining;
- representing data features and characteristics;
- developing analytical theories, models and tools to address the identified issues and problems;
- evaluating the performance and impact of the representation and modeling;
- deploying models and developing applications and problem-solving systems.

Figure 7.8 indicates the main research constituents and tasks in establishing and conducting domain-specific X-analytics. In general, X-analytics involves both (1) the top-down modeling and representation of domain factors and issues in terms of holism (see more discussion on holism in Sect. 5.3.1.2) and (2) bottom-up analytical and learning of domain-specific problems and data in terms of reductionism (see more discussion on reductionism in Sect. 5.3.1.1).

7.7 Summary

Analytics and learning are the kernel stone of data science and are ubiquitous in the era of data science and analytics. The ubiquity is embodied in

- various *stages* of analytics and learning, e.g., historical data analysis, real-time detection and forecasting, and future prediction and alerting;

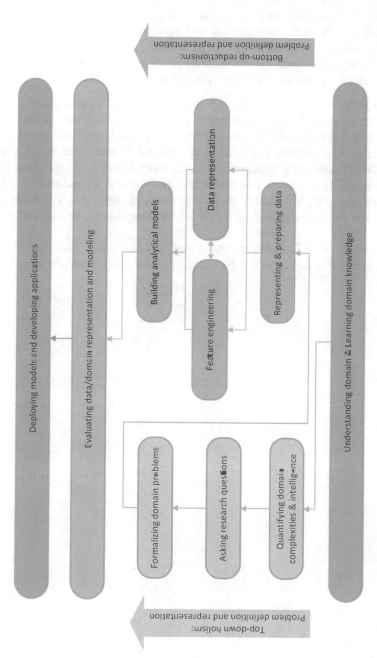

Fig. 7.8 The general X-analytics working mechanism

- diversified *forms* of analytics and learning, e.g., from descriptive analytics to predictive analytics and then to prescriptive analytics;
- a variety of *approaches* to the implementation of analytics and learning, e.g., knowledge-based methods, statistical and probabilistic learning, evolutionary computing-based optimization, and deep networks; and
- different *domains* for undertaking analytics and learning, e.g., financial analytics, government analytics, behavior analytics, and social analytics.

To enable the ubiquitous analytics and learning, different analytics and learning problems, tasks and techniques are required. In book [67], a summary of the family of analytics and learning techniques is provided, following by introduction, categorization and summarization of classic analytics and learning techniques, which cover learning foundations, measurements, and many different classic approaches for conducting diversified analytics and learning problem-solving and tasks. Further, advancements in analytics and learning are presented in the book. The book also introduces the concepts and techniques for inference, optimization and regularization.

The applications of the above discussed analytics and learning are discussed in Chap. 9, which form and drive the data economy and data industrialization, to be discussed in Chap. 8.

Part III
Industrialization and Opportunities

Chapter 8
Data Economy and Industrialization

8.1 Introduction

Data science and big data analytics have led to next-generation economy innovation, competition and productivity [288], as typically shown by the rapidly updated big data landscape [26]. Significant new business opportunities and previously impossible prospects have become possible through the creation of data products, data industrialization and services, which contribute to the so-called *data economy* [106, 220, 279, 472].

In this chapter, we discuss these new opportunities.

- First, we discuss what makes data economy, and how data economy transforms the real economy (i.e., the economy of producing goods and services);
- Second, data industrialization opportunities are discussed, building on both existing industries and new data industries; and
- Lastly, more specific data service opportunities are briefed.

8.2 Data Economy

Data economy is the current new economy. Comparing to the real economy and traditional financial economy, many fundamental questions need to be addressed in discussing data economy, for example:

- What is data economy?
- What are the new data economic models?
- What are the distinguishing features of data economy?
- What is the relationship between data economy and the artificial intelligence renaissance?

© Springer International Publishing AG, part of Springer Nature 2018
L. Cao, *Data Science Thinking*, Data Analytics,
https://doi.org/10.1007/978-3-319-95092-1_8

- How does data economy transform existing real economy and traditional financial economy?

This section discusses these important issues.

8.2.1 What Is Data Economy

What is data economy? *Data economy* is data-driven economy, and the term refers to the design, production, distribution, trading, and consumption of data products (systems, software), services and intelligence. In contrast to traditional core business-driven economy, data economy may not be traded in a given geographical location; rather, it may be conducted in regional or global areas through online, connected networks (mobile, IoT devices, wireless networks, etc.) or office markets.

In the era of data, this new economy tends to involve everything which is data-driven, forming a *datathing*. A *datathing* is a data-driven thing in which

- data is embedded in the thing, and
- the thing is associated with data.

Datathings (and data products) are essentially digital, intelligent, networked, and connected, and are presented in a mixture of real and virtual, and offline and online states.

Figure 8.1 roughly categorizes the data economy family into six categories, covering both the new economy and the translation and transformation of classic economy and businesses.

- Data-enabling technological businesses, which are driven fully by and work fully for data. They are the core of data economy.
- Data-intensive core businesses, which transform the existing core economy to one that is data-oriented, analytical, connected, personalized, optimal and intelligent.
- Classic IT businesses, which can be transformed to new-generation businesses that are smarter, more efficient, and more cost-effective.
- Traditional businesses that are data-poor, or utilize data poorly, and can be transformed by quantification, digitalization, and computerization to make them more effective, cost-effective and intelligent.
- Non-digital businesses, such as legal services, which can be digitalized to make them more effective, cost-effective and intelligent.
- Data service businesses, which are emergent, rely on data, and convert data to services.
- Future economy, which is unknown, and in which unpredictable and exciting new data businesses will increasingly appear in existing and future domains in ways that may be surprising.

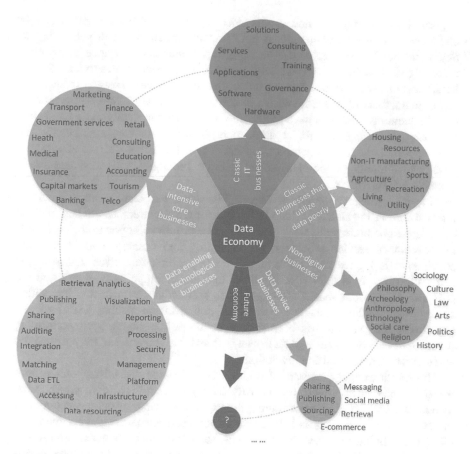

Fig. 8.1 The data economy family

The first category is made up of data-enabling technological businesses. Today, it can safely be said that data-enabling technological businesses are the core of the new-generation data economy and intelligence economy. This *new data economy* is data product-based and data technology-driven. Data-enabling technologies cover the whole spectrum of the data world, including such areas as

- data acquisition: data resourcing, accessing, matching, integration, and auditing;
- data preparation: data extraction, transformation and loading (ETL);
- data sharing and publishing;
- data management: infrastructure, platform, storage, backup, management, retrieval, security and governance;
- data manipulation: processing, reporting, visualization, and analytics.

The second category consists of data-intensive core businesses, which mine data values for enterprise innovation, and create differentiators that increase business opportunities, competition and productivity. Data and IT-aided traditional core businesses have been leading the way in IT transformation to data-driven IT, in particular, financial businesses such as banking, capital markets, insurance, accounting, auditing and consulting services, a large proportion of government services, library, health and medical services, transportation, telecommunication services, education and training, tourism, and retail businesses. These were all early birds in the new era of data economy, and they have played the driving role in datafying, quantifying, analyzing and using data for creating markets and businesses, and beating the competition. They have benefited from Internet technologies and are now taking advantage of mobile technologies, social media technologies, and broad data-driven technologies and data utilization.

The third category comprises classic IT businesses. Classic manufacturing-focused core IT businesses include IT hardware manufacturers, health and medical device manufacturers, software providers, service providers, application developers, IT governance service providers, IT consulting and training, and IT solution providers. They have also all launched corresponding initiatives and strategic actions (e.g., in Intel [374]) for big data, IoT, and/or cloud computing), and are pursuing data product-based transformation, productivity growth and innovation strategies as well as pursuing operational and decision-making strategies and objectives. Data science has been their new innovation engine for transforming device-focused businesses to data-driven soft and hard businesses for productivity and competition upgrade and revolution.

The fourth category consists of traditional businesses that utilize data poorly. Many business sectors were traditionally data-poor or utilized data poorly, such as non-IT manufacturing, agriculture, property and housing, resources, energy and utility services (water, gas and electricity), sports and recreation, environment and climate, and living and entertaining services (such as hiring businesses, retail stores, and restaurants). Business operations and management were not comprehensively computerized, and data were not systematically collected and well utilized. It is encouraging to see that such businesses are also investing in data analytics to transform their productivity and competitive advantage, and converting their businesses to data-enabling enterprises, such as by building their businesses on top of mobile and social media services.

The fifth category consists of traditionally digitization-irrelevant businesses. Non-digital business examples are social fields, such as philosophy, sociology, politics, arts, history, anthropology, law, archeology, culture, ethnology, and social care. Digitization has not been common in these areas; however, they have accumulated rich data and information, usually held in paper-based archives. Computerization and digitization have been adopted in these areas and are fundamentally changing them. The translation to digital data not only brings them into the realm of data-driven operations and management, but also creates new digital services.

The sixth category is made up of the many data service businesses, which are emergent core data businesses. An increasing number of organizations recognize the value of data as a strategic asset and invest in building infrastructure, resources, talent, and teams to convert general, domain-specific or private data to services.

They focus on building, providing and maintaining services data on top of other data, such as messaging, retrieval, social networking, or for core business transformation such as e-commerce services. They are evolving very rapidly, gaining significant market share, and attracting increasing resources and investment. They are evidenced by large data service enterprises such as Facebook, Google, Alibaba and Tencent. Leading Internet-based data-driven businesses [126], such as Google, Facebook, SAS, Alibaba, Baidu and Tencent, have overtaken traditional enterprise giants.

Lastly, unknown future data economies will emerge, connecting the above businesses and/or our work, study, living, travel and entertaining activities, styles, and tools.

8.2.2 Data Economy Example: Smart Taxis and Shared e-Bikes

To illustrate the concept of data economy and the difference between data economy and traditional physical and manufacturing-driven economy, we use the example of the transformation taking place in China's taxi and shared e-bicycle businesses.

As recently as 5 years ago, the difficulty of finding a taxi on Beijing's streets at peak time was a common complaint, while during off-peak periods, taxi drivers had to wait outside hospitals and major hotels in the hope of finding business.

About one year ago, when I needed a taxi, I downloaded mobile car leasing applications and I could choose from several applications to make an immediate call for a taxi or privately leased car. Drivers of taxis and private cars received a bonus fee (RMB5–20 in general) from the mobile application provider when you hired a car. Unemployed or part-time private car owners would download several such applications and then drive into business areas to capture booking requests and receive the bonus. They were not interested in taking full-time jobs. Taxi drivers were seriously offended, and I was told by some taxi drivers that they lost about one third of their business.

Within the last year, new policies have been issued which prevent non-Beijing residential private car owners to do conduct this kind of business, and while Beijing residential private car owners can offer their services, they need to register. All eligible drivers can operate as a taxi service without receiving a bonus, but drivers may still receive a bonus from clients at peak times or in locations where it is difficult to obtain a car. Taxi drivers get many more jobs than they did a year ago but still attract fewer jobs than they did prior to Uber and DiDi-type private taxi services entering the market.

Another surprising and stunning example is the bicycle business. About 2 years ago, every family in China owned two or three bikes. This situation has fundamentally changed. On the streets of Beijing, Wuhan, Changsha and smaller cities, a range of bikes for hire can be found—for example, yellow bikes called *ofo, bluegogo* bikes, and red *Mobikes*. These bikes are everywhere—on campuses, on the street corners in residential areas, and at bus stops and train stations. If you download a mobile App from one of these companies, you can pay a deposit (RMB99 or RMB300), go to street, find a bike, open the App, scan the bicode, and cycle to wherever you want to go. When you reach your destination, you scan the barcode again, lock the bike and drop it off wherever is convenient. The charge is only 0.5RMB-1RMB for half an hour, and if you use a bike on Monday, you may get free use on Tuesday and Wednesday.

This new data-driven biking business model is fundamentally transforming bike ownership and marketing models. Families do not need to buy bikes any more, or to worry that their bikes will be stolen when parked outside. It is common for people to carry their bikes up several floors to their home to avoid theft. Bicycle factories are suffering the loss of traditional individual-driven markets, and personalized bikes, even expensive multi-geared bikes, are disappearing. Solid, simple but intelligent bikes are occupying the bike hiring market, which is rapidly taking over the original individual bicycle market share.

Generation gaps are appearing that are small compared to generation gaps of the past. Young people born after 1990 and 2000 behave very differently from those born in the 1970s and 1980s. The post-1990s generation do not take cash or a wallet with them in China; they use their smart phone to pay for everything, including bike hire and paying the supermarket bill for eggs and milk.

Figure 8.2 illustrates a next-generation conceptual model of smart e-bikes, which integrate intelligent services into bikes.

Fig. 8.2 Conceptual model of a smart e-bike. Note: the bike chosen is simply to illustrate the concept

- GPS location system and intelligent switch: detecting QR code or GPS sensor-based recognition for switching on and off;
- Wireless access and USB port: accessing wireless and USB;
- Entertaining systems: listening to radio, music etc.;
- Lighter/camera for photos and recognition: taking pictures and recognizing QR codes, plate numbers etc. for parking and storage;
- Speed detection and controller: detecting speed, and automatically adjusting speed according to road condition analysis;
- Driver fit detectors: detecting driver's fitness, preferences of speed, acceleration speed, and relations with road conditions;
- Road condition sensors: sensing road conditions, e.g., slope, smoothness, wetness, and hardness etc.;
- Battery and magnetic charger: charging the bike; and
- Bike roadworthiness detectors: diagnosing bike's condition and roadworthiness.

Shared e-bikes and smart taxis are just two examples of datathings. In fact, many existing devices could be equipped with similar functions and services offered by smart e-bikes, e.g., fridges, microwaves, robot cleaners, vacuum cleaners, cars, coffee machines, and fitness equipment. These traditional, physically real items will be quantified as datathings, virtualized and connected in the Internet-based virtual network, and will become connected, sharable and smart.

8.2.3   New Data Economic Model

Data economy is driving the so-called "new economy". What, then, are the new data economic models?

The data-driven economy is fundamentally reshaping and transforming traditional physical objects-based economy. Data economy is making traditional objects smarter, and creating new cyber-physical datathings. Datathings are the core objects in the data economy, but they are different from physical objects in the traditional economy. Datathings are self-datafied, virtualized, and interconnected to form higher levels of datathings and datathing factories. Datathings are essentially intelligent, or at least much smarter than traditional objects.

Below, we summarize the new economic models and forms of the datathing-based new economy. The data economic model is described in terms of economic activity, ownership, stakeholder and customer relationships, pricing and valuation, sales and marketing modes, payment, regulation and risk management, and ethics and trust.

First, *economic activities*: These are the core phases and activities of an economic event, i.e., when the design, manufacture, distribution and consumption of an item have been changed to adapt to be individual-oriented vs organization-oriented, sharing and hiring-based vs private ownership-based, networked (datathings are usually online), intelligence-driven (datathings are intelligent), and active supply-driven vs demand-driven economy.

Second, *ownership*: Datathings follow a service model in which service providers of datathings own the products. They thus have responsibility for distribution, storage, maintenance, quality control, security and safety, etc. Related jobs and services result, such as datathing quality checkers and repairers, and regional agents for storage and maintenance.

Third, *stakeholder relationships*: Supply and demand stakeholders are superseded. In service-based business models, there are service providers, service consumers, and service regulators. Service providers are responsible for supplying services and maintaining the quality of those services. Service consumers are responsible for usage-based payment and maintenance. Third-party-based service regulators ensure service quality and safety, and build new ethical norms for operating service models and services.

Fourth, *sales and marketing*: The new ownership and stakeholder models reshape current sales and marketing models. Service providers sell services rather than actual datathings, thus licensing and hiring datathings replaces product purchase-based sales. Classic advertising channels and service providers lose their markets, because new marketing strategies are adopted, e.g., marketing campaigns that are directly targeted at key business players, clients and consumers such as offering direct bonuses to taxi drivers and private car owners, rather than promoting smart taxi services through traditional advertising channels. Marketing fees are distributed to directly influence clients (business operators) and end consumers, who are rewarded with the benefits of adopting new services. Marketing activities are driven by intelligent and networking technologies, e.g., social media and mobile apps.

Fifth, *pricing and valuation*: New pricing models and valuation models replace the classic systems. As service providers own datathings, a large bulk order replaces individual or small orders, putting service providers in a strong position to substantially reduce prices in their negotiations with designers, manufacturers, and distributors. Consumers are no longer sensitive to product prices since they hire services, and thus are more sensitive to hire fees, service quality, and product performance. Membership-based or hop on/hop off models are widely employed in service usage. Just like tap on/tap off bus systems, smart bike users can tap on to use a bike and then tap off to end the service use. Services of services becomes a new and important business, to maintain the service quality and operations of datathings. Service providers negotiate with them for pricing and business models.

Sixth, *payment*: Service providers have to hire third-party e-payment services to enable frequent, ad hoc service usage by consumers through smart personal devices. E-payment services become a very big enterprise, beyond the scope of traditional banking services. They receive large amounts of deposit fees and service usage fees. For example, if one billion smart bikes are distributed to thousands of Chinese cities, and each service user deposits RMB99 for a bike, then RMB99bn sits in the service provider's account. Compared to the actual service usage fees, managing the value of service deposit fees presents both great opportunity and risk.

Seventh, **regulation and risk management**: New data economic models fundamentally challenge existing regulation systems, including policies, regulation models, stakeholder relationships, and risk management. They have to apply to service providers, including providers of servicing data services, to ensure open, fair, transparent, and efficient data service operations. Regulators are also responsible for protecting the security of consumers' service deposit, ensuring quality is maintained (especially safety) in the services being used, and ensuring a fair relationship exists between service providers and service consumers.

Lastly, *ethics and trust*: Ethics and in particular trust are essential features in operating services and maintaining stakeholder relationships, and in ensuring the success of the service model. Trust must be in the quality of the services must be guaranteed, as must the authentication and credibility of service providers and consumers, and e-payment systems. In addition, as service usage and operations rely on consumer sharing, consumers also have a strong ethical responsibility to follow service rules, respect their peers, maintain the product during usage, and avoid and report illegal or criminal behavior.

Table 8.1 summarizes the main aspects of the comparison between the data economy and real economy.

Table 8.1 Comparison between data economy and real economy

Aspect	Data economy	Real economy
Economic entities	Intelligent datathings and data products	Fundamental function-focused real things and products
Economic activities	Public and group usage-oriented design; outsourcing-based intelligent manufacturing and assembling; networked and online distribution; and active supply-driven consumption	Private and organization-specific design; one-stop manufacturing and assembling; customized distribution; and demand-driven consumption
Ownership	Licensing, sharing and hiring-based services	Private ownership
Stakeholder relationship	Service providers, service consumers, and service regulators, and e-payment service providers	Manufacturer, sales, marketing, and service agencies
Sales and marketing	Selling data services; group consumer-targeted and online marketing and peer-to-peer promotion	Selling products; traditional marketing strategies and channels
Pricing and valuation	Pricing on service licensing models	Pricing on products and maintenance
Payment	Cashless e-payment services	Cash and banking-based payment systems
Regulation and risk management	Regulating services and relevant service stakeholder relationships and responsibilities	Regulating goods and services relationships and responsibilities
Ethics and trust	Trust, social and ethical systems must be formed	Weak involvement in trust and ethics

8.2.4   Distinguishing Characteristics of Data Economy

When we talk about data economy, a fundamental question to ask is, "What are the distinguishing features of data economy?" or, "What makes data economy different from existing economy?"

The smart taxis and bikes discussed above are no longer traditional taxis and bikes—they are smart taxis and smart bikes. When they are online, i.e., connected to clients via smart phone or wireless networks, they become part of the IoT and are partially online and virtual. When they are offline, they are real and just normal taxis and bikes. Smart bikes are not fundamentally different from traditional bikes from the basic transport functionality perspective, but they are very different from the perspective of connected intelligence, which is data-driven. Smart bikes are not normal bikes in the sense that they are unlike traditional family-owned bikes; they cannot be privately owned and operated as normal taxis or family bikes without being fully disconnected from their service systems.

Importantly, data economy may not follow classic economic theories, models, or design. We have to figure things out as we go. Trial and error, guesswork and checking may be the approaches often used in business operations and decision-making. This may result in serious issues, risks and challenges in operations and long-term development, although it can also potentially present surprising opportunities.

The new data economy has distinguishing features that are uniquely enabled by data, intelligence and networked resources and consumers. These data economy features are embodied in

- not being theory-guided: Classic economic theories, methods and systems, including business models, pricing, valuation, marketing and risk management may not be applicable to data economy; there is new ground for development.
- innovation: Datathings are intelligent things, and data economy is intelligent economy. Proposals, operations and management within this new economy rely on research and innovation of new and intelligent datathings, effective and efficient business models, payment systems, and stakeholder management.
- new business model: Data services adopt new business, distribution, and marketing models, such as initial large investments for fast product distribution and market occupation, direct end-user targeting, etc.
- combining physical and virtual economy: Datathings are cyber-physical things, present both real and virtual forms and stay in both the real and virtual worlds. They connect physical businesses with virtual services; e.g., with smart taxis, classic taxi businesses are combined with mobile and wireless network-based businesses, e.g., mobile applications-based economy.
- mixing online and offline businesses: Datathings present in online and offline states and connect online and offline businesses; e.g., when a smart bike is online it is connected to networks and intelligent online services. In retail business, a trend is to combine online shopping with physical offline stores-based delivery and maintenance.

- cashless payment services: Data services require cashless payment services that are connected to smart phones and personal digital devices.
- connected and network intelligence-driven: Business owners (e.g., DiDi), business operators (e.g., taxi drivers and private car leasers) and end users (car hirers) are all connected by mobile applications and services (e.g., DiDi app) with classic mobile telecommunication systems (e.g., mobile call services) and networking services (e.g., wireless networks).
- ubiquity: The provision of data products and data services is ubiquitous, embedded in everything related to our daily life, entertainment, travel, social life, work and study.
- timeliness: Data services build wide and strong market visibility, acceptance and share within a very short period (a few months rather than years), which do not require the years of market development needed by the classic economy.
- uncertainty: Active datathing distribution and market occupation require high initial investment, which will bring high return but also high risk; they may face much stronger competition and faster market shuffling as new business models are proposed by new competitors. Even smarter business operation and marketing strategies may appear quickly, with new and different business operation models for the same business.
- change: New business emerges at a much faster speed than is possible in traditional businesses due to the high level of uncertainty and competition; a data business will experience rapid evolution and significant changes that could exceed business owners' expectations and imagination;
- challenge: Business owners and service providers face significantly faster, bigger and more powerful challenges and risk, but also have greater opportunities during the design, distribution and consumption stages.
- opportunity: Datathings did not previously exist, markets for them did not previously exist, and nor did related business models, operations, marketing and consumer management. Unforeseeable opportunities may therefore exist, depending on the nature of the innovation and enterprise.

Table 8.2 summarizes the main distinguishing characteristics of the data economy.

8.2.5 Intelligent Economy and Intelligent Datathings

Artificial intelligence (AI) is making a return as an increasingly popular topic. We call this generation of AI *advanced AI*. What is the relationship between data economy and the re-emergence of advanced artificial intelligence?

This wave of data economy upgrading and data-driven industry transformation features a revolution in advanced artificial intelligence-enabled technologies and businesses, and the complementary advances in AI and the AI-driven data economy

Table 8.2 Distinguishing characteristics of the data economy

Aspect	Explanation
Hypothesis-free economy	No predefined theories, models and systems governing the design, distribution and operations of data products and services
Innovation	Novel datathings, business models, payment systems, marketing strategies, and stakeholder management
Combining real and virtual economy	Products, resources, and services from both real and virtual economy are involved and combined
Mixing online and offline business	Online and offline resources and services are combined in datathings
Cashless payment services	Data economy relies heavily on e-payment, especially cashless payment services, as economic activities are widespread across daily living, working, studying, traveling and entertainment-related businesses
Connection and networking	Data, datathings and consumers are all connected and networked through IoT and Internet
Ubiquity	Data, datathings, data services and their consumers are ubiquitous, involving and relying on individuals
Timeliness	Timing is critical for creating and supporting data economy; datathings are sensitive to timing
Uncertainty	Many things could be unclear or less certain than in real economy, as there may be no systematic economic theories and risk management systems available; the return, risk, competition, effectiveness and efficiency of operational and marketing models may initially be unclear
Change	Significant to mild changes may occur in many ways in business and service models, business operations, marketing strategies, and stakeholder relationships
Challenge	Significantly faster, bigger and more powerful challenges exist as a result of uncertainty and change
Opportunity	As no fixed theories, models, marketing and operations are available, innovation and an individually-targeted economy indicate significant new opportunities

are largely propelled by data science and analytics. This in turn drives a new era of *intelligent economy*.

The *intelligent economy* is composed of inventing, manufacturing, applying, and commercializing intelligent datathings, such as infrastructure, tools, systems, services, applications, and consultations for managing, discovering, and utilizing shallow and/or deep data intelligence, and synthesizing relevant X-intelligences and X-complexities [64] to create increasingly advanced intelligent systems, such as driverless cars.

Datathings are intelligent things incorporating intelligence that may extend beyond their basic functions. Their design needs to incorporate intelligent technologies, in particular recognition (e.g., biomedical features-based), networking (connected to Internet or wireless networks), positioning (capable of determining locations, e.g., GPS location), sensor ability (e.g., Quick Response (QR) Code), personal digital device-friendly (e.g., smart phone and iPad), e-payment (e.g., Alipay or WeChat Pay), and embedded technologies (all functions are embedded in the device).

The manufacturing of datathings has to incorporate intelligent systems and technologies in basic functions. A large quantity of intelligent products is usually ordered to address active supply-based distribution requirements. The distribution of such intelligent products is driven by market design and targeted client markets, which are actively selected and determined by business models. Responsibility for such as storage, maintenance, and quality control of these intelligent datathings has been redistributed to service providers and professional service agents.

Consumers are manipulated by new business models and smart technologies. Sharing and hiring products replaces traditional private ownership. Consumers have to use smart technologies such as smart phones, GPS-based location services, cashless e-payment services, QR code recognition, and even biomedical recognition systems (e.g., face recognition) with vendor applications installed and e-payment set up in order to use data products.

8.2.6   Translating Real Economy

With the increasing popularity and importance of the data economy, where does this leave traditional economy and real economy? (Here, real economy refers to economy that is built on tangible assets; in contrast, virtual economy is mainly built on virtual assets.) How does data economy transform existing real economy and traditional economy?

The emergence of data economy and its significant development has been ignited by two major mutually improving forces. One is the recognition of the values and potential of data, which has been driven and promoted by the evolution of the so-called *new economy* and the transformation of traditional data-intensive industries, such as large private data enterprise. The other is the advancement of data science and big data analytics and their rapid engagement in business, which is conversely significantly influencing and driving a new wave of data-driven and data-enabled increase in productivity and industry transformation, and the development of a new-generation data economy.

Data science and analytics have been the key drivers for the data analytics-driven industrial transformation. A typical indicator is reflected in the 2010 IBM Global CEO study, from which the resultant report [219] drew the following conclusions: "Yet the emergence of advanced technologies like business analytics can help uncover previously hidden correlations and patterns, and provide greater clarity and certainty when making many business decisions." To manage the increasing complexity, the CEOs in this study believe that "a better handle on information and mastery of analytics to predict consequences of decisions could go a long way in reducing uncertainty and in forging answers that are both swift and right." This leads to their desire to "translate data into insight into action that creates business results" and to "take advantage of the benefits of analytics."

In addition to purely novel datathings such as social media and instant messaging tools such as Twitter, Facebook, WeChat and Skype, most datathings are a smart version of traditional real things. Smart taxis and bikes become typical datathings, and other typical datathings include smart phones, smart home devices, driverless cars, and unmanned aerial vehicles. Such datathings and relevant data economy transform classic real economy in terms of

- economic entities, and their structures and relations; e.g., from landline telephones to smart phones. Smart phones are further connected to many other datathings as well as things and activities related to work, study, travel, entertainment and everyday living.
- economic activities, processes, forms, pricing, valuation, and markets; e.g., the design, functionality, production, distribution, logistics, ownership, marketing, management, and maintenance of smart bikes are totally different from the activities around traditional family-owned bikes.
- payment systems; e.g., smart phone-based cashless payments replace cash-based payments including credit card, online banking, and cheque-based payments.
- ownership, marketing and sales modes, and customer relationship management; e.g., friend recommendation becomes more influential than TV advertising, street advertising boards, and paper media, and free services offering bonus-based promotion are often more attractive and effective than discounted pricing.
- relevant economic norms, rules, policies, and regulations; e.g., in Uber and Didi business, governments need to extend the policies they make for regulating taxi drivers and taxi owning companies to private car owners who are Uber drivers, and may have to change existing policies and regulation of the taxi business.
- social and ethical norms and rules; e.g., smart bikes are shared by many different users who have to follow social and ethical norms but also have to develop higher social and ethical standards in using these shared assets.

The rapid change in China's taxi business model and the replacement of home-owned bikes by hired smart bikes are enabled by data and intelligence. As analyzed above, this data and intelligence-driven economy has fundamentally transformed classic business and economic models, ownership and customer relationships, pricing principles, valuation, marketing, payment, regulation and ethics.

8.3   Data Industry

Data industry refers to the industrialization of data-enabled or data-driven products and systems, called *new data industries*, the transformation of existing data businesses, called *transforming existing data industries*, and the transformation of traditional industries and real economy into data-driven businesses, called *translating traditional industries*.

In this section, these aspects are discussed to outline the opportunities for data industrialization.

8.3.1   Categories of Data Industries

Data industries are taking shape and gaining significance as a driving force in the new global economy. As shown in Fig. 8.3, data industries can be categorized into three families: new data industries and services, transforming existing data industries and services, and transforming traditional core businesses and industries.

First, *new data industries and services*: As shown on the left-hand side of Fig. 8.3, data industrialization directly creates new data-centric business in which companies, organizations and even countries compete over how to best use data to create new data products and data economy. New data industries and services may be created in such areas as data/analytics design, data/analytics content, data/analytics software, data/analytics infrastructure, data/analytics services, and data/analytics education.

Second, *transforming existing data industries and services*: As shown in the middle section of Fig. 8.3, data has a business-related relationship with many activities and opportunities, involving the whole lifecycle and processes from data generation to data usage for decision-making. These may consist of

- data management: data acquisition, storage, management, backup, transportation, communication, logistics, security, and governance;

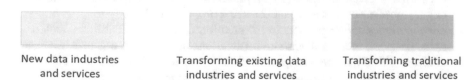

New data industries   Transforming existing data   Transforming traditional
and services   industries and services   industries and services

Fig. 8.3 Categories of data industries

- data manipulation: data processing, preparation, reporting, analytics, visualization, and optimization; and
- data applications: data publishing, sharing, decision-support, and value-added applications.

Lastly, *transforming traditional core businesses and industries*: In recent years, core businesses, including retail business and manufacturing, have given way to a new economy centered on the data industry and digital economy. This is evidenced by the domination of data-enabled companies listed in the top ten global companies, especially the largest data company, Google, and the largest Initial Public Offering, Alibaba. However, core businesses, real economy, and traditional industries have also increasingly recognized the value of data, and have adopted data analytics and data science to transform themselves. This transformation affects many core businesses, such as transport, logistics, manufacturing, e-commerce, tourism, medical and health services, banking, finance, marketing, insurance, accounting, auditing, the public sector, retail business, and education, as well as nature, science, and everyday living, as shown in the right-hand box of Fig. 8.3.

The three categories from left to right in Fig. 8.3 correspond to the three inner to outer rings in Fig. 8.4.

8.3.2   New Data Industries

Section 8.2.1 shows how novel core data businesses are formed by category 1: data-enabling technological businesses, and category 6: emergent data service businesses of data economy. In addition to the areas listed in Fig. 8.1, without loss of generality, the inner circle in Fig. 8.4 illustrates those aspects in which fundamental data technologies and the resultant new areas of data business may grow. The main driving forces of the new data industry come from the following six core areas: data/analytics design, data/analytics content, data/analytics software, data/analytics infrastructure, data/analytics services, and data/analytics education.

- *Data/analytics design* includes the invention of new methods of designing and producing digital and data products, services, business models, engagement models, communication models, pricing modeling, economic forms, value-added data products/services, decision support systems, automation systems and tools;
- *Data/analytics content* includes acquiring, producing, maintaining, publicizing, disseminating, recommending and presenting data-centered content through online, mobile, social media platforms and other channels;
- *Data/analytics software* refers to the creation of software, platforms, architectures, services, tools, systems and applications that acquire, organize, manage, analyze, visualize, use and present data for specific business and scientific purposes, and provide quality assurance to support these aspects;
- *Data/analytics infrastructure* relates to creating infrastructure and devices for data storage, backup, server revenue, data centers, data management and stor-

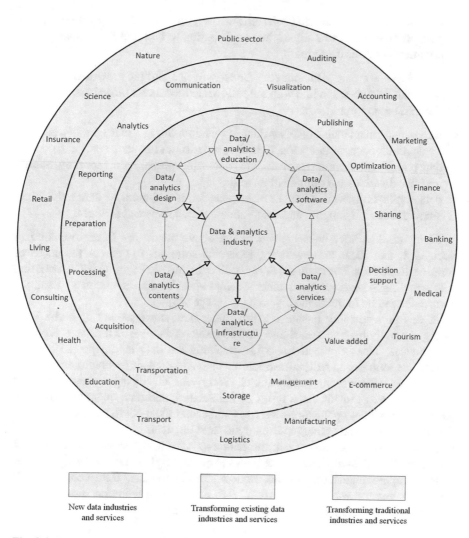

Fig. 8.4 Data and analytics-enabled industries and industrial transformation

age, cloud, distributed and parallel computing infrastructure, high performance
computing infrastructure, networking, communications, and security;

- *Data/analytics services* focus on providing strategic and tactical thinking leadership, technical and practical consulting services, problem-oriented solutions
  and applications, outsourcing, and specific services for data auditing and quality
  enhancement, data collection, extraction, transformation and loading, recommendation, data center/infrastructure hosting, data analytics and more;

- *Data/analytics education* enables the building of corporate competency and
  training, as well as offering online/offline/degree-based courses, workshops,

materials and services that will allow the gaps in the supply of qualified data professionals to be filled, thus contributing to building and enhancing the community of this discipline.

The above six core data/analytics areas will see further growth of new and existing data businesses in line with the following three major directions, illustrated by the middle ring in Fig. 8.4:

- data management which involves data acquisition, storage, management, backup, transportation, communication, logistics, security, and governance;
- data manipulation which involves data processing, preparation, reporting, analytics, visualization, and optimization; and
- data applications: which consist of various forms and areas of data publishing, sharing, decision-support, and value-added applications and services.

Many vendors have been created in the above areas, and those discussed in Sect. 8.2.1, that focus on developing tools and solutions to address the needs of some of these areas. These are the core enablers in the data economy, differentiating data enabling businesses from other business-centric data economies. Example companies can be found in the big data landscape [25].

In this data-enabling business wave, many new start-ups and spin-offs have emerged rapidly in recent years and have focused on data-based business, products and services. This is reflected in a fast-evolving big data landscape [25], which covers data sources and API, infrastructure, analytics, cross-infrastructure/analytics, open source initiatives, and applications. Every year, this changing landscape sees significant, swift growth. As a result of datafication and data quantification, new platforms, products, applications, services and economic models, such as Spark and Cloudera, have quickly emerged in analytics and big data.

Interestingly, as we have seen, the data economy occupies the top ten largest capital entities. The data industry continues to create new business models, products, services, operationalization modes, and workforce models. Data economy will further change the way we live, work, learn and are entertained as new facilities and environments are created in which data plays a critical role.

8.3.3 Transforming Traditional Industries

If data is viewed in the same way as oil, as the new international currency, then clearly the global economy is experiencing a revolutionary transformation from poor data utilization to one that is richly data-explored. Traditional industries which welcome and enable the transformation acquire and collect data, converting themselves from data poor to data rich and data-driven, to increase productivity and become more competitive.

In fact, the respective core data/analytics sectors and core procedures referred to in Sect. 8.3.2 can be developed, implemented and incorporated in any data-related sectors, especially data-intensive domains and sectors such as telecommunication, government, finance, banking, capital markets, lifestyle and education. The application and deployment of advanced data technologies into traditional real economy and core business, including manufacturing and services for transforming our everyday living, will introduce increased opportunities for better collection, management and use of data in traditional industries. Advanced analytics on these domain-specific data can be undertaken to improve productivity, effectiveness, and efficiency, and create new value-added growth in the traditional economy.

Data science and data technologies are driving the transformation of traditional industries,

- from offline real businesses to online virtual businesses and the mixture of real and virtual businesses; e.g., a bike manufacturer who produces and sells both family-owned bikes and smart e-bikes.
- from manufacturing-driven businesses to integrated businesses that connect design, manufacturing, distribution and consumption through online and networked services; e.g., a bike manufacturer who creates online shops and design studios to fulfill global market needs and jobs, produces bikes in response to online orders for global users, and collaborates with global logistics companies to deliver products.
- from demand-driven design, manufacturing, and distribution to data analytics-evidenced innovation-driven design, service and consumption; e.g., a bike manufacturer who collects bike orders and bike designs from global markets, analyzes the preferences in different countries and markets, and then designs suitable products to satisfy these market needs.
- from private ownership-based operations to licensing, hiring and usage-based service operations, e.g.: hiring laptops for fixed period and tap on/tap off smart bikes based on usage; e.g., a bike manufacturer who grows to certain level may license their bikes to smart e-bike service providers and also maintain the bikes for those providers.
- from pay for use to open access: the open access model changes the publishing business model; authors create the work, pay for its publication, and offer free access to readers, to achieve rapid and effective community benefit, and to increase the impact of an author's work.

The core traditional industries that constitute category 2 of the data economy family (Fig. 8.1) have experienced or are experiencing significant changes in creating new growth areas, business models, operational and marketing strategies, and efficiency, competitiveness, and productivity. Examples are given below.

- In *capital markets*, faster infrastructure can be built to process and analyze hundreds of thousands of orders per second and billions of transactions per day in real time, facilitating high frequency orders, trade models, business rules, and real-time market surveillance. In turn, this enables connection to other markets to support real-time cross-market trading and surveillance.

- In *telecommunication*, classic voice call service-focused businesses can be upgraded to multimedia services that support voice and video calls, and instant messaging can be combined with location-based recommendation to expand classic telecommunications businesses.
- In *banking*, cheque and credit card-based banking business can be replaced by mobile-based e-payment, leading to significant changes in retail banking business.
- In *insurance*, new insurance products and policies can be created based on the analysis of insuree preferences, budget, and needs, to provide affordable, on-demand, and personalized insurance products.
- In *manufacturing*, 3D printing and automated, personalized and customized data product manufacturing have fundamentally re-shaped traditional manufacturing businesses.
- In the *food industry*, food design, nutritional configuration, production, logistics, and targeted distribution based on customer preferences replace classic food business operations.
- In *logistics*, proactive planning and operations may be driven by job requirements, cost benefit analysis, and optimization of route planning and selection.
- In *transportation*, the assignment of transport devices, route selection and planning, and pricing and payment methods is optimized based on data analysis.
- In *government services*, evidence-based on-demand and customized policies and services replace population-based service delivery, and active customer engagement and behavior management are enabled based on analyzing government-citizen interactions and customer needs.
- In *tourism*, global tour planning, cost-effective pricing, targeted customer promotion, and personalized care may be offered as products to travelers.
- In *marketing*, traditional marketing channels such as TV, radio and questionnaire-based surveys are replaced by mobile and social media-based end user-targeted promotions and marketing campaigns.
- In *retail business*, goods and pricing are configured according to user needs, payments are cashless, and community-based logistics and delivery are connected to online shopping orders.
- In *healthcare and medical services*, personalized care, medication and treatment are provided based on understanding patient fitness, health, sensitivity to specific medications and treatments, the effectiveness of those medications and treatments, and dynamic physical conditions.
- In *accounting and auditing*, classic routine reports are largely complemented by analytical reports to gain deep insights. Accountants and auditors are required to conduct data analysis of accounting and auditing data to identify insights and invisible issues.
- In *consulting services*, generalized consultations are upgraded to evidence-based, personalized consultations.
- In *education*, courses and lectures are customized according to students' background, preferences, career plans, learning performance, and market needs through online and mobile-based offerings.

8.4   Data Services

Data services are services that are mainly designed and driven by data, and offered and provided principally through data-enabling technologies and networking channels. Data services may be created purely on the basis of data as datathings and data products, or as new service forms built on and transforming existing businesses and services. Figure 8.5 illustrates the directions of developing data services, in alignment with the categories of data industries discussed in Fig. 8.1.

In this section, we discuss data service models, and particularly illustrate data services that can be enabled by data analytics.

8.4.1   Data Service Models

Data services form part of the whole landscape of data and analytics which, as noted above, is changing every aspect of the way we live. Data services can be differentiated from traditional services by the fact that they are not traditional physical material- or energy-oriented services. Data services are information and intelligence-based, networked, and smart.

Data services have important characteristics that differentiate them from classic services.

- Data services act as core businesses of an economy, rather than auxiliary businesses;
- Data-driven production and decision-making emerges as the core function in large organizations for complex decision-making and strategic planning, rather than adjunct facilities;
- Data services are online, mobile and socially based, embedded in our activities and agenda;
- Data business is global and 24/7, offered in any place at any time, on demand or in a supply-driven mode;
- The provision of data services does not require traditional production elements such as intensive workshops, factories and office facilities;
- Data-driven services offer real-time public service data management, high performance processing, analytics and decision-making;
- Data-driven services support full life-cycle analysis, from descriptive, predictive and prescriptive analytics for the prediction, detection, and prevention of risk, to innovation and optimization;
- Data-analytical services are intelligent or can enhance the intelligence of common data and information services;
- Data services enable cross-media, cross-source and cross-organization innovation and practice; and
- Data services demonstrate significant savings and efficiency improvement through the delivery of actionable knowledge/insights.

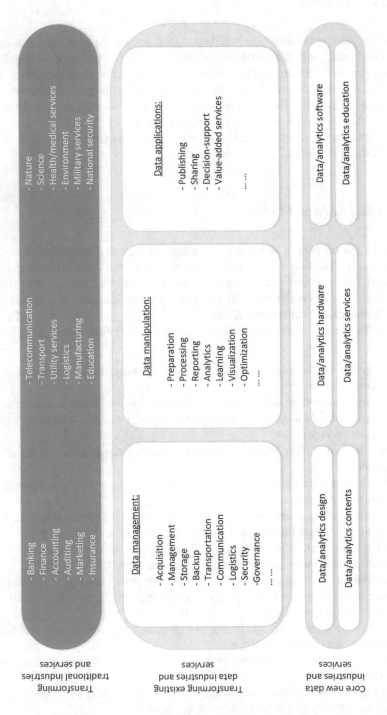

Fig. 8.5 Spectrum of new and translational data services

A major challenge and increasing need in the data industry is to provide global or Internet-based data services for such organizations as multi-national companies and whole-of-government. These services need to

- Define the objectives and benefits of global data analytics;
- Support good data governance, security, privacy and accountability to enable smarter data use and sharing;
- Support data matching and sharing in the context of cross-organizational, cross-platform, cross-format and cross-analytical goals;
- Prepare global and organization-specific local/departmental data;
- Foster global and local analytics capabilities, capacity and competency;
- Enable sharing and collaboration in data and analytics skills, infrastructure, tools, techniques and outcomes;
- Support crowdsourcing, collaborative and parallel analytic tasks and analytic workflow management;
- Support analytic capability and package sharing;
- Support data and data software versioning management and control [30] at a global and collaborative level;
- Support the visualization and dissemination of outcomes to targeted audiences and in personalized preferences.

Data-driven industry and service are forming new trends in data science for business, for instance:

- Advanced analytics is no longer just for analysts [181]; "dummy" analytics is becoming the default setting of management and operational systems;
- Cloud data management, storage and cloud-based analytics are gaining popularity [181] and are replacing traditional management information systems, business support systems, and operational support systems;
- Data science on scale from multiple sources of data is becoming feasible; Internet-based services are a strongly growing area of the new economy;
- Analytics as a service is becoming feasible with appropriate social issue management, as analytics becomes a reality everywhere and is embedded in business, mobile, social and online services;
- Visual analytics is becoming a common language;
- Data services can be mixed with virtual reality and presented in a way that combines physical and virtual worlds, resources and intelligence;
- Services on matched and mixed data are streamlined into a one-world process, addressing both local and global objectives.

8.4.2   Data Analytical Services

It is possible to create new services in the six data/analytics industrialization areas referred to in Sect. 8.3.2: design, content, software, infrastructure, services, and education. As analytics plays the fundamental role in data science, we illustrate the data analytical services below. Data values can be converted to data services as new services by analyzing data, or existing businesses and services can be transformed by undertaking analytics on acquired data.

For example,

- in the *data/analytics design* industry, design studios may be established to design analytical models, systems and tools to product users, with corresponding training and maintenance services.
- in the *data/analytics content* industry, service opportunities can automatically collect, analyze, present and recommend data relevant to a specific need from multiple online sources.
- in the *data/analytics software* industry, many companies and startups focus on inventing, producing, selling and maintaining specific software, tools or online services, e.g., related to cloud analytics or visual analytics.
- in the *data/analytics infrastructure* business, companies may focus on producing data storage and backup solutions, while others may be dedicated to developing data analytics platforms and programming tools.
- in the *data/analytics services* sector, recommendation services, which are becoming increasingly popular, could be customized for news recommendation, travel planning, wine selection, or making restaurant suggestions.
- in the *data/analytics education* sector, corporate training and public/online short courses are in demand and can be offered to enterprises and individuals to upgrade their knowledge, capabilities and competency.

Some typical data services delivered through analytics for both core real business and new economy are listed below as examples:

- *Credit scoring*: to establish the credit worthiness of a customer requesting a loan;
- *Fraud detection*: to identify fraudulent transactions and suspicious behavior;
- *Healthcare*: to detect over-service, under-service, fraud, and events like epidemics;
- *Insurance*: to detect fraudulent claims and assess risk;
- *Manufacturing process analysis*: to identify the causes of manufacturing problems and to optimize the processes;
- *Marketing and sales*: to identify potential customers and establish the effectiveness of campaigns;
- *Portfolio trading*: to optimize a portfolio of financial instruments by maximizing returns and minimizing risk;
- *Surveillance*: to detect intrusion, objects, persons and linkages from multi-sensor data and remote sensing;
- *Understanding customer behaviors*: to model churn, affinities, propensities, and next best actions on intervention behaviors;
- *Web analytics*: to model user preferences from data to devise and provide personalized and targeted services.

Many core and traditional businesses can be converted to data-driven economy and industrial transformation. There are many opportunities for creating data services or converting traditional services to data services. For example,

- in *capital markets*, data-driven financial technologies can enable more efficient trading and surveillance services, e.g., creating high frequency-based algorithms for trading.
- in *telecommunication*, highly important customers in most developed communities can be prioritized for customized multimedia call services by analyzing a customer's profile, usage, preferences and budget.
- in *banking*, low risk home loan scoring can be enabled by deeply analyzing loan applications and borrower data to avoid potential high risk, and to optimize resources for high return but low risk borrowers.
- in *insurance*, by analyzing insurance data, services can be created to determine fraudulent claims or over claims.
- in *manufacturing*, better supply and demand relationships may be modeled by analyzing client, market, and competitor data, and personalized product design and manufacturing may be supported once client requirements and feedback are understood, and market demand and changes have been predicted.
- in *food industry*, significant food waste every year could be substantially reduced worldwide by helping families and food providers to better predict needs and preferences and to design and pack preference-based food options accordingly.
- in *logistics*, the analysis of logistics data, route options and conditions, demand, weather conditions, and associated costs can optimize a logistics company's tasks in dispatching and scheduling, route planning, and achieving a supply-demand balance.
- in *transportation*, data collected and analyzed about road conditions, weather conditions, vehicle accidents, and driver behaviors can help transport authorities to better plan routes, monitor road safety and achieve efficiency.
- in *government services*, matching and analyzing data from government agencies can assist in determining allowance entitlements (e.g., unemployment support), decide who might be receiving excessive payments as a result of under-declaring their income, and establish who genuinely needs government help.
- in *tourism*, more personalized cost- and time-effective tour planning may result from the collection and analysis of tourist profiles, travel histories, feedback, the popularity of scenic locations according to seasons and timing, weather conditions, safety and security, and emergency situation data.
- in *marketing*, targeted and end user-oriented advertising may be made possible by collecting and analyzing user data from sessions, cookies, accessing behaviors, and preferences retrieved from browsers and mobile applications.
- in *retail business*, by analyzing importing, sales, and storage data, and the demand and profitability of goods, stores may be offered optimized goods ordering and servicing.
- in *healthcare and medical services*, after analyzing medical performance by alignment with patient health, medical conditions, treatment history and

feedback, unnecessary services and the potentially negative impact of some medicines on specific patients may be avoided.

- in *accounting and auditing*, accounting fraud and poor performance decisions may be identified by analyzing relevant data, and operational and financial areas to be improved or optimized may be determined.
- in *consulting services*, deep insights into corporate operations, business performance, productivity, and cost benefit may be discovered by analyzing the relevant data of a client.
- in *education*, early identification of students who may experience learning difficulties may be achieved by collecting and analyzing data about students' engagement in class, online learning, library, teamwork, background, learning capabilities, and cultural and family background, and students and education administrators may be advised accordingly.

8.5 Summary

The focus of this chapter has gone beyond the scientific aspect of data science to focus on the economic and industrial opportunities that can be achieved by building on data and converting data to values.

As the discussion in this chapter shows, the foundation of a new economy known as data economy has been established, and is significant for the dramatic way in which it is reshaping our core economy and traditional businesses. This is embodied in terms of new data industries, upgrades to existing data businesses, and the conversion of traditional economy.

In book [67], readers can find discussion on the opportunities and case studies of the data economy and industrialization in many business domains, and the transformation of domain-specific businesses and economy through the application of data science and technologies.

The new data economy has given birth to a new profession, the data science profession, whose purpose it is to develop data science as well as enable data economy. The focus of Chap. 10, is on the data science profession and the potential to create an exciting new generation of data professionals.

Chapter 9
Data Science Applications

9.1 Introduction

In principle, data science can be applied to any application area or business domain. However, data science projects are intrinsically different from ordinary software development and IT projects. The success of data science applications requires many higher level skills, experience and strategies than ordinary IT projects, and involve the applicability of data science techniques, the process of conducting data science tasks, the generalization and specialization of data science algorithms and solutions, and the methodologies for managing data science development tasks. This chapter offers general guidance on these aspects, as well as discussion on success factors in data science projects.

We also discuss the application areas and business domains in which data science may play an important role. We briefly list the data type and possible application and use cases of data science and analytics to advertising, aerospace and astronomy, arts, creative design and humanities, bioinformatics, consulting services, ecology and environment, e-commerce and retail industry, education, engineering, finance, gaming industry, government, healthcare and clinics, living, sports, entertainment and relevant services, manufacturing, marketing and sales, medicine, management, operations and planning, publishing and media, recommender systems, science, security and safety, social science and social problems, physical and virtual society, community, networks, markets and crowds, sustainability, telecommunications and mobile services, tourism and travel, and transportation.

© Springer International Publishing AG, part of Springer Nature 2018   263
L. Cao, *Data Science Thinking*, Data Analytics,
https://doi.org/10.1007/978-3-319-95092-1_9

9.2 Some General Application Guidance

9.2.1 Data Science Application Scenarios

The current buzzy discussion on developing technologies and the commercialization of artificial intelligence, big data, cloud computing, and blockchain all take advantage of data science. Data science has a critical driving role in the next revolution in science, technology and economy, in such scenarios and forms as

- enhancing existing technical systems and tools, business applications and services to more advanced, effective and efficient stages. For example, migrating online banking risk analytics systems to tackle extremely imbalanced and suspicious distributions of fraud (e.g., one fraud in 10M transactions) or optimizing manufacturing processes by identifying and constructing higher-performing processes and workflows.
- transforming traditional technical systems and tools, business applications and services to new generations that have significantly advanced or fundamentally new functionalities built on existing technologies and applications. One example in banking risk management is the upgrade of existing credit card, anti-money laundering and online banking risk management systems by adding sequential behavior and interaction analysis to analyze sequential interactions between customers and the bank, identify possible customer churn, and recommend marketing strategies for retaining customers. This will make existing banking customer engagement more active, personalized, and real-time.
- inventing and producing new technical systems and tools, business applications and services that are driven by data science and technologies. An example in the banking business is to create new cross-product marketing strategies and campaign activities or produce compound/buck/packaging products that provide a more cost-effective and easy-to-manage experience to customers who are likely to purchase multiple products from one bank. In fact, many new businesses and companies have been or will be created by introducing new business models, data tools and applications, and the new economy heavily depends on the invention and adoption of new data-driven economy and the industrialization of data science.

9.2.2 General Data Science Processes

Data science tasks follow certain processes and specifications in any data science application scenario.

- Standard processes in data science work consist of business understanding and requirement analysis, data understanding and problem definition, data preparation and feature engineering, modeling and optimization, evaluation and refinement, and deployment and adaptive enhancement. The specifications for these procedures and their decomposed tasks can be found in such methodologies as CRISP-DM [104], SEMMA [346] and DM-AKD [77], and in book [67].

- It is often said that in conducting classic data analysis and business intelligence projects, 80% of the time is spent on preparing data, while only 20% is spent on modeling and evaluation. Data science projects build on data preparation and exploration, focusing on analytics and learning by spending about 80% of time and resources on advanced analytics (including deep analytics and deep learning) and actionable knowledge and solution delivery. Dedicated efforts are typically made by data scientists on domain-specific algorithm development and customized and personalized solutions and evaluation systems for domain-specific data science problem-solving.

- As discussed in Sect. 10.3.1, data science team members may hold a variety of positions and roles appropriate to the responsibilities and tasks required throughout the project, or in parallel. The data science development process is jointly managed by these different roles. Data scientists focus on data-driven discovery, hence are more specialized in several key tasks of a complete development life cycle, i.e., data exploration, knowledge discovery, and result interpretation and presentation.

9.2.3   General vs. Domain-Specific Algorithms and Vendor-Dependent vs. Independent Solutions

In the market, vendors have developed many analytics and learning tools that include algorithms and/or systems developed and tested on public or domain-specific data. Examples include commercial tools such as the SAS toolset and IBM Waston and PASW toolset, and open source tools such as Apache toolset and Tensorflow. Use of these off-the-shelf tools is often proposed for general purposes or domain needs.

Consulting firms play a key role in assisting enterprise and government to incorporate data science in enterprise innovation and production. General advice and solutions are frequently created and refined for a range of business needs. We have seen that consulting firms often slightly tune the consultancy papers developed for one company to address the needs of another user. This shows the general applicability of their solutions and advice.

End users in business have to consider whether an off-the-shelf toolset fits their domain-specific problem, data characteristics and complexities, and whether vendor-specific solutions provide a perfect match and extensible opportunities for the demands and development of their business. We make the following observations as a result of our practice of enterprise big data analytics:

- No commercial toolset perfectly fits individual requirements without intensive and substantial customization of functionalities, tuning and training of algorithms on the data, and evaluation according to personal criteria. Before paying a bit bill to obtain a tool licence, comprehensive planning, review and expert opinion on requirement analysis and design are a must.

- Independent advice from domain experts and academic leaders with real-world experience may be sought before taking advice and solutions from consulting firms and vendors. Really useful algorithms and solutions are often domain-dependent and require many iterations of tuning and customization to fit the domain-specific data, problem characteristics, and business requirements.
- In a large business, it is critical to hire professional data scientists who fully understand the history, present dynamics and future potential of data science, and have hands-on experience. Holders of these valuable assets will conduct the requirement engineering of data science innovation, and will judge whether a vendor-based solution fits the needs of the business. Starting with something small, tangible, and practical is always a good strategy for newcomers to data science, before they invest in a more powerful toolset and more comprehensive exploration agenda.
- Learning from reflection on failures in other domains and organizations is invaluable. Before we commit to a new customer's projects, we often provide references so that our customers can consult previous clients to enable them to understand the value, process, lessons, and advantages and disadvantages of different options in undertaking enterprise analytics. More experienced end users can share invaluable knowledge to enable informed, tailored and effective/efficient agendas and strategies to be devised for specific businesses.

9.2.4 The Waterfall Model vs. the Agile Model for Data Science Project Management

General IT engineering projects (particularly for software development) are typically managed by the waterfall project management methodology (also called the waterfall model) [458], which is a linear sequential approach consisting of a logical regression of steps from system conception and system engineering, system analysis, system design, system construction and coding, and system testing to system maintenance throughout the system development life cycle. While the structural and iterative waterfall model enables adaptation to shifting teams, forces teams to be structured, and measures development in terms of milestones and schedules, it often involves lengthy and risky investment, and constraints on ad hoc, dynamic and significant running-time design alternations, during-procedure user feedback, and innovative explorations.

Agile methodology [448] is a rapid, iterative and team-based application development methodology that has found favor in both ordinary IT projects and data science projects. A team defines, conducts and evaluates complete functional components by collaborating with customers (end users) within each short term (called a sprint, typical duration 2 weeks). A daily stand-up session (e.g., 15 min) is organized in the morning to update and review progress, and to resolve and determine major issues and priorities. At the end of the sprint, demonstrations of sprint outputs are made and reviewed. Business owners in the team evaluate and prioritize the sprint task's business value and planning (including further development, reprioritization, and termination or completion of the sprint component). The agile development process merges end users into the technical development team, allowing customers to continuously oversee and review progress from an early stage, interact with developers, periodically demonstrate a basic, small but complete version of system components, and make decisions on priorities and changes to timing, resources and schedules. End users own the project and directly guide the formation of project deliverables throughout the project. However, the agile approach may also place a burden on customers in terms of commitment, and as a result of requiring the team's full dedication to the project, as well as synchronization at a single, concentrated location. The approach may also impose a burden because of the risk that significant changes will be required, the uncertainty of scope, priority, resources and deliverables, and the challenges and quality issues induced by the refactoring of system requirements, infrastructure, design and construction in large and complex system developments. The agile model may better fit developments that are small (team and resources), flexible (requirements and fund availability), dedicated (customer involvement), de-composable and less-coupled (systems can be easily decoupled to more independent components).

Both waterfall and agile models were originally proposed for software and system development, not for data science projects. It is important to recognize the differences in nature between data science projects and ordinary software projects. Data science tasks are generally more exploratory (they include understanding data characteristics, selecting and constructing suitable features and algorithms, and evaluating results), analyst capability-dependent (different analysts with different experiences and capabilities may produce different outputs), iterative (spanning the whole process from problem understanding to feature engineering and modeling, and result evaluation), less logical and sequential (while some high-level procedures are applicable to a project, the knowledge discovery is less process-focused and more findings-driven), and may be more uncertain (including uncertainty about the best approach and what values can be discovered from the data).

In data science projects, the choice of methodology depends on many aspects of a project's nature and development. The waterfall model better suits a large team, whereas the agile model prefers a small, dedicated and synchronized team; it requires less customer involvement (only at milestones) than the agile method (throughout the project); works better for clearly scoped and fixed budget contracts, whereas the agile method allows for unclear or exploratory scoping at the cost of additional funding, resources, and schedule; and in a large scale project, achieves milestones and final deliverables as agreed or fails to achieve them, whereas the agile model enables small and partial success from each sprint and has a lower risk of full project failure. If a project or task can be decomposed and the project's scope and features are flexible, the exploratory nature of data science may mean the agile model is better suited to data science projects.

In applying the agile model in data science projects, customers should be educated to understand that "agile" does not mean they can do o change whatever they want in the course of the project. Agreement on upfront constraints may be formed before the project kickoff to avoid significant disputes between developers and customers on cost and funding, team personnel and capabilities, scheduling, and the corresponding deliverables. For larger-scale data science projects which involve multiple big teams and significant resources, and aim for ambitious outcomes, the agile model may be adjusted by adding advantageous elements from the waterfall model, e.g., creating a full picture of the connected milestones, specifying the scope of each sprint upfront, making clear the evaluation criteria and methods, and forcing teams to be structured and collaborative. When the focus is on knowledge, intelligence and wisdom as the outcomes of data science, data science methodologies need to involve these aspects into the development process and management. This may require a project management support and development environment that enables intensive interactions between data scientists, and between domain experts and data scientists, and conceptualizes data-driven findings for user-friendly presentation, visualizing data findings, and enabling and enhancing the interpretability and reproducibility of data findings by business people.

9.2.5   Success Factors for Data Science Projects

Many factors affect and determine the success of a complex data science project. Indeed, many aspects of the data science project itself often affect the project's success, e.g., capabilities, development process, algorithms and solutions, and evaluation. Here, we summarize several such factors which may determine success:

- *Thinking*: Whether the technical and business teams have built the data science thinking, and whether they have built a common understanding of the nature, characteristics and differences between data science projects and more general IT projects as data scientists and engineers.
- *Organization maturity*: Whether the organization has sufficiently mature data science strategies, data, infrastructure, capability, and computing to conduct the tasks.
- *Requirement*: Whether there is a common and in-depth understanding of the requirements, objectives, scope, budget, schedule, outcomes, constraints, uncertainty and risk of the problem and project.
- *Complexity*: Whether the characteristics and complexities of the data science problem (including domain-specific data, problem, knowledge and practice) are sufficiently understood, represented, and/or quantified. It is necessary to have a list of the known and unknown aspects of the problem.

- *Team*: Whether the capabilities of a team and the individuals in it are sufficient for tackling the characteristics and complexities of the data science problem. The team capability set needs to fit the known and unknown aspects of the characteristics and complexities, thus the team needs to be sufficiently experienced to recognize and disclose the significant yet often hidden value from the data. Highly experienced managers and scientists may play a critical role in ensuring the appropriate data understanding, design and evaluation of the problem, as well as communicating with stakeholders.
- *Modeling*: Whether the algorithms and solutions (including the assumptions and constraints held by the algorithms and solutions) suit the characteristics and complexities of the data science problem.
- *Evaluation*: Whether there are quantifiable and commonly agreed evaluation methods, measures and success rates for the data science team and business owners to evaluate and deliver the results, and whether there are effective and business-friendly tools to present and interpret the data findings to business people.
- *Methodology*: Whether there is a customized development methodology that suits the domain-specific nature of a data science project and the specific nature of the project. If the agile model is used, whether the team is synchronized and dedicated, budget and resources are flexible, possible failure is effectively communicated, and other possible uncertainty and risk are recognized by both technical and business people.
- *Business engagement*: Whether sufficient support is available from end users, domain experts, and business executives, and whether business concerns are effectively addressed during the project period and in the project deliverables. Whether the end users are involved throughout the project or at the milestone points of the project, what roles end users have in the project execution period, and whether common understanding and dispute resolution methods are built upfront.
- *Communication*: Whether efficient and effective communications and interactions are supported and undertaken in the team, and between the data science team and executive business owners, and whether business users and owners are sufficiently involved in the development.

9.3 Advertising

Targeted advertising [433] is a popular area for applying data analytics. It requires an understanding of the most receptive audiences for targeted ads. Data science and analytics may assist in recommending

- ads to the individuals most likely to be interested in them;
- the most suitable channels for promoting ads;
- the timing and most appropriate time slots for placing ads;

- the match between ads, goods, products, customers and advertisers; and
- the evaluation of the quality, impact and value of specific ads.

Demographic data may be explored and include a web user's cultural, religious, economic, educational information; psychographic data about the user's values, attitudes, opinions, sentiment, lifestyle, personality and interests; behavioral information about browsing activities, ordering information, and pages visited; and fake, misleading, and falsified data.

9.4   Aerospace and Astronomy

The application of data science and analytics in aerospace forms a new interdisciplinary area: *aerospace data analytics* or *flight data analytics* [176, 267]. Data science can be helpful for such issues in aerospace as

- analyzing airline operational data to optimize maintenance programs, pilot and crew scheduling, to improve overall operational efficiency;
- identifying trends and patterns;
- improving the design, construction, and operation of aircraft;
- generating maintenance alerts, such as warning of the impending failure of an engine component;
- reducing fuel consumption rate, stress levels, temperature, and power usage;
- minimizing turnaround time; optimizing the design of more efficient and cost-effective aircraft;
- enabling improvements across all stages of the aircraft lifecycle, from conceptualization to design, production, after-market support and in-flight operations;
- enabling anomaly detection, incident examination, and fault detection; and
- improving flight safety and life expectancy.

Aerospace-related data may consist of resources on flight conditions, operations, and services; crew information, behaviors and service records; aircraft service and maintenance logs; resource usage; and data related to accidents. Astronomical data has many aspects [367], for example, basic data, cross-identifications and bibliography for astronomical objects; extragalactic databases that include cross-identification, positioning, redshifts and basic data; data related to space missions; and other astronomy information.

9.5   Arts, Creative Design and Humanities

While little data is available for analysis in relation to classical and historical artworks, modern artists and creative designers tend to use more advanced creative design technologies such as digital design, 3D techniques, and virtual reality to create artworks. Data science can be helpful for

- evaluating artworks;
- quantifying and forecasting artwork quality, prices, and changes in value;
- understanding market needs and consumer preferences and interests;
- recommending artworks to the most receptive consumers;
- optimizing the supply and demand relationship; and
- informing design and innovation.

Humanity analytics can also analyze the art market, artistic quality, and public engagement, such as the enjoyment, contemplation, and understanding of artworks.

Arts and humanities and creative design involve data about artwork design and description; comments and commentaries about artworks; artist profiles; the provenance, market valuation, and sales history of artworks; and purchaser data and history.

9.6 Bioinformatics

The marriage of biology with informatics creates the new field of bioinformatics [422]. Bioinformatics is critical for problem resolution and many other purposes. Typical areas are

- understanding biological processes;
- the extraction of useful results from biological data;
- understanding the genetic basis of disease, and unique adaptations, desirable properties, or differences between populations;
- sequencing and annotating genomes and their observed mutations;
- the analysis of gene and protein expression and regulation;
- the discovery of evolutionary aspects of molecular biology;
- the detection of species changes, mutations in genes and cancers (genomes of affected cells);
- the comparative analysis of genes and other genomic features across organisms; and
- the analysis of relationships within biological pathways and networks.

These are conducted on biological (genetic and genomic) and protein data. Other data such as bioimaging data may also be involved.

9.7 Consulting Services

Data science and analytics consulting services are increasingly being offered, but in fact, data science can also improve consulting service quality and performance. Data recorded in relation to consulting services, such as workloads, assignments, work plans, schedules, deliverables, reviews, progress, accidents, team collaborations, meetings and collaborations with clients, issues, and feedback can be used to quantify service quality, performance, potential risk and its causes, and efficiency. Optimization can be conducted to recommend the best possible match between job requirements and consultant capabilities, improve scheduling and planning, and to ensure high-performing team and customer relationship management.

Relevant data may consist of job descriptions, team information, scheduling and job assignment data, progress, review and reflection data, client feedback and complaints, and accident information.

9.8   Ecology and Environment

Environmental and ecological data are rich and comprehensive. They may comprise vegetation development and evolution; biological systems; climate and climate change data; animal profiles and behaviors; and disasters and hazards such as flooding, diseases, land slides, earthquakes.

The interdisciplinary interaction between environment science (water, air and land), ecology, and data science forms a new area called environmental informatics [51, 52, 224], ecological and environmental analytics. Many issues can be deeply explored for possibly better insights and solutions using ecological and environmental data. Examples consist of

- investigating the effect of various environmental factors on animal behavior (e.g., turtle movement and nesting, bird migration, salamander migration, and other amphibian activities), and terrestrial and submerged aquatic vegetation growth;
- improving the quality of environmental management by assessing risk and analyzing management effectiveness in all phases of environmental hazards;
- understanding the wide range of environmental risks (e.g., flooding, natural disasters, animal diseases, chemical spills, and high winds);
- prioritizing risk management actions;
- quantifying the relationships between environmental risks and influencing factors;
- quantifying the impact of environmental risks on a population and potential secondary impacts;
- optimizing environmental risk management w.r.t. influencing factors and social, economic, political and technological trends and costs;
- predicting the possible consequences of a situation; and
- recommending improvements for other risk assessment and management methods in given scenarios.

9.9   E-Commerce and Retail

Data mining and analytics have been widely used in e-commerce [257, 336] and retail industries [258] for better shopping experience. Typical applications and data analytics tasks conducted on e-commerce and retail data include:

- optimizing query search and product ranking by responding to user queries with the most relevant products, or listing the products searched in order of relevance;
- understanding query intent for more personalized responses;
- product recommendation to identify and recommend similar products;
- fraud detection, such as detecting a seller's misuse of system facilities and falsification of product categories, or a customer's use of stolen credit cards;
- understanding business operation, inventory and sales matching performance and detecting problems such as predicting inventory and sale trends, and inventory patterns;
- categorization, such as mapping products to the correct product category and building an inventory classification for the most relevant product;
- seller, retailer and customer profiling and behavior patterns, service usage, and preferences;
- service quality evaluation by using session information to understand visitor/customer profiles, browsing and viewing behaviors, and interest in specific pages, products and product categories; and
- user categorization of new users, readers, competitors, browsers, repeaters, heavy spenders, and original referers.

Further customer-centric analytical work can be conducted by

- segmenting customers based on their value, potential and engagement;
- modeling customer propensity to pursue certain purchase pathways, product categories, price ranges etc.;
- analyzing and predicting customer visiting frequency, value growth, and cross-category purchase habits;
- evaluating customer satisfaction and the effectiveness of product catalogue and market campaigns; and
- building personalized communications with high value and high potential customers.

Analyzing e-commerce and retail may involve data such as user (seller, buyer) demographic data; product attributes, listing price, price ranges, and product categories; in-shop and online browsing behaviors and click-through rates, user purchase transactions in POS systems, and product sell-through rates; user behavioral data from a query to a product page view, and all the way to the purchase event; query logs, other users' searches, and the semantics of query terms; data on loyalty programs, email campaigns, and marketing activities.

9.10 Education

Educational data mining and learning analytics have long been an area of data science in education. Issues that benefit from data analytics include

- benchmarking the entrance score for a course;
- understanding student admission preferences and designing effective strategies to attract students;
- analyzing and predicting student attrition, driving factors, and prevention strategies;
- predicting student risk in respect of completing assignments and passing subjects, and suggesting early intervention strategies;
- understanding the high-performing behavior of students on campus, online, and in social media;
- converting low-performing students to high-performing students by identifying and changing their behaviors;
- quantifying and optimizing the effectiveness and utility of campus and online services and facilities;
- optimizing a student's learning path and progression by configuring subjects that satisfy the student's interest, background and career objective;
- suggesting student career plans and actions to fulfill these plans on campus;
- quantifying teaching performance at different granularities and identifying unsatisfactory and low-performing teaching methods, including possible changes to subject content; and
- identifying students in financial or emotional difficulty and recommending strategies for intervention.

Educational data may consist of student entrance data, class engagement activities, academic performance, access to online, library and social media-based resources and materials, interactions with lecturers and tutors, student survey data, data on a student's family situation, and socio-economic data about the suburb in which a student lives.

9.11 Engineering

Engineering is a substantial area for data science and analytics, and involves many areas of engineering, such as software engineering, civil engineering, energy engineering, food engineering, mining engineering, city engineering, and industrial engineering, as well as engineering design, manufacturing, and logistics engineering of products [162]. For example, mining software engineering data [468] has been a typical application for diagnosing problems in software design, coding, debugging, testing, and use. Analyzing engineering data can address many engineering-related issues, such as

- benchmarking the effectiveness and efficiency of engineering processes, systems, resource usage, and project management;
- optimizing the design, configuration, manufacture, and planning of devices;
- benchmarking and optimizing the balance between consumption, demand, supply, and logistics;
- detecting, alerting and preventing gaps, exceptions or even accidents in engineering systems, processes, and workplaces;
- monitoring the condition of infrastructure and devices;
- optimizing human resource configuration, high-performing work patterns, effective teamwork and collaborations;
- enhancing people satisfaction, performance, professional development, leave management, and career planning;
- quantifying and improving quality, reliability, and status of engineering systems; and
- quantifying and improving management, governance and organization.

Depending on the specific engineering area, the engineering data involved may relate to infrastructure, people, devices, products, processes, management, manufacturing, implementation, usage, consumption, workplace, and environment-related configurations, parameters, variables, costs, sensor data, accidents, conditions, and status, and relevant external data such as contextual information.

9.12 Finance and Economy

Financial and economic problems consist of risk and portfolio management, financial analysis, pricing, investment, trading, surveillance, accounting, auditing, and consulting. Analyzing financial and economic data can address many potential problems and issues, and can improve financial services and economy in ways that cannot be achieved without data-driven discovery.

Financial and economic data analytics applications may include

- recommending and optimizing portfolio management by portfolio analysis to suggest trading rules, products, timing, and positioning in a market;
- analyzing price movements and market data for designing and optimizing high frequency financial trading algorithms for algorithmic trading;
- identifying and forecasting trends and exceptional market movements;
- benchmarking the price of financial products;
- corporate finance-related analysis and management;
- bitcoin and virtual money-related analysis and regulation;
- analyzing dependency across products, markets, and portfolios to create or improve trading strategies and optimize return on investment;
- benchmarking risk factors, risk areas, and mitigation strategies for financial risk management;

- detecting anomaly, fraud and problems in accounting, auditing and finance, uncovering evidence for auditing, and supporting anti-money laundering;
- risk analytics in Internet insurance and Internet wealth management;
- optimizing financial supply chains;
- analyzing opportunities and risk in mobile payment and e-payment;
- discovering market manipulation strategies and evidence;
- discovering insider trading behaviors;
- predicting links between market movement, trading strategies, announcements, and sentiment in social media.

Both macro and micro financial and economic data can be combined to understand the relationships and interactions between micro-level products and trader behaviors and macro-level markets and company dynamics, and their influence. Financial data may consist of tick by tick trading behaviors and transactions, market data, investor information, account information, payment information, announcements, news, commentaries, and social media data, accounting data, auditing data, market surveillance business rules, regulation policies, cross-market data, and external data, such as government policies, political events, and natural accidents. Economic data describes the economy and related information, such as unemployment rates, GDP, microeconomic and macroeconomic data.

9.13 Gaming Industry

The gaming industry has developed rapidly, especially in respect of the wide acceptance of online and mobile games. Gaming businesses accumulate large amounts of data about game players, playing behaviors, multiple game player interactions, game results, and incentive or payment information. In addition, the gaming industry also engages in marketing, sales, advertising, and the supply and demand of game products.

Analyzing gaming data can

- model game players to understand user behaviors, preferences, sentiment, motivation and intent, and satisfaction with specific games;
- improve and optimize game design, implementation, operation and management;
- design dynamic, evolving, and personalized games;
- detect, predict and prevent illegal gaming behaviors, gambling, money laundering and fraud; and
- design and optimize game design and production for positive learning and entertaining objectives.

Additional analytics can be undertaken to analyze game sale data, to make personalized game recommendations, and to optimize game marketing and productivity.

9.14   Government

The government sector has been the driving force in initiating, applying and promoting data analytics and applications. Government data analytics can be applied to

- government services to quantify service quality, improve service objectives, detect and prevent problems in servicing;
- improving interactions and engagement efficiency and effectiveness between government, clients and citizens;
- improving and optimizing the effectiveness and utility of government services;
- identifying and preventing fraud, compliance, and risk;
- detecting, preventing and enhancing national security, cyber security, and homeland security;
- understanding the sentiment of clients and citizens;
- optimizing government policies, strategic planning, and strategy design for industrial transformation, economy, productivity and increasing competitiveness;
- discovering and preventing national risk such as financial crisis; and
- detecting and intervening in natural disasters and social emergencies.

Different government departments and agencies different data resources. For example, social security services accumulate such data as service allowance policies, recipient information, recipient data from other government agencies such as healthcare, taxation, bureau of statistics, immigration and customs, as well as from banks, and feedback data such as over-the-counter services and call center data, and overpayment-related information such as debt amount and duration. Data matching and open data have been enabled in some countries, which results in the incorporation of data from other government departments and agencies. External data should also be involved, such as social media data, news, and blogs.

9.15   Healthcare and Clinics

Healthcare data analytics [218, 427] can play a critical role in enhancing the quality of healthcare services and resource effectiveness. Typical issues that could be resolved by analyzing healthcare data are

- clinical analysis, including benchmarking and optimizing healthcare services, and clinical quality of care;
- conducting effectiveness analysis of healthcare services to provide personalized healthcare and clinical services;
- analyzing and enhancing patient safety and detecting and preventing medical errors;
- detecting and preventing under-servicing, over-servicing, and the misuse of healthcare and clinical services;

- analyzing and optimizing healthcare, clinical resources, and service quality and effectiveness, as well as optimizing physician profiles and clinical performance;
- benchmarking and determining healthcare resources and services, including the pricing of medicines;
- conducting financial analysis to optimize operational and administrative costs;
- detecting fraud and abuse in public and healthcare services;
- optimizing the supply vs demand of healthcare services;
- evaluating the quality and competency of service providers for improved patient care;
- preventing disease and optimizing disease management; and
- creating educational and training opportunities for service providers to improve career planning and service quality for better patient wellness.

Healthcare analytics involve a variety of data sources, such as transactions from hospital information systems; clinical data collected from electronic medical records (EHRs); patient behavioral, preference, and sentiment data; visit data of patients from healthcare service providers; pharmaceutical data; claims and cost data; retail purchase data; hospital and clinic operations data; and external data about patients' social relationships, activities, emotional wellbeing, etc. Health and clinical data may also be connected to genomics, imaging, pharmaceutical and diagnostic data for comprehensive targeted and personalized health service provision, remote service provision, and telehealth.

9.16   Living, Sports, Entertainment, and Relevant Services

Mobile services and social media are deeply embedded in our daily living, entertaining and services. They create a significant amount of fast-growing data every day, examples of which are (1) daily life-related activities and services; (2) entertaining-related data, such as restaurants, coffee bars, theaters, arts, and online services such as music, video, and games on demand; (3) lifestyle service data, such as cleaning services and delivery services; (4) sports and fitness-related data; and (5) classic, online, mobile and social media data.

Data science is transforming our lifestyles [365], entertainment [37], sports [439], and relevant services. Examples include

- quantifying and improving urban, community and city livability;
- understanding residents' preferences, needs and satisfaction;
- predicting and recommending services based on mobility and service demand and preferences;
- recommending physical entertainment and sports services;
- recommending online, digital and mobile application-based media content and entertainment services;
- detecting and optimizing lifestyle, health and emotional wellbeing;

- analyzing and optimizing service quality, effectiveness, and customer satisfaction;
- customizing and recommending personalized services;
- predicting and optimizing athletic performance by analyzing data on athletes and coaches, as well as the profiles, sentiment, and behaviors of other staff, and playing conditions prior to and during sports participation; and
- detecting and preventing illegal, unhealthy, and unsafe activities and events related to daily life, entertaining, and services (e.g., terrorist activities, fraudulent activities, money laundering and illegal payments).

9.17 Management, Operations and Planning

Management, operations and planning (MOP) form part of any business or enterprise, and generate massive amounts of data. This data consists of information about MOP-related (1) teams, staff and managers; (2) activities, resources, and records; (3) performance data; (4) events, exceptions and accidents; and (5) policies, rules and regulations. MOP data may also be connected to production, marketing and sales data in business.

Data obtained from management, operations and planning sources can be used to explore many MOP-related issues and to gain insight into improving MOP. Examples are

- quantifying high-performing MOP individuals, groups, and management teams, and identifying key factors that may improve MOP human resource value and performance;
- analyzing MOP-related activities to discover effective or ineffective MOP patterns, factors associated with ineffective MOP activities, and opportunities to transform MOP effectiveness;
- understanding the distribution and drivers of MOP performance, benchmarking MOP performance indicators in MOP scenarios, and promoting positive MOP activities;
- detecting MOP accidents and the reasons they occur, and preventing MOP exceptions; and
- analyzing and optimizing the applicability and effect of MOP policies.

In addition, opportunities exist to analyze the enterprise-wide impact of MOP on product engineering, marketing, sales and customer satisfaction.

9.18 Manufacturing

Manufacturing industry accumulates a plentiful amount of data which consists of rich information about the manufacturing lifecycle, relevant people, processes, and supply/demand. It involves (1) product design, production, and implementation; (2) product logistics and supply chain; (3) product raw materials and resources; (4) production processes, conditions, and status; (5) information on manufacturers, the workplace and performance; (6) product market value, demand, and supply; and (7) product sales data, such as real-time shop-floor data.

Data science can transform traditional manufacturing to another level: intelligent and personalized manufacturing, and analytics can play an important role in manufacturing [18, 459]. Possible analyses of manufacturing data include

- diagnosing, preventing and reducing process flaws, quality issues, and accidents in manufacturing procedures by collecting and analyzing production compliance and product life traceability data;
- detecting and reducing waste and variability in production processes to improve product quality and yield;
- analyzing designs, specifications and configurations to optimize product competitiveness and productivity;
- analyzing operational and managerial data to optimize manufacturing effectiveness and efficiency to make time and money savings;
- optimizing manufacturing processes and production flow by analyzing process data, worker behaviors, capacity, efficiency, performance, and problems;
- measuring and comparing the effect of production inputs, processes, and designs on yield;
- benchmarking work patterns, identifying high-performing work patterns and converting low-performing work behaviors to high-performing activities for better workforce efficiency and performance;
- analyzing and improving process smoothness and team collaborations during the production process, including management, engineering, machine operation, and quality control;
- quantifying the relationships between demand, production, sales, storage, and delivery and distribution to optimize and speed up product turnover and throughput, and reduce storage and distribution costs;
- quantifying the cost-effectiveness and profitability of material usage;
- optimizing product pricing and return on investment;
- analyzing and improving customer satisfaction by collecting and analyzing product quality and market feedback; and
- quantifying and improving product warranty periods, processes and customer experience.

9.19 Marketing and Sales

Marketing and sales are highly data intensive sectors. Marketing and sales-related data consists of (1) marketing information, including marketing activities, campaign activities, focus groups and customers, promoted products, product information (including prices, volume, etc.), marketing channels, surveys, and information on campaign effectiveness; (2) sales information, including sales channels, product pricing and discounting, sales volume and revenue; (3) competition data including competitor product pricing, focus groups and customers, market share and revenue, marketing channels, strategies and activities; (4) customer data, including customer demographics, relationships, and socio-economic status; (5) customer purchase history, transactions and behaviors; (6) customer feedback and satisfaction data, including survey feedback, complaints, product reviews, qualitative surveys, focus groups, interviews, and call center data; (7) customer online data, including product browsing, clickthrough and sellthrough activities on shopping website; and (8) external data, such as social media feedback and comments on products.

Marketing and sales data analytics can address many issues, such as

- product pricing;
- forecasting product and market demand, size, revenue and profitability, and predicting market trends and growth potential;
- predicting unmet market trends and unsatisfied customer needs;
- quantifying and improving brand perception;
- identifying and predicting future market development and new customers;
- analyzing customer preferences and recommending preferred products;
- analyzing the effect of marketing on customers (for example, price sensitivity), and suggesting relevant marketing activities and potential product recommendations;
- analyzing and predicting the effectiveness and efficiency of marketing and sales channels;
- predicting and preventing customer churn and value dynamics, and forecasting customer lifetime value;
- comparing and analyzing competitor pricing, marketing and sales strategies and their effectiveness;
- estimating inventory capacity needs;
- estimating brand value;
- estimating and improving the accuracy of prospect lists, and prospect engagement;
- detecting non-customers and creating marketing and sales strategies to convert them to customers; and
- detecting and preventing fake customer activity or manipulation, and quantifying its impact.

9.20 Medicine

Data science is fundamentally transforming both Western and Chinese medicine [349]. Medical data science is conducted on a wide range of data, including DNA, protein, and metabolite-related data to cells, tissues, organs, organisms, and ecosystems; disease-specific information; medical and clinical information about patients such as clinical service and treatment data, physicians' notes, hospital admissions and discharge data from hospitals and GP surgeries; radiology images and notes, pathology samples and results, and hospital medical images; and government-held medical records. Biological data may be used to enable genomic and genetic analytics related to bioinformatics (Sect. 9.6) and healthcare data (Sect. 9.15) and may also be combined with medical data for analysis. Data recorded in wearable devices and health fitness-based mobile applications are also important for complementing private health and medical data.

Medical data analytics may be undertaken for such purposes as

- understanding the biology of disease;
- analyzing and optimizing medicine and treatment effectiveness;
- evaluating and personalizing treatments and medicine for specific patients;
- prescribing personalized drugs and selecting personalized treatment methods based on data-driven evidence;
- matching diseases and clinical conditions with medicines and treatment methods;
- evaluating and reducing the risk of methodological flaws;
- detecting medical accidents and intervention;
- detecting and preventing medical fraud;
- optimizing the effectiveness and efficiency of medical services;
- estimating and preventing re-admissions;
- pricing medicines and medical services; and
- forecasting hospitalization demand, costs and trends.

By incorporating external data from mobile health applications and wearable devices, lifestyle and entertainment activities, and patient behaviors can be combined with healthcare and medical data to understand and detect behavior patterns for healthy living, rapidly recovering matching patterns between diseases, treatments and patient behaviors. The combined analysis of biological, clinical, and pharmaceutical data can identify the most effective drugs or combined treatments for specific diseases, and assist in recommending the most effective or cost-effective medication.

9.21 Physical and Virtual Society, Community, Networks, Markets and Crowds

Human physical society is expressed through the existence and activities of communities, networks, markets and crowds, e.g., social networks, social media, residential communities, country fairs, collective gatherings such as exhibitions, infectious disease networks, and computer networks. Physical world forms and activities have been accompanied by virtual societies since the development of social networks, social media, and mobile messaging and communications services.

Both physical and virtual societies shape our life, work, study, and economic activities, communications, and friendships.

The many elements of physical and virtual societies interact with and complement one another, jointly generating massive amounts of data related to our daily activities, communications, and personal affairs. For simplicity, we will refer to these societies, networks, media, markets and crowds as *communities*. Community data reflects the activities and characteristics of community members, or participants, including the relationships between participants in a group and across groups, communications between participants (e.g., making audio/visual calls and communicating via messaging systems), information about activities organized within and across communities, including contextual information about those events and activities, and the administration, management and regulation of communities.

Physical and virtual communities seamlessly and deeply express the anytime, anywhere, anybody, and anything of our lived economic, social, and cultural activities. Analyzing community data is thus critical and increasingly challenging. Community analytics is a highly popular topic in data science, and is an important business driver of the data economy. Many topics, areas and problems have been widely analyzed, some of which are listed below.

- *Complex system modeling* [34, 348] or *complex network modeling* [100], which describes, simulates, represents and formalizes the formation, intrinsic working mechanisms, evolution, and dynamics of communities/networks. It explores networking characteristics and system complexities in terms of metrics such as connections, distributions, structures, and relationships, and each of these aspects is extended as a research direction.

- *Network and community formation*, which studies how networks and communities are formed. It covers, for example, evolutionary paths, growth, distribution and features, and the emergence of small-world networks and scale-free degree distribution.

- *Mechanism modeling*, which explores the form and evolution of collaboration, conflict, competition, confrontation, influence, and consensus building in community interactions, as well as transitions, evolution, and conflict resolution.

- *Transmission modeling*, which models the transmission mechanisms, modes, information protocols, and materials exchange within and between communities; the capacity for random failures and targeted attacks in transmission, and the robustness against such vulnerabilities; and failure resolutions. Different systems may have different mechanisms, e.g., disease transmission and crisis contagion may work differently.

- *Influence modeling*, which studies the types of influence and their processes and impact on conformity, socialization, peer pressure, obedience, leadership, and persuasion [275, 429]. Influence may be built on information, materials, or authority, leading to different antecedent-consequent models, or on more specific issues such as estimating the role and significance of important community members, or the working mechanisms and impact of events (e.g., circulating

an announcement) on community members, community structures and relation-ships.
- *Structure modeling*, which captures and represents the type, form, scale, hierar-chy, and features of topological, informational, social, and influence structures.
- *Interaction modeling* or *relation learning*, which characterizes the parties involved in community interactions, and the form, strength, frequency, direction, scale, and hierarchy of interactions (or relations) between community members or sub-communities.
- *Information content modeling*, which measures the richness of information con-tent in a community, such as its volume, heterogeneity, importance, popularity, and business value.
- *Controllability modeling*, which measures the degree of community stabilization, observability and manipulability of community states, members, interactions, behaviors, input, and output.
- *Trust modeling*, which characterizes trust [423] (and reputation, which is also related to influence) in terms of metrics such as quality, significance, and impact, and models trust uncertainty, evolution, transition, and transferability;
- *Relational learning* or *statistical learning* [207, 431], which learns latent rela-tions and interactions, often based on dependencies and uncertainty, between community members and between communities.

More specific applications in social and information networks include

- *Social network analysis* [353, 430], which studies the social structures, relation-ships, types, graphs, and features of social networks;
- *Linkage analysis* [426], which focuses on modeling, representing, detecting and predicting relationships, connections, and interactions between community members and categories, on specific topics such as modeling linkages and *link prediction*;
- *Community detection* [380], which splits a community cohort into sub-communities in a static or dynamic setting and often involves social relationships, member information and contextual information. Typical topics and issues are *hierarchical community detection*, *subgroup discovery*, and infinite community detection;
- *Heterogeneous information network analysis* [357], which represents and models heterogeneities in multiple information networks;
- *Visualization* [254], which presents communities and networks and their internal working mechanisms in terms of diagrams, images or animations.

9.22 Publishing and Media

The publishing and media sectors produce massive amounts of data on a daily basis. This data is spread through traditional print materials such as books and newspapers, as well as via digital media such as online news feeds, mobile news applications, digital libraries, and social media.

Analyzing publishing and media data can address many issues, as shown in the following examples:

- understanding the efficiency of publishing systems from the perspective of design, manufacturing, logistics and distribution;
- optimizing the relationships between publication production, inventory, and market sales;
- predicting future topics of interest and market demand for specific content, and devising corresponding publishing plans and actions;
- estimating and optimizing publication costs and pricing;
- analyzing competitor data and the potential for competitive advantage;
- detecting changes in readers' interests;
- identifying valuable readers and improving customer satisfaction;
- analyzing and optimizing marketing strategies for focus readers and groups; and
- optimizing distribution channels and modes by matching them with preferred reader channels and modes, such as online media, mobile applications, digital publication, and open access.

9.23   Recommendation Services

Recommendation has become a fundamental service which has evolved rapidly into many aspects of business, work, study, entertainment, and daily life in general. Data science has extensively driven the emergence and development of recommender systems and services, to the extent that recommender systems are changing our ways of servicing and doing business with customers. Recommendation involves customers (individuals, groups, organizations), products or product packaging, and the background domain of customers and products. Recommended products can be anything from physical products and services to digital products and services.

As a large, rapidly evolving sector, recommendation-oriented data analytics cover a wide range of situations and opportunities. Typical areas include

- recommending similar new or existing products to customers;
- recommending existing or new products to similar customers;
- recommending products to groups of customers or organizations;
- recommending products from other categories, domains or regions to existing customers in existing domains;
- recommending products to satisfy personalized preferences, interests, and needs;
- understanding and addressing customer responses, feedback, sentiment, and genuine and dynamic demand;
- estimating customers' future demands, including changed interests for corresponding recommendations; and

- reasoning about and inferring the potential needs of customers by analyzing customer behaviors, status, needs and intent through the analysis of external data, such as data from smart home and smart city-oriented sensors, sensor networks, online and social media, and mobile applications.

9.24   Science

Science is experiencing a paradigm shift from hypothesis, theory and experimentation-driven discovery to data-driven discovery [95, 209, 305]. This applies to broad domain-specific scientific areas, including physical, chemical, medical, earth, environmental, biomedical, and neural sciences. Specific scientific problems and new opportunities can be addressed or better handled from a data-driven perspective, using data-driven approaches. Examples are the data-driven biomedical analysis of genomic and genetic data for evidence and practice-driven medicine, medication, and treatment.

Areas that can benefit from data-driven scientific discovery, which may complement existing scientific paradigms, may include but are not limited to

- new hypothesis and hypothesis-free discovery from big data and significant amount of practice, e.g., discovering effective disease-medicine matches that are sensitive to specific patient circumstances yet are not recognized in existing treatment protocols;
- addressing open scientific questions by iteratively exploring and refining indications and hints from observations and practice, and then capturing general and intrinsic working mechanisms, principles, and theories; for example, exploring how neurons in the human brain communicate with one another;
- proposing new scientific questions from data and practice observations that cannot be addressed by existing knowledge and theories; in data analytics, the results cannot always be explained by or linked to existing knowledge and theories; e.g., hidden coupling relationships may take place in social networks that cannot be represented by existing relation learning theories and mathematical and statistical theories;
- identifying limitations and gaps in existing scientific theories, approaches, and tools by analyzing large scale observations and practices; e.g., recognizing the limitations of assumptions and constraints in applying existing theories that have not been broadly considered;
- fusing multiple cross-science data and knowledge to discover more general and comprehensive domain-independent scientific principles and findings, as well as cross-disciplinary new scientific fields, e.g., data-driven behavior science to quantify and compute behaviors;
- generating new scientific theories and systems that are inspired by observation and practice-based discovery, verified by scientific methods, and converted to theoretical systems;

- representing unknown problems in terms of scientific characterization and formalization to expand the scientific body of knowledge.

9.25 Security and Safety

Security has evolved to a broad field of general issues related to society, finance, economy, nation and homeland, and information and network systems. This has generated a need for a wide range of security applications to handle such problems as homeland security, cyber security, network and computer security, economic security, political security, and social security. Security issues can be complicated; some are illegal and criminal, others are abnormal and usual. In all cases, the relevant domain-related data is rich and important for enhancing security and safety.

Data science has the capacity to address a substantial number of issues in security and safety, since security and safety problems are often exceptional, implicit, sophisticated and difficult to predict. Data science can enhance security and safety by such means as

- detecting and predicting exceptional security-related threats, attacks, fraud, violations, events, and accidents, and their underlying targets, e.g., potential terrorist activities;
- discovering the causes, formation, and driving forces of security events, e.g., identifying reasons why information hacking is triggered, and attacks on financial systems;
- estimating and predicting the potential consequences and severity of security violations;
- capturing behavior patterns, exceptional means, and information on networks, groups, and organizations that might violate security and surveillance policies, regulations and systems;
- describing and identifying the profiles, behaviors, and circumstances of policy violators;
- analyzing and improving the effectiveness of security detection, assurance and enhancement strategies, policies, and methods;
- monitoring the development of suspicious people, events, scenarios and circumstances, and suggesting the most effective intervention actions;
- simulating and quantifying the contagion process, pathways, channels, and effect of high risk and exceptional security violations to relevant businesses and systems;
- analyzing the connections between security accidents and suspects and other aspects, e.g., the instructors and manipulators behind the scenes, other violation plans, and criminal activities.

9.26   Social Sciences and Social Problems

Social sciences and social problem-solving are increasingly quantified, with the increasing and widespread datafication and quantification of relevant social areas and problems. This involves areas such as both data rich and data poor areas; for example, (1) traditionally data poor and highly qualitative and empirical areas such as law, sociology, psychology, humanities, public policy, public health, safety, social security, and national security; and (2) areas that mix both qualitative and quantitative methods, such as education, behavior science, economics, and marketing.

Data science can deepen the understanding of formation, evolution, dynamics, driving forces, causality, and emergency of social problems. Examples are to

- discover citizen's sentiment on government policies for social security and taxation from analyzing call center feedback data and social media data;
- quantify the effectiveness of public health policies by analyzing healthcare and medical data;
- detecting social safety, emergency and accidents by analyzing relevant events, accidents and information;
- complementing empirical research by identifying quantitative indication and supporting evidence to quantify social impact and effect of social policies and mechanism designs, e.g., market regulation rules and trading rules;
- identifying new opportunities for improving and renovating social science research by providing quantitative methods and systems;
- optimizing the use, allocation and usability of natural and social resources;
- modeling the behaviors and dynamics of complex social systems; and
- optimizing other public goods.

9.27   Sustainability

Sustainability [432] involves broad natural and human systems and problems, including natural ecosystems and environments, and human economics, politics, city and land, lifestyle, and culture. Sustainability areas and issues involve comprehensive data and information in the relevant domains and systems. For example, ecosystems and environments consist of data related to biological systems, vegetation change, animal behaviors, climate change, natural disasters, society, community, and human activities.

Data science can enhance sustainability and assist in achieving sustainable development and strong resilience. By analyzing sustainability-related data, we can address problems and aspects such as

- analyzing and detecting the sustainability change and major driving factors;

- simulating and describing sustainability factor interactions and consequence evolution;
- estimating the capacity of target systems in given context;
- characterizing and predicting the dynamics of potential sustainability degradation;
- detecting and predicting extreme sustainability accidents;
- quantifying impact and influence of external disturbance, human activities, policies, and exceptional events on target system's sustainability;
- analyzing the effectiveness and efficiency of sustainability intervention strategies, policies, methods at multiple scales;
- suggesting intervention actions on enhancing sustainability; and
- quantifying and benchmarking sustainable and resilient economy, environment, society and community development at hierarchical scales and within certain contexts.

Additional issues that can benefit from sustainability data science include modeling the relationships between increasing human socialization in aboriginal regions and environments, between economic activities and ecosystems; and modeling the recovery dynamics of destroyed ecosystems.

9.28 Telecommunications and Mobile Services

Telecommunications are industry sector that have rich data and information, and have applied data analytics in various telecommunications and mobile businesses. Telecommunications data consists of customer demographic information (e.g., cultural and educational background); social relationships (e.g., previous employer, education institutions); economic circumstances (e.g., subscription service level, payment methods); call detail records (e.g., call frequency, duration, destination); network and communication data (e.g., equipment, configurations, resources, services, fault, accidents, etc.);

Data analytics can address many problems and issues in telecommunications [284]. Typical issues consist of

- categorizing customer demand, preference and interest and their changes;
- detecting and predicting customer churning and suggesting customer care and retention;
- estimating prices, market valuation, and demand dynamics of new products and services;
- analyzing competitor's advantages, strategies and opportunities;
- predicting market, technology, and demand trend on specified products and services;
- detecting fraud and suspicious activities in telecommunications;
- quantifying customer value and identifying and retaining high value customers;
- quantifying and identifying high value products and services;

- recommending cross-products and services to high value and high profit customers;
- designing and optimizing new mobile services and mobile applications; and
- predicting future directions of telecommunication technologies, markets, services, channels, and business models.

9.29   Tourism and Travel

Tourism and travel are a data rich sector [466], which connect to various relevant services and businesses, such as transport facilities including airline, accommodation, entertainment, tourism places, and social networks and social media. Much of tourism and travel data, such as private data related to hotel and flight booking, itinerary and scheduling, and public data such as comments made in social media on scenery places and restaurants, photos shared in social networks, is available through service providers and online data sharing. Analyzing such data is helpful for tourists and tourism administration.

Analyzing tourism and travel data can assist in smarter planning, administration, resource effectiveness, and safety and security. Typical applications of data science in tourism and travel businesses include

- optimizing itinerary and schedule planning;
- recommending cost-effective and personally preferred hotel and transport combinations;
- recommending attractions and places of most interest to travelers;
- suggesting friends or group-based opportunities by predicting and recommending tourists with similar interests and intent;
- predicting risk, security and safety-related accidents and suggesting intervention strategies and arrangements;
- early warning of significant health, safety, and risk exceptions and emergencies during traveling and mitigation for managing the exceptions and emergencies;
- quantifying and benchmarking satisfaction rate of tourism places;
- quantifying and alerting health, safety and security levels of tourism places;
- designing new tourism products and services to satisfy tourist's expectations and create positive surprise;
- detecting specific on-the-fly travel risk and accidents while traveling, and suggesting resolution arrangements;
- evaluating the effectiveness of tourism marketing activities and suggesting effective marketing strategies; and
- forecasting the economic effect of marketing strategies and tourism products.

9.30   Transportation

Transportation is another area accumulating rich data. Transport data involves many aspects related to road, air, and water transport. Relevant data involves the operations of trains, flights, vehicles, and fares; road conditions; weather conditions; and accidents. Other relevant information is stored in road and transport control and command systems at headquarter offices; road surveillance systems; smart cards, GPS vehicle locations, cell phone Call Detailed Records, and mobility tracking apps; and related insurances, such as travel insurance and car insurance providers. These aspects of information are highly helpful for intelligent transport management, planning, optimization, and risk management.

Data science has played a fundamental role in intelligent transport systems [292, 428]. Typical roles consist of

- estimating and predicting traffic flow, volume and demand w.r.t. timing, period, and regions, such as by analyzing telecommunication services usage and their alignment with travel, and evaluating travel demand management performance;
- estimating transport system effectiveness and efficiency, service reliabilities, as well as risk;
- optimizing traffic capacity and network efficiency;
- early warning and prediction of traffic overcrowding and congestion w.r.t. timing, period and locations;
- predicting the likelihood of traffic accidents and analyzing factors driving incidents and events;
- modeling and visualizing the dynamic public mobility and implementing mobility-based management, dispatching, and planning;
- exploring travel behavioral regularities;
- quantifying travelers' satisfaction and sentiment;
- building cross air-road-water transport management, monitoring and optimizing systems; and
- making traffic control plans and alternative arrangements for large-scale special events.

Analyzing transport data, combined with other relevant external data, such as telecommunications data, location data, mobility data, entertaining services, can also provide value-added services, such as (1) estimating and optimizing route planning and journey time for GPS; (2) recommending living and entertaining services to travelers; (3) sending alerting messages about potential traffic overcrowd, jamming, and emergencies.

9.31 Summary

There are many more aspects of data science applications that could be discussed, but they are well beyond the scope of this book. Instead, readers can find relevant books and references about practice and case studies in the application and/or industry tracks of major data science conferences, such as KDD and DSAA, and application and programming-focused books on data analytics, big data computing, cloud computing, and machine learning. The book: Data Science: Techniques and Applications [67] by this author consists of case studies of analytics and learning, and discussion on the methodologies, processes, and project management of data science in the real world.

Chapter 10
Data Profession

10.1 Introduction

We are lucky to be living in the age of analytics, data science, and big data. These three fields represent probably the most promising areas and future direction in the current Information and Communications Technology (ICT) and Science, Engineering and Technology (SET) sectors and disciplines. While a bright future seems to lie ahead of us, the responsibilities and qualifications of a data scientist, or more broadly, data professional, are often not clearly defined, despite the fact that the role of data scientist has been described as the sexiest job in the twenty-first century [121]. At times, it seems that everyone on the street currently lays claim to being a data scientist.

This is a clear indication that data science is driving a new profession, i.e., the *data science profession*, or simply *data profession*. Many emergent *data science roles* (or for short *data roles*) are being created to form a special community of *data science professionals* (or *data professionals*). These roles will surpass those of traditional business intelligence professionals (BI professionals) [345] and data administrators, as well as new roles such as big data experts and data scientists.

So, what will make up the next generation of data scientists? This fundamental question addresses a wide-ranging topic, that is, what forms the data profession; what are the definitions and qualifications of the various data roles and data science competencies; what constitutes data capability maturity; what is the core knowledge base and skill set for data professionals; and what are the competency, maturity, and ownership requirements for a data-enabled organization? It specifically demands a comprehensive definition of the responsibilities and qualifications of data scientists.

To answer this broad question, it is essential to conduct an in-depth exploration of the intrinsic characteristics, agenda, future directions and gaps in the data profession and data science field, as well as developing a disciplinary roadmap, body of knowledge, set of competency standards, code of conduct, and self-regulation mechanism for the data profession. This may be jointly undertaken by relevant

© Springer International Publishing AG, part of Springer Nature 2018
L. Cao, *Data Science Thinking*, Data Analytics,
https://doi.org/10.1007/978-3-319-95092-1_10

funding bodies, policy makers, professional bodies, and experts already working in this field.

This chapter discusses the formation of the data profession and considers the definition of data roles, the required core knowledge and skills for data professionals, the model of data science maturity, and the responsibilities and qualifications of data scientists and data engineers.

10.2   Data Profession Formation

The growth and recognition of an emerging field can be effectively measured by the breadth, depth and speed of the formation of its professional communities. The data science and analytics communities have grown incredibly quickly over the last decade. The term *data profession* refers to those areas in which a critical mass of professionals essentially handle data. The data profession has emerged as a professional area that significantly involves and drives data-specific science, engineering, and administration.

This section discusses the formation of the data profession in terms of its disciplinary significance indicators, research progress, and community, professional, and socio-economic development.

10.2.1   Disciplinary Significance Indicator

The formation of the data profession has been marked by key disciplinary significance indicators, as shown in Fig. 10.1. They include:

- scientific development—significant data science research progress, reaching a critical mass of dedicated scientific publication venues;
- scientific community—significant global data science communities with a critical mass of disciplinary scientists;
- professional development—significant number of global professional organizations with corresponding technical progress, and
- socio-economic impact—significant demonstrated data economic productivity and social impact.

10.2.2   Significant Data Science Research

The first significance indicator is the emergence of dedicated publication venues for reporting the significant scientific progress made in data science. Several journals dedicated to data science have been established in addition to many existing journals

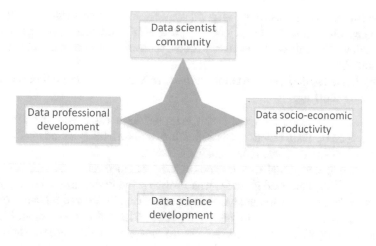

Fig. 10.1 Disciplinary significance indicators for data profession

and magazines that play the dominant role in reporting relevant progress, especially in the areas of artificial intelligence, data mining, machine learning and statistics; for example, the Machine Learning Journal and the Journal of Machine Learning Research, which report on progress in machine learning. These dedicated journals include the Journal of Data Science [235], launched in 2002, which is devoted to applications of statistical methods at large; the electronic Data Science Journal [140] relaunched by CODATA in 2014; EPJ Data Science [151] launched in 2012; the International Journal of Data Science and Analytics (JDSA) [236] in 2015 by Springer; IEEE Transactions on Big Data [385] in 2015; and the Springer Series on Data Science [369] and Data Analytics Book Series [113].

Other publications are in development by various regional and domain-specific publishers and groups. Some examples are the International Journal of Data Science [226], Data Science and Engineering [139] published on behalf of the China Computer Federation (CCF) [139], the International Journal of Research on Data Science [227], and the Journal of Finance and Data Science [237].

These dedicated publication venues report the views, debate and disciplinary discussions on data science, theoretical and technological progress, and technical and general practice and lessons learned in developing applied data science practice.

10.2.3   Global Data Scientific Communities

The second significance indicator can be found in the creation of global data scientific communities and the emergence of a critical mass of disciplinary scientists. The evolution of data science communities has been significantly enabled and boosted by related conferences, workshops and forums dedicated to the promotion of data science and analytics. There are also many well-established venues which either focus on specific areas of data science, such as KDD and ICML, or have adjusted their previous non-data and/or analytics focus, such as the traditional AI conferences IJCAI and AAAI.

More dedicated data science communities are emerging through focused efforts, such as the following:

- The first conference to adopt "data science" as a topic was the 1996 IFCS Conference on Data Science, Classification, and Related Methods [225], which included papers on general data analysis issues.
- The IEEE International Conference on Data Science and Advanced Analytics (DSAA) [135] launched in 2014, was probably the first conference series dedicated to data science and analytics in an interdisciplinary and business-oriented way. Co-sponsored by ACM SIGKDD, IEEE CIS and the American Statistics Association (ASA), DSAA attracts wide and significant interest from major disciplines including statistics, informatics, and computing, representatives from government and business, and professional bodies and policy-makers. The IEEE Conference on Big Data is an event dedicated to broad areas of big data.
- Several other domain-specific and regional initiatives have emerged, such as the three initiatives in India, i.e., the Indian Conference on Data Sciences, the International Conference on Big Data Analytics, and the International Conference on Data Science and Engineering, and several Chinese conferences on big data technologies and applications sponsored by the China Computer Foundation and other associations.
- A number of other conference series have been renamed and repositioned from their original focus on topics such as software engineering and service-based computing to connect with big data and data science, drawing mainly on key topics of interest and participants from their original areas.
- Data analytics, machine learning, and big data have eclipsed the original topics of interest in many traditionally non-data and/or analytics conferences, such as IJCAI, AAAI, VLDB, SIGMOD and ICDE. Not surprisingly, more than 50% of papers at some of these venues now frequently cover data science matters.
- Many professional interest groups have been set up in social media, including Google groups, LinkedIn, Facebook and Twitter, and are among the most attractive and popular venues through which big data, data science and analytics professionals can share and network.

A small group of data science researchers play a pioneering role in the formation of communities dedicated to data science research and practice. Of the sizeable cohort who claim to be data scientists, only a very small proportion are original thought-leaders and intrinsic data science research initiators who truly contemplate, research and practice data science-related issues, and drive the disciplinary development of data science. The data science era requires a large group of disciplinary thought-leaders and frontier researchers.

A majority of practitioners who claim to be data scientists are (1) simply continuing what they did in their previous domain, such as researchers in statistics, database management, information retrieval, data mining, machine learning, computer vision, and signal processing; (2) refocusing their interest on data-related areas; (3) incorporating data technologies into their specific areas; or (4) applying data technologies without research and innovation.

10.2.4   Significant Data Professional Development

The third indicator is the growth and development of professional (online) communities and organizations, and the public or private establishment of educational activities designed to build and train data science professionals, promote big data, analytics and data science research, practice and education, and promote interdisciplinary communication. For example:

- The IEEE Big Data Initiative [223] aims to "provide a framework for collaboration throughout IEEE", and states that "Plans are under way to capture all the different perspectives via in depth discussions, and to drive to a set of results which will define the scope and the direction for the initiative."
- The IEEE Task Force on Data Science and Advanced Analytics (TF-DSAA) [382] was launched in 2013 to promote relevant activities and community building, including the annual IEEE Conference on Data Science and Advanced Analytics.
- The China Computer Federation Task Force on Big Data [85] consists of a network of representatives from academia, industry and government, and organizes its annual big data conference with participants from industry and government.
- Several groups and initiatives promote dedicated analytics and data science activities. For instance, Datasciences.org [118] collects information about data science research, courses, funding opportunities, professional activities, and platforms for collaboration and partnership. The Data Science Community [137] claims to be the European Knowledge Hub for Bigdata and Datascience. Data Science Central [138] aims to be the industry's online resource for big data practitioners. The Data Science Association [133] aims to be a "professional group offering education, professional certification, conferences and meetups" [172], and even offers a "Data Science Code of Professional Conduct."
- Many existing consulting and servicing organizations have adjusted their scope to cover analytics, where they previously focused on other disciplinary matters. Interdisciplinary efforts have been made to promote cross-domain and cross-disciplinary activities and growth opportunities. Examples include INFORMS [230], Gartner, McKinsey, Deloitte, PricewaterhouseCoopers, KPMG, and Bloomberg.

Data science education and training play a key role in forming the data profession and training a generation of qualified data professionals, and many relevant data science courses are offered through academic institutions and online course providers. These courses train data scientists at Bachelor, Master and Doctoral levels. More discussion and information about data science related courses can be found in Sect. 11 and at www.datasciences.info.

10.2.5 Significant Socio-Economic Development

The last criterion is to evaluate the socio-economic and cultural impact, and other benefits, resulting from the developments in data science research and technologies, and the development of data science communities and professionals.

Multinational vendors, online/new economy giants, and service providers play a critical driving role in community outreach in this respect. Many relevant initiatives have been launched by these bodies, such as those by SAS [347], IBM [220], Google [190] and Facebook [156]. These major global data economy drivers have either re-structured and transformed their classic business focus to a data-oriented one, or have established themselves as data-driven businesses. The classic IT giants such as Apple, Microsoft, IBM, Samsung and Huawei, analytics-focused companies such as SAS, and data-driven service providers such as Alphabet, Facebook, Twitter, Alibaba, and Tencent have substantially reshaped the global economic map and now dominate the new economy market. Their economic impact is huge, as indicated by their market value-based ranking in the global economic structure [168].

Although their share of the market and impact are huge, however, these giants are few in number. The impact of the data profession is increasingly sparked by the exponential scale of the emergence and growth of startups and spinoffs in the data area. This trend is well reflected by companies that are listed as entering the big data landscape [26] and the fast growing data-driven IPO (initial public offering) companies in recent years such as Airbnb, Uber, Snapchat, and Alipay. These companies are typical individual-user based data services, where data plays an essential role in enabling people's connections, data sharing, business activities and e-payment.

10.3 Data Science Roles

Although vendors and consulting firms have disclosed that there will be a shortage of hundreds of thousands of data scientists in the coming years for any one country, data science roles have not been specified in terms of a well-designed structure or capability set. Building an effective *data science team* for a large organization involves far more sophisticated skill requirements and complementary capabilities than can be achieved by just hiring a pool of data scientists [322].

10.3.1 Data Science Team

Data science roles are those professional positions whose job holders undertake and fulfill a wide range of data-oriented tasks and responsibilities according to the nature of the company or organisation, its focus and objectives.

Figure 10.2 illustrates some of the major roles that may be required in an enterprise data science team. We categorize these data-related or data-enabling roles into six categories: *architects, engineers, data scientists, decision strategists, administrators*, and *executives*, which between them cover the holistic and comprehensive capabilities and competency requirements for undertaking corporate data science.

- *Architects*: Architects are people who develop, implement, and manage the following: corporate data infrastructures and architectures; database, data warehouse and cloud storage and computing systems, platforms, and network; enterprise application integration; distributed and high performance analytics systems and clusters; corporate data and analytics solutions; corporate solutions for data governance, sharing, publishing, privacy, and security.
- *Engineers*: These are people who focus on engineering and developing the architect's designs and solutions. Engineering roles include data engineer, software engineer, application engineer, system engineer, network engineer, security engineer, programmer, developer, data publisher, and reporting officer.
- *Data scientists*: Data scientists focus on the scientific discovery and exploration of data and data-driven problem-solving. Their roles include data modeler, data miner, machine learning analyst, statistical analyst, optimization analyst, forecasting analyst, predictive analyst, visualization analyst, quantitative analyst, and insight analyst.
- *Decision strategists*: The role of the decision strategist is to focus on enabling and executing decisions and strategy design and implementation, and may include business analyst, business intelligence analyst, model operator, deployment strategist, business behavior strategist, and decision strategist.
- *Administrators*: Administrators manage requirements, data, solutions, deliverables, and processes; for example, requirement manager, project manager, quality manager, network manager, data manager, stakeholder manager, deployment manager, solution manager, and librarian.
- *Executives*: These are senior leaders who lead the data science team and work at various levels within an enterprise to build and manage hierarchical or global holistic and systematic planning, strategies, governance and assurance. These particularly include chief officers, such as lead data scientist, chief data officer, chief analytics officer, chief information officer, and chief technology officer.

Note that not all the roles in Fig. 10.2 may be required in an organization. Roles are often merged into more general and "fuzzy" roles such as "engineer", "data scientist", "analyst', "modeler" and "strategist", and these job holders then frequently take on multiple responsibilities and tasks. Enterprise data science teams tend to have more specific roles than smaller and less data-intensive organizations.

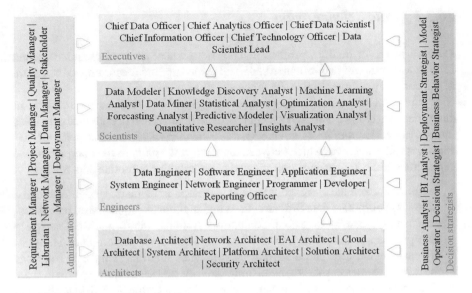

Fig. 10.2 The data profession family

A team is often expanded with more tightly defined roles when it reaches a professional level in which focused specialists play a critical role in major tasks. Interestingly, it is common to find that different titles may be used to denote responsibilities that are essentially similar.

10.3.2   Data Science Positions

Several skill streams are discussed below to expand on the *data science roles* categories in Fig. 10.2. Major data science roles in a corporate organization may consist of analytical architects, data engineers, business analysts, data modelers, system engineers, knowledge discovery scientists, model deployment managers, network engineers, model operators, decision strategists, and business behavioral strategists.

The responsibilities for the above positions are briefly explained below.

- *Analytical architects* determine and produce the architecture and infrastructure to undertake, operationalize and upgrade data engineering, analytics, communication, and decision-support;
- *Data engineers* acquire, connect, prepare and manage data, especially for large-scale, multi-source, and real-time data management and preparation;
- *Business analysts* seek to understand business and its relationship to data, define analytical goals and milestones with data modelers and data scientists, and decide and verify evaluation metrics for business satisfaction;

- *Data modelers* who aim to understand and explore data insights by creating data models to make sense of data through explicit analytics and regular analytics;
- *System engineers* who program and develop system prototypes and platforms by incorporating data models and analytical logics into enterprise information space and decision-support systems;
- *Knowledge discovery scientists* who discover implicit and hidden insights and present outcomes in terms of knowledge representation and deliverables that fit business needs and operations;
- *Model deployment managers* who manage deployed solutions, monitor and refine model fleet management, and conduct scheduling and planning;
- *Network engineers* who enable the communication, transport and sharing of data outcomes across an organization or enterprise;
- *Model operators* who use dashboard, visual to on-the-fly analytics and self-service analytics to analyze data and produce results for operations, and who escalate outcomes for routine management;
- *Decision strategists* who determine what and how should be produced for decision makers, and assist them to reach their decisions;
- *Business behavioral strategists* who create, select and optimize treatments and next best actions to modify the attitudes and behaviors of target populations, following the recommendations of the decision strategist.

These roles are not distributed equally across the areas listed in Fig. 10.2, since greater capacity in data engineering, system engineering, and network engineering is usually required in data-oriented organizations and projects, rather than relying on data scientists and strategists.

10.4   Core Data Science Knowledge and Skills

Data scientists, and data professionals more broadly, need relevant data science knowledge and skills to undertake their responsibilities. This section thus discusses major knowledge and capability set requirements, and highlights the importance of communication skills for data professionals.

10.4.1   Data Science Knowledge and Capability Set

The core knowledge base and skill sets for individual data science roles can be divided according to the particular responsibilities of that role. Holders of data science roles may be required or expected to have a comprehensive knowledge of the field and a broad capability set, although the focus of their role may be quite narrow. For example, a data scientist may focus on predictive modeling, but they will also usually be required to have a deep understanding and ability to undertake descriptive, diagnostic, predictive and prescriptive levels of analytics and model development.

Figure 10.3 illustrates the core categories of comprehensive knowledge and skill in data innovation and services that may be expected in both a data science team and a senior data science role. The data science skill set includes *data science thinking*, *theoretical foundation*, *technical skills*, *work practice*, *communication*, *management* and *leadership*.

- *Data science thinking*, to support the data science objective to "think with data", which requires thinking the habits, knowledge and skills to enable creative thinking, critical thinking, cognitive thinking, imaginary thinking, inferential thinking, reduction, abstraction, and summarization, as well as research methods and decision sciences. Refer to Chap. 3 for more discussion on data science thinking, in particular creative thinking and critical thinking in data science.
- *Theoretical foundation*, consisting of knowledge of relevant theories in disciplines and areas that include statistics, mathematics, understanding data characteristics, data representation and modeling, similarity and metric learning, algorithms and models, qualitative analysis, quantitative analysis, computing (computational) science, complexity analysis, evaluation methods and enhancement, and meta-analysis and meta-synthesis.
- *Technical skills*, composed of skills and techniques for data preparation, data exploration, data mining, document analysis, machine learning, pattern recognition, information retrieval, data management, data engineering, analytics programming, high performance computing, networking and communication, operations research, human-machine interaction, visualization and graphics, software engineering, and system analysis and design.
- *Practices*, including practical components for building real-life data infrastructure and architecture; data management, retrieval, storage, and backup; data manipulation and processing, various levels of business-oriented analytics, and optimization; experimental design, simulation, and evaluation; project development, and management; case studies, applications, prototyping and deployment.
- *Communication*, consisting of presentation, story-telling, reporting, documentation, group collaboration, teamwork, seminars and workshops, reflection, and refinement to various stakeholders, such as end users, business operators, project owners, and decision-makers.
- *Management*, encompassing governance, organization, projects, resources, data quality, roles, responsibilities, risk, impact, privacy, security, social and professional issues, and deployment and decisions.
- *Leadership*, which requires senior executives to make decisions on such strategic questions as: What competitive and business value, benefit and impact can be gained from undertaking a data science project? What essential elements are required to effectively and efficiently undertake data science projects? How should a data science team be set up? Where is the potential risk, and how can it be mitigated? How could data science practice be improved? Where are the gaps in staffing that will further work and career development in data science? These

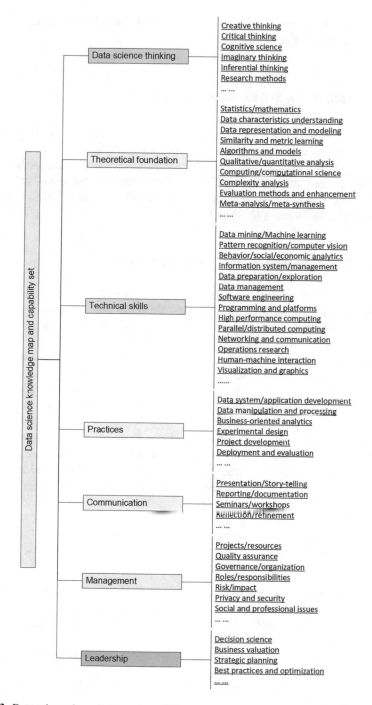

Fig. 10.3 Data science knowledge and capability set

issues involve high level decision science, strategic planning, business evaluation, best practice, and optimization.

10.4.2   Data Science Communication Skills

It is said "if we cannot communicate effectively we cannot do a good job" [46]. Effective communications are essential for the success of data science.

This section extends the discussion on data science communications. Additional discussion on communications with stakeholders can be found in Sect. 2.6.3.6. Many references to developing general skills for effective and critical communications can also be found in [323] and [119].

10.4.2.1   General vs. Specific Communication Skills

The communication skills required to be a good data scientist are multiple, and are both general and highly specific. General skills are those that are applicable to any occasion or project; specific skills are those required to address the particular features and challenges of data science teams and projects.

Commonly recognized general skills that enable effective communication include being a good listener, using nonverbal communication means such as eye contact, body language, hand gestures and tone, and in written communications, being clear and concise, showing confidence through justification, being open-minded, friendly, empathic and respectful.

In communicating with business stakeholders, it is crucial for data scientists to be able to provide a clear and fluent translation of data science findings to non-technical teams and senior executives who may not be in the technical domain, such as marketing and sales personnel, chief executive officer, chief operating officer, and chief financial officer. A data scientist needs to enable decision-making within the business by arming colleagues with domain problem-specific strategic plans, business evaluations, evidence-based and quantified insights, and clear recommendations and solutions in respect of data science projects.

General skills for effective communications may consist of:

- actively learning and understanding the domain and the business, including understanding business goals, concerns, and expectations;
- learning presentation skills that are to the point, cutting edge, confident, engaging, evidence-based, and on-schedule;
- developing business writing which is professional, well-worded and well-grounded, output-driven, performance-based, and evidence-validated, and developing the ability to provide brief, actionable and insightful executive summaries;

- showing respect towards peers and avoiding unscientific thinking and inappropriate team collaborations;
- fitting-in to a corporate and organizational culture by being respectful, open, and engaging;
- using a range of communication channels and media types, e.g., by emails, analytical reports, and executive reports;
- incorporating indicators such as significance, likelihood, severity, risk level, and priorities for informative, balanced suggestions between whole vs. partial, global vs. local, and past vs. present vs. future actions, plans, benefits and purposes into communications.

Communication skills are data role and responsibility-sensitive, but are also independent and involve interactions with business stakeholders. Specific data communications skills and methods of enabling effective communications with business stakeholders may consist of:

- explaining things with graphs and charts rather than by numbers and percentages, and providing summary reports with key factors, evidence, major findings, trends, and highlights for attention, with appropriate justification;
- translating analytical results from the data in response to business questions in a way that tells a story about the formation of a business problem and the solutions to address the problem;
- presenting data findings in terms of knowledge, insights, and recommendations;
- providing data-driven and evidence-based presentations and justification for recommendations;
- converting quantitative results to business rules and intervention actions;
- providing business users and owners with evidence-based recommendations for them to make decisions, and explaining to them how, when and what actions to take;
- ensuring that communications are output-focused and performance-based by showing what outcomes or business impact could be achieved if the findings are implemented.

Figure 10.4 summarizes some of the key general and specific communication skills required by data scientists.

10.4.2.2   More vs. Less Effective Communications

The Burtch Works survey [465] categorizes communication skills for quantitative professionals into four levels: very strong, good, okay, and could be better.

- Very Strong: I can translate technical findings to a non-techie audience with the greatest of ease!
- Good: I haven't confused anyone yet, but there's always room for improvement.
- Okay: They're not bad, but they're probably not my greatest strength either.
- Could Be Better: Hey, I'm a numbers person!

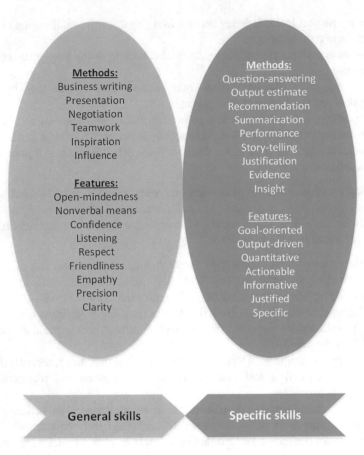

Fig. 10.4 General and specific communication skills for data scientists

This survey also shows that although more than half the respondents believed they had very strong communication skills, their bosses had divided views. So, what communication skills do senior executives expect of data scientists?

- The ability to translate data findings into different versions to suit both technical and business audiences;
- The expertise to convert data results into actionable insights that their managers, marketing team, or even CEO can easily understand, are convinced by, and can act on;
- The willingness to advise what strategic decisions and actions should be taken by appending supporting evidence to justify the analysis of any given scenario, and the confidence to also outline the conditions and constraints inherent in taking those actions;
- The capacity to cater for hierarchical peer-to-peer communications, e.g., enabling the Chief Analytics Officer and Chief Data Officer to communicate with the Chief

Information Officer, Chief Technology Officer, Chief Financial Officer and Chief Executive Officer for enterprise and strategic decision-making;
- Being prepared to communicate more broadly with others such as marketers, journalists, and policy-makers, to interpret social values and business impact.

What makes a data scientist less effective or ineffective in terms of communications? In [272], the following five dysfunctions of a team are cited:

- the absence of trust,
- the fear of conflict,
- the lack of commitment,
- avoidance of accountability
- inattention to results.

In addition, there are other aspects that downgrade or disable effective communications in a data science team. We detail those dysfunctional communication issues below; they consist of:

- being too detail-oriented, preferring numbers and figures;
- lacking a global picture and focusing only on low-level, local and partial aspects;
- being too determinate and expecting the data findings to perfectly fit the business scenario (without realizing the exploratory nature of business findings, and the implicit and invisible characteristics, system complexities and X-intelligence in business problems and data);
- being too quantitative and relying on data alone, without the involvement of domain knowledge and domain experts;
- being too technical, and failing to present the social, economic, business impact and value estimation;

To leverage the shortcomings in data role-based communications, data science team leaders need to effectively gain understanding and support of business owners to include hierarchical non-technical stakeholders in the team and to appropriately define the business goals, milestones, evaluation metrics, and delivery and deployment plan. Business experts and domain specialists may play important data roles as business analysts and business strategists, to leverage the limitations in knowledge and communication skills usually faced by data scientists.

Table 10.1 summarizes the less effective vs. more effective communication skills and attributes in a data science team.

10.5 Data Science Maturity

How do we know whether an individual or organization is mature in terms of data science? To answer this question, we propose a data science maturity model and discuss the concepts of data maturity, capability maturity, and organizational maturity in relation to data science.

Table 10.1 Less vs. more effective communications in data science team

Less effective	More effective
Data-driven exploration	Goal-oriented interpretation
Details with numbers and tables	Summary with graphs and visual diagrams
Low-level, local and partial picture	High-level, global and comprehensive picture
Determinate methods and results	Exploratory and dynamic methods and results
Quantitative results	Answering questions
Data findings	Story telling
Identified patterns and findings	Actionable insights
Empirical observations and arguments	Solid evidence and justification
Technical significance	Social and economic impact and benefits

10.5.1   Data Science Maturity Model

Foundational and applied data science skills and capabilities are required to conduct data and analytics research, innovation and industrialization, and to become a qualified data science professional.

Maturity refers to the level (stage or degree) of capability and the capacity for the full and optimal development of data science. Models are created to measure the maturity of data science in an organization. A *maturity model* provides a structured benchmark for measuring the maturity of data science. At least two structural layers exist in a maturity model:

• The lower layer describes the distinct characteristics or assessment items;
• The upper layer clusters general views.

In software development, the Capability Maturity Model [325] is a tool that objectively assesses the ability of a contractor's processes to implement a contracted software project. The model is based on the Process Maturity Framework. More information about the Business Process Maturity Framework is available from the Object Management Group [40]. The concept of *capability maturity* refers to the "degree of formality and optimization of processes, from ad hoc practices, to formally defined steps, to managed result metrics, to active optimization of the processes" [441].

The current data science community faces significant shortcomings in terms of data, analytics and organizational capability maturity. We propose the concept of *data science maturity*, which measures the stages of advancement of data science resources, capabilities, processes, and results in an organization. Here *resources* refers to the relevant data used for data science; *capabilities* refers to the knowledge and ability for conducting data science, *processes* refers to the procedures and behaviors of undertaking data science, and *results* are the data science outcomes. These are fundamentally supported and conducted on data according to the relevant capabilities of an organization.

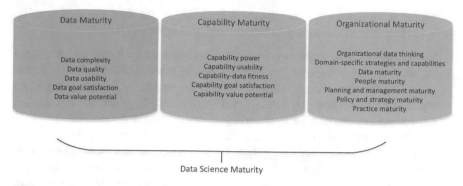

Fig. 10.5 Data science maturity model

We define data science maturity by the following three aspects: *data maturity*, *capability maturity*, and *organizational maturity*. In the following sections, we define and discuss the models for measuring these three aspects of data science.

Shortcomings (gaps) and immature behavioral stages, practices and processes within an organization are common in current data science development, for example:

- At the level of *data maturity*, gaps are seen in local and specific data management and use, in departmental data management and use, and in enterprise-wide data engagement and use.
- At the level of *capability maturity*, such as in analytics, gaps appear in explicit and descriptive analytics, predictive modeling, and actionable knowledge discovery and delivery. They also feature in simple applications of off-the-shelf algorithms and tools for specific problems to more sophisticated explorations of how and why business innovation can be driven by advanced analytics.
- At the level of *organizational maturity* [105], gaps exist in the skill sets required for ad hoc and general use of off-the-shelf or vendor-motivated case studies by individuals, to gaps in planned functional innovation and organizational transformation through the creation of data-driven approaches, and the systematic planning and execution of enterprise differentiation driven by data insights.

Figure 10.5 summarizes the main definitions and aspects of the data science maturity model. It is important for us to further define, quantify and evaluate data science maturity, and make the respective effort required to increase that maturity.

10.5.2   Data Maturity

Data maturity refers to how successfully the data of an organization can effectively fulfill its anticipated purpose and deliver the expected outcomes. Data maturity is measured according to five criteria: the complexity of the data, the quality of the data, the usability of the data, goal satisfaction, and value potential.

- Complexity of data: *data complexities* are represented by intrinsic data characteristics and their challenges, which are usually embodied through data factors and such aspects as dimensionality, volume, heterogeneity, linearity, variability, velocity, certainty, sparsity, balance, and various coupling relationships (including correlation, association, dependency, causality, and latent relations.)
- Quality of data: *data quality* refers to the intrinsic condition of the data, i.e., how well the data fits the purpose, fulfills requirements, or accurately portrays the scenarios in a given context. Typical data quality issues may include accuracy, precision, completeness, validity, veracity, conformity, integrity, consistency, reliability, and trustfulness.
- Usability of data: *data usability* refers to the ease of use and learnability of the data to achieve specified goals in a given context. Data usability may be measured in terms of availability, acceptability, observability (visibility), readiness, effectiveness, flexibility, efficiency, learnability, utility, and standardization. Data usability is closely related to data quality, but focuses on the user friendliness and learnability of data for achieving expected objectives.
- Goal satisfaction: *data goal satisfaction* refers to the extent to which the data can achieve the specified objectives. Data goal satisfaction may be measured and represented in terms of the richness of data, the fit level between data and goals w.r.t. expected goals, and performance in achieving expectations.
- Value potential: *data value potential* refers to which opportunities, abilities and unrealized values the data can lead to and generate, and to what extent. Data value potential may be modeled in terms of *theoretical values* for creating new theories, *technical values* for contributing to new technologies or techniques, *business values* for productivity, and *social values* for the benefit of society and community.

Maturity can be quantified in terms of multiple levels, e.g., low, medium and high, which are defined according to specific aspects within a specified context. The definitions for these levels may be specified for complexity, quality, usability, satisfaction, and potential. Here, we categorize data maturity in terms of the following levels: unorganized, partially organized, sufficiently organized, and perfectly organized.

- *Unorganized data* is very ad hoc, chaotic and undocumented. It has no specifications, protocols, structure or statements, and cannot be used. The data needs to be fully organized per analytical requirements before it can be used.
- *Partially organized data* follows certain structures and specifications that are not sufficiently arranged and documented, and cannot be used directly. Such data needs to be re-arranged for use according to specified goals.

- *Sufficiently organized data* conforms to certain specifications, standardization and documentation, and is ready for use for specified goals, or can be used with limited re-arrangement.
- *Perfectly organized data* is very well organized and documented, and does not require additional re-arrangement to fully achieve specified goals.

10.5.3 Capability Maturity

Capability maturity in data science refers to the level of knowledge and ability, and the practices and experience required to handle data science problems, to achieve optimal outcomes within a given context. Capability maturity can be measured in terms of the power of capabilities, usability of capabilities, fitness between capabilities and data, and satisfaction of capabilities for achieving goals.

- Power of capabilities: *Capability power* refers to the strength of capabilities in terms of addressing given data complexities and quality, and achieving the expected goals in a given context. Capability power may be objectively evaluated in alignment with data complexities and quality, and subjectively evaluated with regard to goal satisfaction and value potential.
- Usability of capabilities: *Capability usability* refers to how easily the capabilities can be used in data handling. This may be measured in terms of readiness, acceptability, flexibility, friendliness, learnability, and standardization.
- Fitness between capabilities and data: *Capability-data fitness* refers to how well capabilities fit data complexities, which may be specified and evaluated in terms of fit level, scope, aspect, and performance. As discussed in Fig. 3.8, the capability and data fit may fall into one of five scenarios: perfect fit, inadequate data fit, inadequate model fit, limited fit, and failed fit. In addition, the capability fit to data may also be measured by capability generality and the flexibility of fitting specified data characteristics and complexities.
- Capability satisfaction of goals: *Capability goal satisfaction* refers to the extent to which the capabilities can achieve the expected goals. This may be represented and measured by the power of capabilities to achieve specified goals, the fit level between capabilities and goals w.r.t. specific data, and the performance of capabilities in addressing goals.
- Capability value potential: *Capability value potential* refers to the level of opportunity that capabilities can bring about. This may be evaluated in terms of theoretical values, technical values, business values, and social values, as for data value potential. In addition, value potential may also be justified in terms of area, scope and level of unresolved data complexities, performance, and/or applications.

10.5.4   Organizational Maturity

Organizational maturity in data science refers to the level of maturity of organizational data thinking, data maturity, people maturity, management maturity, organizational policy and strategy, and organizational experience and practices. These may be embodied in *horizontal* aspects, such as enabling, handling and transforming the organization to data-driven, evidence-based design, operations, productivity, decision, and management; and *vertical* aspects, which are the domain-specific data issues and organizational strategies for addressing these issues.

- Organizational data thinking: refers to the strategic level of organizational thinking in which data is utilized for the transformation of an organization's vision, strategic planning, operations, productivity, competition, and evolution.
- Domain-specific organizational strategies and capabilities: refers to the maturity of the strategies, capabilities and practices that are specific to the domain of which an organization is part.
- Data maturity: refers to the maturity level of organizational data, which is embodied in the maturity level of both the general and domain-specific data infrastructure, management, analysis, applications, and value-added applications and services.
- People maturity: refers to the maturity level of general and domain-specific data professionals, and the data science team in an organization, including the data science thinking, leadership, technical capabilities, and best practices of the data team and its professionals.
- Planning and management maturity: refers to the maturity of planning and management capabilities, experience and performance of data-related strategies, policies, profession, implementation, optimization and development.
- Organizational policy and strategy maturity: refers to the maturity of organizational capabilities in making, executing and refining policies, standards, specifications, and strategies for data-related business and services.
- Organizational practice maturity: refers to the level of performance in an organization in undertaking data science and engineering, which may be embodied through such aspects as processes (procedures), performance (such as effectiveness and efficiency), and the impact of conducting data science within an organization.

The data science maturity of an organization requires the development, management and maintenance of appropriate strategies, people, knowledge, capabilities, data, technologies, practices, and outputs. This requires an organization to leverage all available capabilities to manipulate data and undertake data science tasks to achieve business outcomes by considering the organizational plan, data and business relevancy, operational constraints, value measurement, actionable insights, and performance management.

Standard building, governance, policy making and collaborations with external resources and authorities are critical components in *organizational maturity*. They support the building, training, sharing and evaluation of data science standards, repositories, data, skills and tools, the oversight of data science activities, recommendations on effective and efficient strategies, policies, business models, organizational structures, enterprise architectures and best practice, and collaborations with thought leaders and professional bodies.

Organizational data science maturity is further determined by the level of data science competency. *Organizational data science competency* is evident in enterprise motivation and the intent to use data science as a service, as well as the facilities to support communication, facilitate interactions, and forge mutual understanding across technical, business and decision-making divisions. An organization's data science competency ranges from ad hoc uptake and technical adoption of data science to the adoption of data science at a business and enterprise level.

There are different levels of *competency ownership* in an organization.

- *Individual owners*: Where individuals own one or more data science projects and competencies. Typically, this applies to an organization in which only a small team is committed to data science and the team leader sponsors and owns projects related to specific and ad hoc business needs.
- *Team owners*: Where data science projects and competencies are not owned by individuals, or a team leader, but by a team within a department. The particular focus of the team is on specific business problems and data science goals.
- *Departmental owners*: Where a department in an organization owns data science competencies and projects. Departmental owners form a functional plan for enhancement through innovation, productivity and decision-making.
- *Enterprise owners*: Where an organization recognizes and owns data science projects and competencies. A systematic and strategic view of data science is taken in enterprise-wide innovation and productivity.

10.6 Data Scientists

It is possible that anyone who holds a data-related role could claim to be a data scientist, but what truly constitutes a data scientist in the context of today's data reality? To address this important and challenging question, this section discusses the definition of *data scientist*, what a data scientist does, what qualifications a data scientist might hold, and the difference between data scientists and holders of related roles, in particular, BI professionals.

10.6.1 Who Are Data Scientists

Data scientists are those people whose major roles and responsibilities are very much centered on data. What are the actual roles and responsibilities of a data scientist? In this section, we summarize the statements taken from several documents on relevant government initiatives.

- In [310], the US National Science Board defines data scientists as "the information and computer scientists, database and software engineers and programmers, disciplinary experts, curators and expert annotators, librarians, archivists, and others, who are crucial to the successful management of a digital data collection."
- In a report [107] from the US Committee on Science of the National Science and Technology Council, data scientists are defined as "scientists who come from information or computer science backgrounds but learn a subject area and may become scientific data curators in disciplines and advance the art of data science. Focus on all parts of the data life cycle."
- In [376], the Joint Information Systems Committee defines data scientists as "people who work where the research is carried out, or, in the case of data center personnel, in close collaboration with the creators of the data and may be involved in creative inquiry and analysis, enabling others to work with digital data, and developments in data base technology."

10.6.2 Chief Data Scientists

In 2015, the White House appointed the first US Chief Data Scientist, saying that the role "will shape policies and practices to help the U.S. remain a leader in technology and innovation, foster partnerships to help responsibly maximize the nation's return on its investment in data, and help to recruit and retain the best minds in data science to join us in serving the public." [421]

More and more industry and government organizations recognize the value of data for decision-making and have set up general and specific data scientist roles to support data science and engineering, e.g., data modelers and data miners, in addition to data engineers and business analysts. A critical data role is that of chief data scientist, a position that may be termed as such, or as Chief Data Officer or Chief Analytics Officer. This chief data scientist takes executive responsibility for data and data-related decisions, strategies, planning, and review, and for the data science team within an organization.

A significant shortage of hundreds of thousands of data scientists is predicted in coming years, yet data science roles have not been defined in terms of a structured capability set. Building an effective *data science team* for a large organization involves far more sophisticated skill requirements and complementary capabilities than simply hiring a pool of data scientists [322].

10.6.3 What Data Scientists Do

The position statements from recruiting organizations indicate that they require customized capabilities to suit their particular purposes. General definitions about data scientists and their roles and responsibilities can be obtained from national scientific committees and professional bodies. In business, immense interest has been expressed by multinational service and product providers, social media and online communities, such as in [156, 222, 247, 276, 358], concerning the roles and responsibilities of data scientists, and what makes a good data scientist.

For instance, a data scientist metromap was created in Chandrasekaran [88]. The metromap covers ten technical areas: Fundamentals, Statistics, Programming, Machine Learning, Text Mining/Natural Language Processing, Data Visualization, Big Data, Data Ingestion, Data Munging and Toolbox. Based on the idea of a railway network, each area/domain is represented as a "metro line", on which the stations depict the topics to be progressively learned and understood. Figure 10.6 is the metromap of activities and tools for data scientists, created by Chandrasekaran [88].

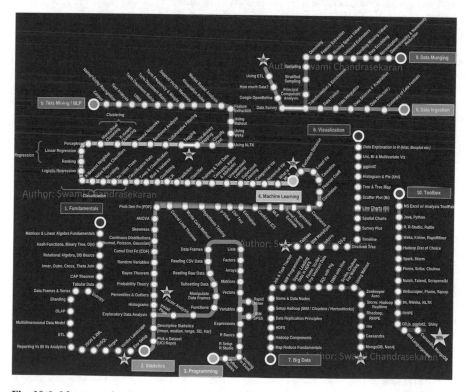

Fig. 10.6 Metromap for data scientists (created by and copied from [88])

In contrast, INFORMS summarizes the following seven job tasks and processes for data scientists [229]: Business Problem (Question) Framing, Analytics Problem Framing, Data, Methodology (Approach) Selection, Model Building, Deployment, and Model Lifecycle Management.

An increasing number of academic and research institutions are working on defining the certification and accreditation of next-generation data scientists. This is reflected in general and domain-specific data science curricula for Masters and PhD qualifications, such as a PhD in Analytics [408] and Master's degree in SCM Predictive Analytics [229].

Typical domain-free and problem-neutral responsibilities and requirements for jobs that have been announced in social media channels (e.g., Google Groups, Facebook and LinkedIn), plus what has been seen over the past 15 years in large governmental and business organizations, can be summarized as follows: learn and understand the business problem domain; understand domain-specific and social restrictions and constraints; understand data characteristics and complexities; set up engineering and analytical processes; transform business problems into analytical tasks; develop data project plans; undertaken analytics task; set up engineering and analytical processes; select and construct features; conduct discovery; discover insights; manage projects; operationalize data exploration and discovery; communicate the results to end users and decision-makers; and 'sell' the solutions.

We explain the demands of these responsibilities and requirements below.

- *Learn and understand the business problem domain*: talk to business experts and decision-makers to understand the business objectives, requirements and preferences, issues and constraints facing an organization; exercise organizational maturity and identify, specify and define the problems, boundaries and environment, as well as the challenges; generate a business understanding report.
- *Extract domain-specific and social restrictions and constraints*: identify and specify social and ethical issues such as privacy and security; develop an ethical reasoning plan to address social and ethical issues.
- *Understand data characteristics and complexities*: identify the problems and constraints imposed by the data; develop a data understanding report.
- *Transform business problems into analytical tasks*: define the analytical objectives, learning models and milestones, development plan, implementation plan, evaluation systems, deployment method, project management methods, risk management, and so on, according to the business understanding results, and by considering domain-specific constraints and data complexities.
- *Build engineering and analytical projects and processes*: devise analytical goals for turning business and data into information, information into insight, and insight into business decision-making actions, by developing technical plans for the discovery, upgrade and deployment of relevant data intelligence.
- *Undertake analytical tasks*: conduct advanced analytics by developing corresponding techniques, models, methods, algorithms, tools and systems; engage in the experimental design and evaluation of data science; generate best practice

experiences, perform descriptive, predictive and prescriptive analytics, conduct survey research, and support visualization and presentation.

- *Develop data usage plans*: identify the problems and constraints of the data; develop a data understanding report; specify and scope analytical goals and milestones by developing project plans to set up an agenda and create governance and management plans.
- *Prepare data*: pre-process and explore the data to make it ready for further discovery (by understanding and describing data characteristics and problems through the use of descriptive methods and tools).
- *Select and construct features*: based on the understanding of data characteristics and complexities, extract, analyze, construct, mine and select discriminative features; continually optimize and innovate new variables for best possible problem representation and modeling; when necessary, conduct data quality enhancement [208].
- *Discover knowledge*: transform business problems to data discovery problems, including analytical, learning and visualization tasks, and conduct advanced analytics by developing fundamental data science methods, techniques, and best practices; perform predictive analytics, algorithm development, experimental design, visualization, and survey research.
- *Extract insights*: combine analytical, statistical, algorithmic, engineering and technical skills to mine relevant data by involving contextual information; invent novel and effective models; constantly improve modeling techniques to optimize and boost model performance, and seek to achieve best practice.
- *Communicate the results to end users and decision-makers*: convert data findings to business-friendly presentations and visualization; conduct frequent interactions with clients during the whole lifecycle; tell clear and concise stories and draw simple conclusions from complex data or algorithms; provide clients with situational analyses and deep insights into areas that would benefit from improvement; translate into business-enhancing actions in the final deployment.
- *Operationalize data exploration and discovery*: develop corresponding services, solutions and products or modules to be fed into a system package on top of user-specified programming languages, frameworks and infrastructure, or open source tools and frameworks; maintain the privacy and security of data and deliverables.
- *Market the solutions*: publicize solutions by writing reports and making presentations to specialists and non-specialists at public events; present executive summaries with precise and evidence-based recommendations and risk management, especially to current and potential end users, business sponsors, business owners, policy makers, and decision-makers.
- *Manage projects*: maintain, manage and refine data projects and milestones, and handle their processes, deliverables, evaluation, risk and reporting to build active lifecycle management.

Some of the above responsibilities may be undertaken by data engineers, or jointly conducted with data engineers or software engineers.

10.6.4 *Qualifications of Data Scientists*

Data scientist candidates need to achieve certain qualifications to satisfy the above responsibilities and job requirements. Although different surveys [150] and reports [121] claim that data scientists enjoy attractive, highly paid jobs, the expectations placed on them in terms of qualifications are also high.

In addition to the analytic skills that are the foundation of the data scientist role, the following qualifications and abilities may be required:

- *Thinking, mindset and ability* to think analytically, creatively, critically and inquisitively;
- *Master's or PhD degree* in computer science, statistics, mathematics, analytics, data science, informatics, engineering, physics, operations research, pattern recognition, artificial intelligence, visualization, or related fields;
- *Methodologies and knowledge of complex systems and approaches* for conducting both top-down and bottom-up problem-solving;
- A *solid theoretical foundation* with a deep understanding and comprehensive knowledge of common statistics, data mining and machine learning methodologies and models; as well as human-computer interactions, visualization and knowledge representation and management;
- *Technical abilities* to implement, maintain, and troubleshoot big data infrastructure, such as cloud computing, high performance computing infrastructure, distributed processing paradigms, stream processing and databases;
- *Background in software engineering* including systems design and analysis, and quality assurance;
- *Interdisciplinary interest and knowledge* in multi-disciplinary and trans-disciplinary studies and methods in scientific, technical, and social and life sciences;
- *Substantial computing experience* with state-of-the-art analytics-oriented scripting, data structures, programming languages, and development platforms in a Linux, cloud or distributed environment;
- *Evaluation capabilities* including theoretical background and domain knowledge for the evaluation of the technical and business merits of analytic findings;
- *Communication and organizational skills* including excellent written and verbal communication [283] and organizational skills, with the ability to write and edit analytical materials and reports for different audiences, and the capacity to transform analytical concepts and outcomes into business-friendly interpretations; the ability to communicate actionable insights to non-technical audiences, and experience in data-driven decision making;
- *Experience and practices* in working with complex data, including large datasets, and mixed data types and sources in a networked and distributed environment; in data extraction and processing, feature understanding and relation analysis.

The above observations reflect the inter-, cross- and trans-disciplinary responsibilities of data scientists. Most data scientists may, as a minimum, be expected to hold degrees in statistical and mathematical science, intelligence science, and information science, to enable them to fulfill their fundamental responsibilities.

10.6.5 Data Scientists vs. BI Professionals

Business intelligence (BI) professionals are those who interact with business stakeholders to understand their reporting needs, collect business requirements, and design, build and maintain BI solutions and reports to address business needs.

While there may be an overlap between the roles of *data professionals* and *BI professionals* [345], a number of research works show that data science professionals are generally much more knowledgeable about data and analytics than business-oriented, with most holding a Master's or PhD degree in statistics or computer science. By contrast, BI professionals may hold qualifications in business, management, marketing, or accounting, in addition to computer science.

An EMC data science community survey [150] shows that

- data scientists can open up new possibilities;
- compared to 37% of BI professionals trained in business, 24% of data science professionals are in computer science, 17% are in engineering and 11% are in hard science;
- compared to BI toolkits, data science toolkits are more technically sophisticated and more diversified;
- the number of data scientists undertaking data experiments is almost double that of BI professionals;
- data science professionals more frequently interact with diverse technical and business roles in an organization (such as data scientists, strategic planners, statisticians, marketing staff, sales people, graphic designers, business management and IT administration, programmers, and HR personnel) than BI professionals;
- compared to working on normal data, big data manipulators tend to tackle more sophisticated data complexities; and
- data science professionals spend almost double the time on big data manipulation (e.g., data parsing, organization, mining, algorithms, visualization, story-telling, dynamics, and decisions) than they spend on normal data.

BI professionals often manipulate tools with skills such as ETL, reporting, OLAP, cubes, business object design, dashboard, web intelligence, and relevant services and platforms.

While centering on data, a good data scientist is also expected to know the underlying domain and business well. Without an in-depth understanding of the domain, the actionability by data scientists of the data deliverables and products may be low. However, a data scientist is no substitute for a domain expert in complex data science problem solving [64]. Similar to other disciplinary specialists, data scientists work more effectively when they collaborate with domain-specific specialists and subject matter experts to achieve a broader impact. This is similar to the requirements of domain-driven, actionable knowledge discovery [77].

10.6.6   Data Scientist Job Survey

The role of the data scientist was recognized more than 10 years ago. In 2004, Dr Usama Fayyad was appointed as Chief Data Officer of Yahoo, which opened the door to a new career possibility: the *data science professional* [206, 281] or more specifically *data scientist*, for people whose role very much centers on data.

As noted earlier in this chapter, the White House subsequently appointed the first US Chief Data Scientist in 2015 [421]. Today, the role of data scientist [322] is regarded as "the sexiest job of the twenty-first century" [121], and next-generation data scientists will be mostly welcomed in (1) the increasingly important data economy, especially ubiquitous data services offered through mobile Internet, wireless networks, the Internet of Things, wearable devices and the social media-oriented economy, and (2) data-driven cross-disciplinary sciences, such as precision medical science.

It is reported that data scientists earn much higher salaries than those in other data-related jobs, with a median salary of US$120k for data scientists and US$160k for managers, according to the 2014 Burtch Works survey [47]. This is attributed to the fact that 88% of respondents in this survey have at least a Master's degree, while 46% also hold a Doctorate, compared to only 20% of other Big Data professionals. In the 2015 O'Reilly survey [253], 23% were found to hold a doctorate, while another 44% had a Master's degree. The median annual base salary of this survey sample was US$91,000 globally, and among US respondents was US$104,000, compared to US$150k for "upper management" (i.e., higher than project and product managers).

10.7   Data Engineers

It is common for the responsibilities of data scientists to be confused with those of data engineers. Put another way, the role of a data scientist is so broad that every data-relevant professional may consider themselves to be doing data science. In this section, we discuss the responsibilities and qualifications of data engineers, to distinguish their role from that of data scientists, who conduct more data-driven scientific work. The relevance of the roles of data scientists, data engineers and other IT professional roles is also discussed to differentiate between their specific responsibilities. If data science and the data economy require a distinct data profession, it may be necessary to impose a more structured distributions of data roles and to different them from classic IT.

10.7.1 Who Are Data Engineers

Data engineers are those data professionals who can prepare, design, construct, manage, integrate, present, monitor and optimize data (including big data) from various sources, and can manipulate data-oriented architectures and solutions for further analytics, discovery, and other applications by data scientists and other data professionals.

Compared to data scientists, who focus on discovering data values and insights, data engineers focus on data manipulation including data preparation, management, integration, presentation, and quality control. *Data architects* are those data professionals who determine which infrastructure and architecture are appropriate for data and analytics.

Data engineers are also different from software engineers. *Software engineers* focus on software requirement analysis, system analysis and design, and testing and quality assurance. Data engineers may sometimes adopt the roles of BI professionals to an extent when they commit to reporting or interacting with business stakeholders.

A data science project is a special IT project. It requires the input of a range of professionals to form a data science team, such as data architects, data scientists, data engineers, BI professionals, system architects, and software engineers. The distribution of their various roles is broadly as shown in Table 10.2.

- System architects are responsible for appropriate system infrastructure and architecture, including their design, implementation and maintenance;
- Software engineers are responsible for system preparation (including for data manipulation and discovery), and their function incorporates requirement analysis, system analysis, design and implementation;
- Data architects are responsible for ensuring the appropriate infrastructure and architecture for data, data discovery, and data products;
- Data engineers are responsible for data preparation, including making the data available, clean and usable for whatever purpose;
- Data scientists are responsible for discovering knowledge and insights from data;
- BI professionals are responsible for converting data to business-ready reports, and for presenting those reports.

A data engineer may take on the responsibilities of the classic database administrator for data management. Given the wide range of applications of big data and distributed computing, big data administrators now work more on distributed and high performance infrastructure, storage, sharing, distribution, programming, and computing.

Data engineers may hold qualifications (knowledge, skills and capabilities) in applied statistics, applied mathematics, information systems (including database management, warehousing, retrieval and query), distributed computing, high performance computing, computer infrastructure, intelligent data processing, discrete

Table 10.2 Related professional roles

Role	Description
System architect	Responsible for appropriate system infrastructure and architecture
Software engineer	Responsible for system preparation (including systems for data management, manipulation and discovery)
Data architect	Responsible for appropriate infrastructure and architecture for data, data discovery, and data products
Data scientist	Responsible for discovering knowledge and insights from data
Data engineer	Responsible for data preparation (making it available, clean and usable for whatever data-driven purpose)
BI professional	Responsible for projecting data to business-ready reports

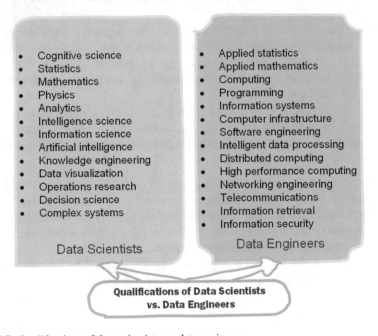

Fig. 10.7 Qualifications of data scientists vs. data engineers

mathematics, networking engineering, telecommunications, programming, computing, and information security.

A comparison of the qualifications that may be held by data scientists and data engineers is illustrated in Fig. 10.7. Data scientists are usually expected to hold degrees in disciplines that enable them to understand data exploration and discovery work, while data engineers are often encouraged to have the knowledge and capability to undertake data-related technical and engineering tasks.

10.7.2 *What Data Engineers Do*

Typical roles and responsibilities of data engineers include: designing, implementing, maintaining, and optimizing data infrastructure and architecture, data extraction, data transformation, data loading, data warehousing, data management, data integration, data matching, analytical infrastructure, analytical programming, analytical project management, data sharing, data security, and data and analytics version control.

Accordingly, a *data engineer* typically engages in the following data engineering tasks and responsibilities:

- Select, install, maintain and manage accurate data systems that meet business requirements and objectives. *Data systems* are any data-related systems for data, infrastructure, analytics, management and decision-support, including management information systems, data warehouse systems, unstructured (e.g., Web and multimedia) data management systems, ETL systems, analytical platforms, data/analytics programming systems, visualization systems, presentation systems, and decision-support systems.
- Acquire, extract and integrate relevant, often multi-source and heterogeneous, data to wrangle, including transactional ACID data and non-ACID sources such as column stores, partitioned rows, documents, graphs and multimedia data;
- Enhance data quality to resolve data quality issues, including cleaning and denoising data, removing redundancy, filling missing data, and removing false information.
- Transform and prepare data for discovery (get data ready for analytics, learning and visualization);
- Explore data to understand data characteristics and complexities, especially guided by or collaborating with data scientists and domain experts;
- Check and enhance data competency in terms of availability, reliability, accessibility, efficiency and power for achieving data potential;
- Program data analytical, learning and computational methods, models, and algorithms for data scientists on samples or at scale to code models, and convert design-time models to run-time executable codes;
- Develop data discovery systems, prototypes and products;
- Dispatch data discovery processes, scheduling, prototyping and jobs;
- Enable and maintain the data discovery performance of data systems including the robustness, trustfulness, scalability, and consistency of results;
- Enable and maintain the computational performance of data systems including system efficiency and scalability;
- Handle data-related social issues, including supporting safe and privacy-protecting data sharing, matching, processing, and usage;
- Protect data systems and manage risk related to such issues as security, privacy and anonymity, robustness, and disaster recovery.

Typically, data engineers may also engage with big data infrastructure and architecture tools; data ETL tools and platforms; enterprise application integration services and solutions; data management, storage, distribution tools; cloud platforms; distributed computing platforms and languages; data visualization and presentation tools; programming language and platforms; system performance management and monitoring platforms; and data security solutions. Some of these associated roles may be undertaken or shared with software engineers and system engineers.

Data engineers may therefore be required to use tools such as Hadoop, MapReduce, data streaming technologies, NoSQL, New SQL, Hive and Pig. In contrast, data scientists may be required to use analytical tools and languages, including analytical tools for statistics, data mining, and machine learning such as SPSS, SPSS Modeler, SAS and SAS Miner, open source languages such as R and Python, and distributed tools such as Apache Spark, Hadoop, and Scala. More discussion on tools for data roles can be found in Sect. 10.8.

Figure 10.8 compares the main responsibilities between data scientists and data engineers. Data scientists focus on scientific research, innovation, the development and management of data, discovery and related systems. Data engineers concentrate on engineering and technical tasks related to requirements, analysis, design, development, maintenance, management, and the enhancement of data, discovery, and solutions.

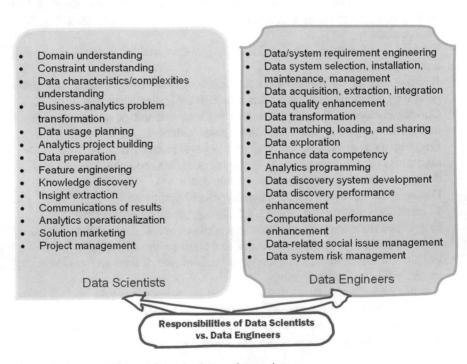

Fig. 10.8 Responsibilities of data scientists vs. data engineers

10.8 Tools for Data Professionals

In this section, we discuss tools that may be used by data scientists, data engineers, and general data professionals to address the above tasks and responsibilities. Tools are categorized in terms of cloud infrastructure, data and application integration, data preparation and processing, analytics, visualization, programming, master data management, high performance processing, business intelligence reporting, and project management. A data scientist may use one or more of these tools on demand for data science problem-solving.

- Statistics tools: Such as R, SPSS, Matlab, SAS, Stata, Statistica, and Systat.
- Cloud infrastructure tools: Such as Apache Hadoop, Spark, Cloudera, Amazon Web Services, Unix shell/awk/gawk, 1010data, Hortonworks, Pivotal, and MapR. Most traditional IT vendors have migrated their services and platforms to support cloud.
- Data/application integration tools: Including Ab Initio, Informatica, IBM InfoSphere DataStage, Oracle Data Integrator, SAP Data Integrator, Apatar, CloverETL, Information Builders, Jitterbit, Adeptia Integration Suite, DMExpress Syncsort, Pentaho Data Integration, and Talend [340].
- Master data management tools: Typical software and platforms include IBM InfoSphere Master Data Management Server, Informatica MDM, Microsoft Master Data Services, Oracle Master Data Management Suite, SAPNetWeaver Master Data Management tool, Teradata Warehousing, TIBCO MDM, Talend MDM, Black Watch Data.
- Data preparation and processing tools: In Today [386], 29 data preparation tools and platforms were listed, such as Platfora, Paxata, Teradata Loom, IBM SPSS, Informatica Rev, Omniscope, Alpine Chorus, Knime, and Wrangler Enterprise and Wrangler.
- Data analytics and machine learning tools: In addition to well-recognized commercial tools including SAS Enterprise Miner, IBM SPSS Modeler and SPSS Statistics, MatLab and Rapidminer [337], many new tools have been created, such as TensorFlow [41], DataRobot [117], BigML [32], MLBase [264], IBM Watson, Apache toolset, and APIs including Google Cloud Prediction API [191].
- Deep learning tools: Caffe, Keras, Apache MXNet, Torch, PyTorch, TensorFlow, Microsoft Cognitive Toolkit, DeepLearnToolbox, DeepLearningKit, ConvNetJS, Neural Designer, and Theano.
- Natural language processing tools: such as Natural Language Toolkit (NLTK), Apache OpenNLP, and General Architecture for Text Engineering (GATE).
- Visualization tools: Many free and commercial software is listed in KDnuggets [246] for visualization, such as Interactive Data Language, IRIS Explorer, Miner3D, NETMAP, Panopticon, ScienceGL, Quadrigram, and VisuMap.
- Programming tools: In addition to the main languages R, SAS, SQL, Python and Java, many others are used for analytics, including Scala, JavaScript, .net, NodeJS, Obj-C, PHP, Ruby, and Go [122].

- High performance processing tools: In Wikipedia [434], about 40 computer cluster software items are listed and compared in terms of their technical performance, such as Stacki, Kubernetes, Moab Cluster Suite, and Platform Cluster Manager.
- Business intelligence reporting tools: There are many reporting tools available [82, 437], typical of which are Excel, IBM Cognos, MicroStrategy, SAS Business Intelligence, and SAP Crystal Reports.
- Project management tools: In Capterra [81], more than 500 software and tools are listed for project management, including Microsoft Project, Atlassian, Podio, Wrike, Basecamp, and Teamwork. The agile methodology is often customized for analytics and learning projects.
- Social network analysis tools: In Desale [125], 30 tools are listed for SNA and visualization, such as Centrifuge, Commetrix, Cuttlefish, Cytoscape, EgoNet, InFlow, JUNG, Keynetiq, NetMiner, Network Workbench, NodeXL, and Soc-NetV (Social Networks Visualizer).
- Other tools: Increasing numbers of tools have been developed and are under development for domain-specific and problem-specific data science, such as Alteryx and Tableau for tablets; SuggestGrid and Mortar Recommendation Engine for recommender systems [183]; OptumHealth, Verisk Analytics, Mede-Analytics, McKesson and Truven Health Analytics [381] for healthcare analytics; BLAST, EMBOSS, Staden, THREADER, PHD and RasMol for bioinformatics.

10.9   Summary

The data profession has ushered in a revolutionary era in human professional development history. The principal feature of this new profession is data-driven everything, which includes discovery, innovation, development, engineering, technologies, real and virtual economy, and real and virtual society. To formalize and standardize the data profession, many new and updated data roles will be required, and we conclude this chapter with a summary in Table 10.3 of major data roles and functions. In Chap. 11, we discuss the educational revolution required to train and educate data professionals to satisfy the necessary qualifications for these data roles.

Table 10.3 Major data professional roles

Roles	Description
Data executive	Senior leaders who lead the data science team and work at different levels in an enterprise to build and manage hierarchical or global holistic and systematic planning, strategies, governance and assurance. These especially involve chief officers, such as data scientist lead, chief data officer, chief analytics officer, chief information officer, and chief technology officer
Decision strategist	Determines why, how and what should be designed, planned, produced and optimized for decision makers, and assists decision makers to make their decisions and achieve their goals
Data architect	Determines and produces the architecture and infrastructure for undertaking, operationalizing and upgrading data engineering, analytics, treatments and decision-support
Data scientist	Conducts scientific discovery and exploration on data and data-driven problem-solving; role is often specified as data modeler, data miner, machine learning analyst, statistical analyst, optimization analyst, forecasting analyst, predictive analyst, visualization analyst, quantitative analyst, and insight analyst
Data engineer	Acquires, connects, prepares and manages data, especially for large-scale, multi-source, and real-time data management and preparation
Business analyst	Understands business and data, defines analytical goals and milestones with data modelers and data scientists, and decides and verifies evaluation metrics for business satisfaction
Data modeler	Understands and explores data insights by creating data models to make sense of data through explicit analytics and regular analytics
Data system engineer	Programs and develops system prototypes and platforms by incorporating data models and analytical logics into enterprise information space and decision-support systems
Knowledge discovery scientist	Discovers implicit and hidden insights and produces outcomes that are presented in terms of knowledge representation and deliverables that fit business needs and operations
Model deployment manager	Manages deployed solutions, monitors and refines model fleet management, scheduling and planning, etc.
Network engineer	Enables the communication, transport and sharing of data outcomes across an organization or enterprise
Model operator	Uses dashboard, visual to on-the-fly analytics and self-service analytics to analyze data and produce results for operations, and to escalate the outcomes for routine management
Business behavioral strategist	Creates, selects and optimizes treatments and next best actions to modify the attitudes and behaviors of target populations following the recommendations of the decision strategist
Data administrator	Manages requirements, data, solutions, deliverables, and processes; role is often specified as requirement manager, project manager, quality manager, network manager, data manager, stakeholder manager, deployment manager, solution manager, and librarian

Chapter 11
Data Science Education

11.1 Introduction

An increasing number of data science courses are available from research institutions and professional course providers. However, most such courses may look like "old wine in new bottles", i.e., they are a re-labeling and combination of existing subjects in statistics, business and IT. A fair summary of the current template for creating so-called data science courses is:

- the hybridization of a number of existing subjects in statistics, computing and informatics, and business, or
- in a more specific sense, the combination of statistical analysis, data mining, machine learning, programming, and capstone assignments.

Is this sufficient to constitute a data science course that will train the next generation of data scientists? Data innovation and economy is highly dependent on students being empowered with a level of readiness in the knowledge, capabilities, competencies and experiences required to handle the related data problems and applications. To achieve these goals, educators should address the significant gaps in the data profession market, the current education and training systems, the body of knowledge provided to students, and the level of organizational maturity [105, 325] and competency.

To train our next-generation data engineers, data professionals, data scientists, and data executives, a systematic educational framework should enable our students to "think with data", "manage data", "compute with data", "mine on data", "communicate with data", "deliver with data", and "take action on data" [65]. This section discusses how data science education should address these important matters. A summary of awarded and online courses in the market, a data science education framework of data science courses, and an overview of undergraduate to postgraduate and Doctoral qualification courses are provided.

11.2   Data Science Course Review

In this section, we give an overview of the data science education currently offered in both research institutions and commercial entities, and analyze the gaps in data science education.

11.2.1   Overview of Existing Courses

Researchers and scientists play a driving role in the data science agenda. In Borne et al. [36], the authors highlight the need to train the next generation of specialists and non-specialists to derive intelligent understanding from the increased vastness of data from the Universe, "with data science as an essential competency" in astronomy education "within two contexts: formal education and lifelong learners." The aim is to manage "a growing gap between our awareness of that information and our understanding of it." In several researches [5, 20, 23, 48, 327, 343], the discussion focuses on the needs, variations and addenda of data science-oriented subjects for undergraduate and postgraduate students majoring in mathematics and computing. Case studies of relevant subjects at seven institutions are introduced in Hardin [205], with the syllabi collected in Hardin [204].

In addition to the promotion activities in core analytics disciplines such as statistics, mathematics, computing and artificial intelligence, the extended recognition and undertaking of domain-specific data science seems to repeat the evolutionary history of the computer and computer-based applications. Data science has been warmly embraced by more and more disciplines in which it was traditionally irrelevant, such as law, history and even nursing [93]. Its core driving forces come from data-intensive and data-rich areas such as astronomy [36], neurobiology [127], climate change [157], research assessment [359], media and entertainment [186], supply chain management (SCM) [208] and SCM predictive analytics [350], advanced hierarchical/multiscale materials [200, 242], and cyber-infrastructure [311]. The era of data science presents significant interdisciplinary opportunities [343], as evidenced by the transformation from traditional statistics and computing-independent research to cross-disciplinary data-driven discovery which combines statistics, mathematics, computing, informatics, sociology and management. Data science drives the disciplinary shift of artificial intelligence (AI) from its origins in logics, reasoning and planning-driven machine intelligence to meta-synthesizing ubiquitous X-intelligence-enabled complex intelligent systems and services [62, 70, 333, 335].

A very typical inter-, multi- and cross-disciplinary evolutionary trend is the adoption and adaptation of data-driven discovery and science in classic disciplines from an informatics perspective. This has resulted in the phenomenon of *X-informatics* for transforming and reforming the body of knowledge. Typical examples include astroinformatics, behavior informatics [55, 72], bioinformatics, biostatistics, brain informatics, health informatics, medical informatics, and social informatics, to name a few [436]. Hence, it is not surprising to see courses and subjects being offered in specific areas such as biomedical informatics, healthcare informatics, and even urban informatics.

Following the creation of the world first coursework Master of Science in Analytics [303] at North Carolina State University in 2007, and the world first Master of Analytics by Research and PhD in Analytics launched at the University of Technology Sydney in 2011 [408, 460], more than 150 universities and institutions have now either created or are planning courses in data science, big data and analytics [360]. The majority of these course initiatives focus on training postgraduate specialists and certificate-based trainees in business disciplines, followed by the disciplines of computer science and statistics.

Several repositories [94, 136, 182, 360, 406] collect information about courses and subjects related to analytics, data science, information systems and management, statistics, and decision science. For example, according to DSC [136] and Github [182], there are currently about 500 general or specific subjects or courses that relate to data analytics, information processing, data mining and machine learning, of which 78% are offered in the US. Seventy-two percent are offered at Master's level, with only 7% at Bachelor level, and 3.6% at Doctoral level. About 30% are online courses. From the disciplinary perspective, some 43% of courses specifically encompass "Analytics", compared to 18% on "Data Science" and only 9% on "Statistics". Approximately 40% focus on business and social science aspects. In Classcentral [94], 138 courses and subjects are available. A number of US programs are listed in USDSC [406], most being created in business and management disciplines.

More than 85% of courses [136] cover a broad scope of big data, analytics, and data science and engineering. Some courses only offer training in very specific technical skills, capabilities and technologies, such as artificial intelligence, data mining, predictive analytics, machine learning, visualization, business intelligence, computational modeling, cloud computing, information quality, and analytics practices. It is very rare to find courses that are dedicated to analytics project management and communication skill training [159], although several courses on decision science are offered.

11.2.2   Disciplines Offering Courses

A summary of data science courses is provided in terms of course-offering disciplines, body of knowledge, institutions, channels, audience, and regions.

With regard to *course-offering disciplines*, many more courses are available today in data science and analytics compared to 5 years ago. This is in addition to related courses offered in classic disciplines, in particular statistics, informatics and computing. As shown in references [94, 136, 182, 360, 406], the following observations are gained from the course review.

- Core relevant disciplines offering data science-related courses are statistics and mathematics, IT (in particular, informatics and computing), business and management. Interestingly, more courses are offered in business and management than in science, engineering and technology (in particular, IT and statistics) [360]. As a result, such courses usually focus on offering a specific discipline-based body of knowledge.
- An increasing number of initiatives have been created or are in development in seemingly "irrelevant" disciplines, such as environment science, geographical science, physics, health and medical science, finance and economics, and even agricultural science.
- Most courses are offered by a single faculty and in an individual discipline. Joint efforts across multiple relevant disciplines are significantly recognized, and need to be encouraged to create data science courses that extend beyond a narrow focus and satisfy holistic requirements. In interdisciplinary terms, a simple solution often adopted by course providers is to create courses by mixing components offered in both statistics and IT/computing.

11.2.3 Course Body of Knowledge

The *body of knowledge* delivered in existing data science courses has the following features.

- The core knowledge and capabilities offered in most courses consist of focused or comprehensive knowledge modules from statistics, data mining, machine learning, and programming.
- Additional knowledge and capabilities offered in some courses include a capstone project component or case studies, visualization, communication, and decision science.
- Some courses, in particular those at Master's and Doctoral levels, have prerequisite course requirements. For example, candidates must have completed a Bachelor degree in IT, statistics, computing science, informatics, physics, or engineering.
- In courses that are not focused on statistics and IT, disciplinary knowledge and capabilities are instead offered as the main course content. Examples are biomedical courses, management courses, and business analytics, which incorporate some components of machine learning, data analytics, and programming.

11.2.4 Course-Offering Organizations

The *institutions* offering data science courses are multiple and comprise the following.

- Research and academic institutions: In addition to awarded courses, short courses are also offered in some universities and colleges.
- Vendors: Multinational vendors, especially data product and service providers in data-related businesses, offer courses which are frequently product-specific and provide training in vendor solutions.
- Professional course providers: An increasing number of professional and commercial course providers who collaborate with academic institutions or individual researchers and professionals have emerged in the education and training markets.
- Joint provision: In contrast to single provider-based courses, some courses involve multiple stakeholders, such as those in which vendors collaborate with universities. This is becoming more common in data science than in other areas, probably as a result of the practice-based nature and significant economic value of data science and analytics.

11.2.5   Course-Offering Channels

Data science courses are generally offered through the following *channels*.

- On-campus courses: Most awarded courses and short courses are offered on campus, and are based on classic teaching and learning modes, which include lectures, tutorials, and lab practice.
- Off-campus courses: Online courses and in-house corporate training are the main types of off-campus courses, which are usually offered to professionals and people who do not seek a degree qualification and a lengthy and comprehensive knowledge advancement.
- Mixed channel courses: More data science courses are now offered jointly by on-campus and off-campus modes, and in other disciplines, than has previously been the case.

11.2.6   Online Courses

Online data science courses significantly complement traditional education and typically offer a successful Internet-based data business model.

About 30% of courses are offered online, and research institutions, vendors, professional organizations, and commercial bodies contribute to the online course offerings.

The corporate training market has seen increasing competition as vendors and universities invest more resources in this area; the SAS training courses are one such example. Online courses, such as those offered as a massive open online course (MOOC) and by open universities, are quickly feeding the market.

Today, an increasing number of courses are offered in the MOOC mode [39, 169], such as those promoted by Class Central [94], Coursera [103], edX [147], Udacity [393] and Udemy [394]. The MOOC model is fundamentally changing the way courses are offered by utilizing online, distributed and open data, curriculum development resources and expertise, and delivery channels and services. Course development technologies such as Google Course Builder [192] and Open edX [318] are used to create online courses and their operations.

In contrast to the awarded courses offered in research institutions, which tend to focus on systematic and fundamental subjects, online course providers take advantage of off-the-shelf software, platforms, languages and tools in particular R programming, and cloud infrastructure MapReduce and Hadoop. This online course structure, in conjunction with offers by university lecturers, creates a much larger pool of subjects than are available in the awarded curricula of universities.

An increasing number of courses are created to address domain-specific demands, such as courses that incorporate statistics, business analytics, web and social network analytics into SCM predictive analytics [350]. However, most online courses focus on classic subjects, in particular statistics, data mining, machine learning, prediction, business intelligence, information management, and database management. New programming languages including R and Python, and cloud infrastructure MapReduce and Hadoop are highlights in these courses. Techniques related to off-the-shelf software and tools are often emphasized. Very few subjects are specified for systematic training of data science and advanced analytics, real-life analytics practices, communication, project management, and decision-support.

11.2.7   Gap Analysis of Existing Courses

Reviewing different course options prompts the following observations for possible gap-filling and future development in data science education.

Data science thinking: Typical cognitive elements, including data-driven discovery, intelligence-driven exploration, data analytical thinking, hypothesis-free exploration, and evidence-based decision-making, are often overlooked in teaching and learning, and in practice. However, *data science thinking* is the keystone of what makes data science a science, and differentiates data science from information science, computing, and statistics.

Cross-disciplinary body of knowledge and capabilities in awarded courses: Most awarded courses are offered within one department or faculty and provide limited access to the interdisciplinary knowledge and skills which are essential for data science. How to bridge the gaps between disciplines and generate "whole-of-discipline" (i.e., involving all necessary disciplines) data science courses are critical challenges for data science course providers and educators.

Old wine in new bottles: It is not surprising that many so-called data science courses are essentially a newly labeled "miscellaneous platter" of several existing subjects offered in different curricula. This is common in most undergraduate courses and Masters degrees by coursework. Because of the re-combination and shuffling of existing subjects, the logical relationships between subjects are often very weakly drawn and do not form a logical, reasonable concept map and knowledge base for training a data scientist.

Advancement of course content: A typical phenomenon in existing data science courses is the advancement level of knowledge and capabilities incorporated into the data science class. Most subjects only address material that has been part of the curriculum for a considerable time, such as the fundamentals of statistics and an introduction to data mining and machine learning, which are two foundation subjects that often feature in data science courses. There are significant gaps in course outlines, which offer only basic and standard content that is widely available in online materials and textbooks and fail to address such critical topics as state-of-the-art data science discovery, development outcomes and best practice.

Practice and work-ready practicality: Data science is a practice-driven and problem-inspired new science in which work practices and case studies are essential components and driving factors. Unfortunately, these components are usually unavailable or are replaced by so-called capstone projects that are essentially trivial and toy example-based exercises, far removed from real-life problems and challenges. The development of pragmatic modules by hands-on data scientists and engineers, such as connecting mainstream data science tools with typical real-life data and problems, is critical. Most academics cannot do this well, thus it is advantageous for industrial players and vendors to collaborate with academics to generate general course materials that will make data science courses operable for training work-ready data scientists.

Problem-solving capabilities and practices: Data science courses should teach people how to solve real-life data and business problems through data-driven discovery, and should provide training in corresponding data-related capabilities and practices so that next-generation data scientists can identify, solve and evaluate new problems and innovations. This involves a mix of data-driven, model-based, domain-specific, and evidence-enabled problem-solving.

Anytime learning: In the age of the Internet and big data, a major advantage can be gained by utilizing and building Internet and data-driven services that provide data science courses at any time via a variety of channels, including social media, online and mobile applications and services. Given today's hectic lifestyle, many people are reluctant to access classroom-based lectures; providing anytime course attendance to a learner's location is therefore essential and will attract a wider audience.

Learning interaction: Data-related course offerings have both similar and different features and requirements from classic courses, but the overall aim is to seek a more interactive, human-centered, and virtual reality-based teaching and learning experience. This goes beyond visualization and interaction design courses, and requires an interactive learning environment in which learners can be part of the learning process, interact with peers, and communicate with teachers in a cloud and virtual reality environment.

Positive learning experience: The active, positive and engaging learning experience of a dedicated data science course is more rewarding than an adapted classic course, because data science learning is data-driven and more intelligence-enabled. Maximizing the discovery, demonstration and reward of ubiquitous X-intelligence [64], new insights, and novel findings is aligned with the nature and characteristics of data science.

Personalization and customization: Existing courses are general purpose and of a standardized design that reflects the educator's thoughts, i.e., what they want to offer, in accordance with the general requirements and qualifications that are usually offered by such a course. A typical problem is that these courses do not cater for or satisfy those who already have some knowledge and background in this area and are looking for something that can be customized to suit personal needs and career planning. Online courses are easier to combine; however, the modules are pre-defined and are usually oriented toward specific topics.

Anytime support and crowdsourcing: Although online materials and references are widely available, the study of data science often triggers specific problems and needs that are not addressed in standardized repositories and common help-desks. Anytime support and crowdsourcing, such as the Linux product development mode, are essential for customized and personalized help and support.

Quality control and assurance: As data science is at an early stage of development, the simple combination and replacement of existing subjects into a package labeled "data science" is unavoidable but disappointing. It is necessary to ensure the quality of teaching, learning and outcomes, including aspects of teaching quality, lecture content, interaction, student engagement, practice, satisfaction, evaluation, and feedback.

Competency: This involves the establishment of courses at introductory, intermediate and advanced levels to cover a complete body of knowledge and skill set, as well as ensuring the credibility and qualification of subjects at each level, the hierarchical structure from professional training to undergraduate, postgraduate and Doctoral courses, and the coverage from low-level hands-on engineering practices to business applications, management and senior executive courses.

Life-long learning: The changing and increasingly complex nature of the data world determines the need for data scientists to engage in life-long learning. There are currently no such course systems in data science to support this. The challenges are multi-fold, such as satisfying the needs of learners at different stages of their career in a range of domains and disciplines, through different channels, and for different purposes.

Actionability and decision science: Increasing research and practice in data science encourages the delivery of actionable insights and findings. This should be applicable to data science courses; attendance in data science courses should enable learners to transform themselves to another capability level and career stage. Another expectation for data science course delivery is to enable learners to become data-driven decision-makers and problem solvers; this requires building data-driven decision science into course materials and course delivery.

11.3   Data Science Education Framework

The discussion on course structures, experiences, and syllabi in relevant initiatives (e.g, [5, 20, 23, 48, 204, 205, 327, 343]) show the need to develop a data science course framework. We thus discuss the nature of a feasible data science education framework in this section.

11.3.1   Data Science Course Structure

The purpose of data science and analytics education is to create and enable the data science profession [281, 415], to train and generate the necessary data and analytics know-how and proficiency in managing capability and capacity gaps, and to achieve the goals of data science innovation and the data economy. Accordingly, different levels of education and training are necessary, from attendance in public courses, corporate training, and undergraduate courses, to undertaking a Master of Data Science and/or PhD in Data Science.

Public courses are designed for general communities to improve their understanding, skill level, professional approach and specialism in data science through the study of multi-level short courses. Typical public courses focus on "know what" and "know how" education, which largely consists of transferring knowledge to audience, and helping them to understand what data science is and how to conduct data science projects.

The scope of such courses may range from a basic level to intermediate and advanced levels. The knowledge map often consists of such components as data science, data mining, machine learning, statistics, data management, computing, programming, system analysis and design, and modules related to case studies, hands-on practice, project management, communication, and decision support. Specific and often highly popular public courses are about tool usage and applications, such as R programming, Python programming, and how to use Spark and Tensorflow.

Corporate training and workshops can be customized to upgrade and foster corporation thinking, knowledge, capability and practice for an entire enterprise, and is a necessary form of training to encourage innovation and raise productivity. This requires courses and workshops to be offered to the entire workforce, from senior executives, business owners, and business analysts to data modelers, data scientists, data engineers, and deployment and enterprise strategists. The topics for such courses include data science, data engineering, analytics science, decision science, data and analytics software engineering, project management, communications, and case management.

Corporate training workshops may be customized to build whole-of-organization hierarchical data science competencies and data science teams, and to transform existing capabilities to data-driven operations, management, production, and decisions. Typical objectives and issues that may be addressed in corporate data science education may assist corporations to understand

- What values, benefit and impact can be gained from undertaking data science and analytics?
- What is data science and analytics?
- What essential elements are required to engage in data analytics projects?
- What better data analytics practice could be implemented in the organization? and
- Where are the gaps for staff in relation to further work and career development in analytics?

Bachelor's Degree in Data Science may be offered on either a general data science basis that focuses on building foundations of data science and computing of data and analytics, or in specific areas such as data engineering, predictive modeling, and visualization. Double degrees or majors may be offered to train professionals who will acquire knowledge and capabilities across disciplines, such as business and analytics, or statistics and computing. In Sect. 11.3.2, the objectives, knowledge, competencies, and compulsory and elective subjects for the degree of Bachelor in Data Science are discussed.

Master of Data Science and Analytics aims to train specialists and talented individuals who have the ability to conduct a deep understanding of data and undertake analytics tasks. Core disciplines and areas should cover statistics, mathematical modeling, data mining, knowledge discovery, and machine learning-based advanced analytics. Interdisciplinary experts may be trained, such as those who have a solid foundation in statistics, business, social science or other discipline and are able to integrate data-driven exploration technologies with disciplinary expertise and techniques. A critical area in which data science and analytics should be incorporated is the classic Master of Business Administration course. This is where new generation business leaders can be trained for the new economy and a global view of economic growth. In Sect. 11.3.3, extended discussion can be found on the objectives, knowledge, competencies, and compulsory and elective subjects for the degree of Master in Data Science.

PhD in Data Science and Analytics aims to train high level talented individuals and specialists who are capable of independent thinking, leadership, research, best practice, and theoretical innovation to manage the current significant knowledge and capability gaps, and to achieve substantial theoretical breakthroughs, economic innovation, and productivity elevation. Interdisciplinary research is encouraged to train leaders who have a systematic and strategic understanding of all aspects of data and economic innovation. The objectives, knowledge and competencies for training Doctoral students in data science are discussed in Sect. 11.3.4.

Figure 11.1 summarizes the hierarchical level of data science education and training, from public courses and corporate training, to undergraduate, postgraduate and Doctoral training, and their respective objectives, capability sets and outcomes.

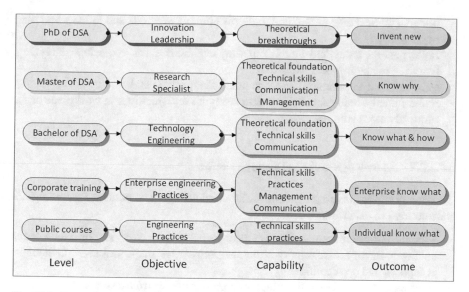

Fig. 11.1 Data science course framework

In the following Sects. 11.3.2, 11.3.3 and 11.3.4, we expand on the structure in Fig. 11.1 to discuss the objectives, knowledge, competencies, and subjects for Bachelor, Master, and PhD degrees in data science.

11.3.2   Bachelor in Data Science

This section shares views about the vision and objectives, knowledge and competencies, and compulsory and elective subjects for undergraduates in data science. The discussion shared in this section considers the resources, experiences, visions and observations shared by many educators and lecturers in this area, such as in [5, 23, 136, 205, 303, 360, 406, 412].

11.3.2.1   Vision and Objectives

The vision and objectives of offering undergraduate courses in data science are to

- train next-generation data professionals, especially data engineers and data analysts (generally called fundamental data professionals);
- foster a basic and general understanding of data characteristics and complexities;
- form mathematical, statistical and computing foundations for data understanding and exploration;

- understand the basic theoretical underpinnings of methods and models and their applicability;
- build general analytical and computing capabilities for acquiring, analyzing, learning, managing, and using data;
- gain hands-on experience and learn practices for data; and
- make students ready for real-life data analytics and practices in workplace or for more advanced studies.

11.3.2.2   Knowledge and Skill Competencies

The blueprint for the integrated curriculum and pedagogical content for undergraduate courses in data science consists of

- Module 1: Introduction to data science: to provide an overview of data science concepts, history, disciplinary structure, knowledge map, tool sets, challenges, prospects, and applications.
- Module 2: Data science thinking: to create thinking traits, styles and habits in conducting data science tasks. Knowledge modules consist of scientific thinking, critical and creative thinking, analytical thinking, mathematical thinking, statistical thinking, and computational thinking.
- Module 3: Data science foundations: to establish foundations for undertaking data science tasks. These cover relevant fundamentals in statistics and mathematics (e.g., mathematical modeling, matrix algebra, time series, probability theory, derivatives, and stochastic processes), and information and intelligence science (e.g., information theory, artificial intelligence and intelligent systems, and software methodologies).
- Module 4: Data engineering technologies: to acquire, wrangle, curate, clean, transform, load, and manage data. These involve data and application integration, data ETL, SQL, data warehousing, and data management.
- Module 5: Data processing technologies: to extract, construct and select features, and to sample, simulate, explore, and process data to make data clean and ready for further exploration. These tasks include feature engineering, data representation, data pre-processing, sampling methods, and statistical analytics, and using relevant tools such as SQL, SPSS, and SAS.
- Module 6: Data discovery technologies: to analyze, learn, make sense of, think with, and assess data. This involves algorithm design, model building, statistical analytics, pattern recognition, data mining, machine learning, optimization, visualization, theoretical analysis, and assessment and validation. This may involve existing analytical software [248], such as R, Python, TensorFlow, RapidMiner, PASW, various deep learning boxes [123], and visualization software [246].
- Module 7: Data computing and programming: to implement computing infrastructure and architectures, and use analytics programming languages, platform and tool sets to program analytics algorithms and models. This may involve computing, cluster, Hadoop, MapReduce, NoSQL, or other analytics programming

systems, and the various analytical and statistical software (e.g., as in the list of statistical software in [445]) and programming tools such as Python, R, Spark, Java, SAS, Matlab, and TensorFlow [41].

- Module 8: Data science practices: to manipulate real-life data and address real-life problems using data-driven discovery and the data science tools available in the marketplace. This may be achieved by undertaking capstone projects or internship work, or by joining a research project in a research center, working on data analytics algorithms, their modeling and real-life applications.
- Module 9: Social methods and ethics: to understand social thinking, evaluation, and skills in social science, behavior science, management, and communications for data science. This may involve the study of data ethics, social methods, social problem-solving, social evaluation, data project and process management, data science input and output management, communication skills, and responsibilities.
- Module 10: New data science advances: to instantly learn the latest advancements, trends and directions in data science theories, technologies, tools, and applications. This may be achieved by organizing and attending seminars, workshops, conferences, tutorials, and winter or summer vocational courses.
- Module 11: Interdisciplinary knowledge and skills: to learn and understand relevant knowledge and skills in other disciplines. This may include gaining domain knowledge and skills in business management, innovation methods, networking and communications, finance and economics, and software analysis and design.
- Module 12: Data economy: to foster the capability and experience to produce data products and services, and to industrialize data applications for decision and production.

11.3.2.3   Compulsory and Elective Subjects

Modules 1–8 above aim to build the compulsory competencies for a qualified data engineer or data professional. They may be further segmented into 16–24 major subjects, to be delivered within a period of approximately 2 years, and may include:

- Introduction to data science
- Data science thinking
- Mathematical foundations
- Probability and statistics
- Mathematical modeling
- Artificial intelligence and intelligent systems
- Data preparation and exploration
- Feature engineering
- Data representation
- Data quality enhancement

- Knowledge discovery and machine learning
- Algorithm and model design
- Analytics programming
- Model selection and assessment
- Optimization methods
- Visualization
- System analysis and design
- Data structure and management
- Enterprise application integration
- Data science project management
- Capstone projects
- Data system design and building

Modules 9–12 aim to foster ancillary knowledge, interdisciplinary skills, and social capabilities to diversify the competencies of data professionals. They may be customized as 6–12 elective subjects, for example:

- Social methods and evaluation
- Data science ethics
- Data project management
- Communication skills
- Data science advancements
- Business management
- Decision methods and systems
- Data economy and industrialization
- Innovation methods

Figure 11.2 summarizes the main knowledge map and subjects to be considered in training qualified Bachelors in data science. The courses for Bachelor of Data Science can be broken down into three stages:

- Stage 1: Data science fundamentals, namely, modules on the introduction to data science, data science thinking, and data science foundations;

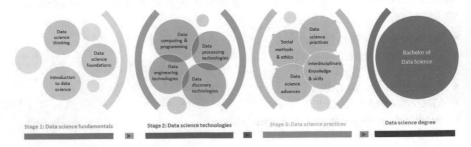

Fig. 11.2 Course structure of Bachelor of Data Science

- Stage 2: Data science technologies, namely, modules on data engineering technologies, data processing technologies, data computing and programming, and data discovery technologies;
- Stage 3: Data science practices, including modules on data science practices, interdisciplinary knowledge and skills, social methods and ethics, and advances in data science.

11.3.3   Master in Data Science

As some surveys show, some 70% of qualified data scientists hold at least a Master's degree in data science related disciplines. Below, the necessary objectives and competencies for a Master's degree in Data Science are outlined.

11.3.3.1   Vision and Objectives

The vision and learning objectives of a Master's program in data science are to

- train next-generation data science researchers, professionals, and entrepreneurs;
- foster a sound understanding of data characteristics and complexities;
- foster the ability to create mathematical, statistical and computing methods and models for sound and sophisticated data discovery and exploration;
- enable students to create, explain and evaluate the theoretical underpinnings of new methods and models and their applicability;
- teach students to conduct and manage sophisticated real-life data science tasks and projects; and
- prepare students for challenging real-life data problem-solving and better practices, entrepreneurial startup, or more specialized deep studies.

11.3.3.2   Knowledge and Skill Competencies

The integrated curriculum and pedagogical content of Masters courses in data science may cover some of the following capability and knowledge components.

- Module 1: Data science thinking: to foster thinking and philosophical traits, styles and habits in conducting data science tasks. Knowledge and capability training modules cover advanced skills, traits, and methods for scientific, critical, creative, analytical, mathematical, statistical, and computational thinking.
- Module 2: Advanced data science foundations: to establish foundations for undertaking advanced and challenging data science tasks. This may involve the study of relevant advancements in statistics and mathematics (e.g., advanced mathematical modeling, matrix algebra, probability theory, derivatives, stochastic processes), and information and intelligence science (e.g., latest advances in

information theory, artificial intelligence and intelligent systems, and software methodologies).

- Module 3: Interdisciplinary methods: to understand and create new knowledge and methods by integrating multi-disciplinary methods, capabilities, and knowledge from disciplines and areas such as statistics and mathematics, physics, biology, finance, economics, marketing, management, decision science, social science, behavior science, information technology, and intelligence science.
- Module 4: Advanced data discovery: to analyze, learn, rationalize, think with, and assess complex data and problems. This involves the research and development of advanced algorithm design, model building, statistical analytics, pattern recognition, data mining, machine learning, optimization, visualization, theoretical analysis, and the assessment and validation methodologies and tools of complex data and models.
- Module 5: Advanced data science practices: to manipulate real-life complex data and address real-life complex problems by inventing data-driven discovery and data science methods and tools that are unavailable in the marketplace. The relevant programs may be sponsored by industry and government organizations, with partnership being formed for students to directly work with business on innovative and in-depth data understanding and problem-solving.
- Module 6: Data innovation and enterprise: to foster innovative and entrepreneurial capabilities, and the experience of producing data products and services, as well as industrializing data discovery technologies and systems. This includes exposure to relevant resources, policies, capabilities and experiences for setting up startups and spin-off laboratory works, including innovation methods, commercialization, marketing and valuation, business operations and management, and entrepreneurial ventures.

11.3.3.3 Compulsory and Elective Subjects

In the modules suggested for Masters students in Sect. 11.3.3.2, Modules 1–3 ensure the compulsory foundations and competencies for becoming a qualified data scientist. Accordingly, the above modules may be segmented into 4–8 major subjects, to be delivered within 1 year or so. These subjects may cover:

- Data science thinking
- Interdisciplinary methods
- Introduction to data science advancements
- Advanced mathematical foundations
- Advanced probability and statistics
- Advanced artificial intelligence
- Social and behavioral science
- Quantitative finance and economics

Module 4 in Sect. 11.3.3.2 focuses on training students in research and innovation capabilities, and in delivering outcomes in big data analytics and data science innovation. Subjects such as the following may be included.

- Advanced mathematical modeling
- Advances in data representation and characterization
- Advanced machine learning and knowledge discovery
- Advanced optimization theories and methods
- Advanced algorithm and model design

Modules 5–6 in Sect. 11.3.3.2 are designed to train students who are ready for the workplace as senior data scientists and leaders, and as entrepreneurs prepared to commercialize their research outcomes. The knowledge and skills required to convert students into senior real-life problem solvers and entrepreneurs may be delivered through the following elective subjects.

- Data profession, economy and industrialization
- Innovation methods
- Commercialization, startup and enterprising
- Business and corporate management
- Investment and risk
- Leadership and management
- Capstone projects (partnered with industry)

Figure 11.3 summarizes the main knowledge map and subjects to be considered in training qualified Masters in data science. The Masters courses in Data Science can be broken down into three stages:

- Stage 1: Advanced data science foundations, covering the modules of data science thinking, advanced data science foundations, and interdisciplinary methods;

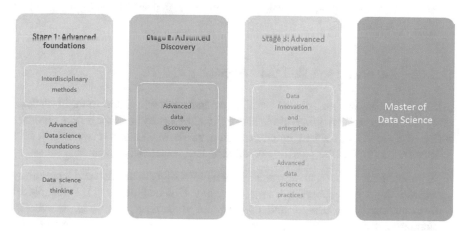

Fig. 11.3 Course structure of Master of Data Science

- Stage 2: Advanced data science discovery; and
- Stage 3: Advanced data science innovation, consisting of modules of data innovation and enterprise and advanced data science practices.

11.3.4   PhD in Data Science

In major data science enterprises and innovation-driven businesses, senior data scientists and data science executives often hold Doctoral qualifications. This section discusses the objectives, knowledge, and competencies of Doctoral students in data science.

11.3.4.1   Vision and Objectives

The vision and learning objectives of a Doctoral program in data science prepare students to

- be next-generation senior data scientists and data science leaders in a variety of institutional settings;
- foster a profound and theoretically rigorous understanding of complex data characteristics and complexities;
- invent innovative mathematical, statistical and computing theories, methods and models for high quality data discovery and exploration;
- invent solid theoretical or empirical underpinnings to new theories, methods and models and explain their applicability;
- excel in sophisticated and challenging real-life data science problem-solving with best practices; and
- lead a team and engage in services and outreach that will have high impact or enhance scholarship.

11.3.4.2   Knowledge and Skill Competencies

Doctoral students in data science are data science specialists and thought leaders. They are trained to solve real-life scientific challenges by inventing new methods with new foundations for new outcomes. To achieve this, Doctoral students in data science should further their ability and experience in such aspects as

- Philosophical thinking: to train and foster a philosophy of science and technology and develop a corresponding way of thinking. This may cover (1) philosophical theories, methodologies, foundations, thought patterns, methods, structures, standards, values, assessment, and implications of data science; and (2) the principles, methods, processes, and instantiation of reductionism, holism, and systematism in data science.

- Scientific paradigms: to understand, invent and instantiate scientific paradigms. These may include (1) principles, methods, processes and instances of experimentation, theory, computation, and data-driven discovery; (2) methodologies, processes, and thought patterns in data science; and (3) paradigms in artificial intelligence, cognitive science, data analytics, machine learning, and statistics, for instance, symbolism (e.g., symbolic algebra, rules, logic, and fuzzy logic), connectionism (e.g., neural networks), nature-inspired methods (e.g., evolutionary learning), structuralism (e.g., topological methods, graphic models, tree methods, network models), analogy (e.g., matching methods, pairwise methods, contrast analysis, and mapping), probabilistic research (e.g., probabilistic graphic models, and statistical models), and hybridization.
- Research leadership: to grasp and train the research culture, methods, and skills to independently undertake high quality and high impact research in data science. These may include research methods, thought traits and patterns, processes and skills for independently (1) recognizing significant yet open scientific challenges and opportunities, (2) creating new and significant scientific theories and systems, and (3) assessing and evaluating scientific outcomes in significant data problems and applications. In addition, students are trained to have the vision of becoming next-generation data science research leaders, through participating in leading data science communities, professional activities, debates, and policy making, supervising junior researchers, and developing scientific arguments for national and disciplinary data science initiatives and research agenda.
- Knowledge advancements: to invent and develop a thesis that significantly advances the knowledge base, fills knowledge gaps in the literature, and demonstrates great scientific and/or practical potential in solving open data science problems. Students may dedicate their doctoral study to original theoretical or practical problems, particular data science foundation building, specific but systematic theories, and solid and reproducible tools and assessment.

As a result, a graduated PhD in Data Science may be empowered with the philosophical attributes and advancements shown in Fig. 11.4.

11.4 Summary

There are many data science-related courses that are offered as awarded courses, corporate training and public courses, and online courses. Compared to the situation 5 years ago, many more courses have appeared on the market, and this trend seems likely to be unstoppable in the coming years. In China, the Ministry of Education recently gave approval to more than 30 universities to offer undergraduate courses in data science and big data technologies. In developed markets, we are seeing the boom in data science education reshaping the education and job markets.

A systematically developed data science discipline, education framework, and course structure and content are not yet available. This chapter has attempted to

Fig. 11.4 Course structure of
PhD in Data Science

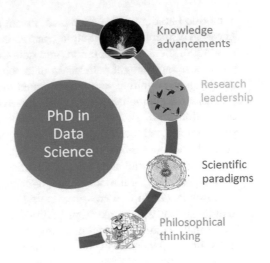

review the current market offerings, understand the gaps, and paint a high-level picture of hierarchical data science education. Much more effort is required to develop handbooks, textbooks, case studies, and content deliverables for training the next-generation data engineers, data professionals, data scientists, and data executives. This is critical for quantifying data science as science, transforming industry and government, and stimulating new data economy.

Chapter 12
Prospects and Opportunities in Data Science

12.1 Introduction

There is continuing debate about how data science will evolve in the next 50 years and what it will ultimately look like. In thinking of the picture of data science in 50 years of time, many questions emerge, for example,

- Why is data science highly important and promising as a new science?
- What contributes to the future of data science?
- How would data science look like by then?

Questions like the above are hard to be addressed with firm answers, as the nature of this type of questions is essentially fast evolving. However, as data is the essential object for data science research, we can still make pretty optimistic forecasting of data science potential and high-level development directions. This is owing to some fundamental drivers that determine the highly promising future of data science.

- Data is becoming ever bigger and increasing ever faster, more diversified and complex;
- Data storage capacity continues to expand to fit increasingly larger and faster growing data;
- Data processing capabilities (in particular, semiconductor technology for processors, distributed computing, and intelligent analytics algorithms) continue to expand to handle more diversified, larger, faster and more complex data;
- Data values continue to be recognized, discovered and utilized.

The last 50 years since the proposal of the concept "data science" has contributed to the progressive and now widespread acceptance of the need for a new science and its initial conceptualization through its transition and transformation from statistics to the merger with existing disciplines and fields. The next 50 years of data science will extend beyond statistics to identify, discover, explore, and define specific foundational scientific problems and grand challenges. With the

© Springer International Publishing AG, part of Springer Nature 2018
L. Cao, *Data Science Thinking*, Data Analytics,
https://doi.org/10.1007/978-3-319-95092-1_12

joint efforts to be made by the entire scientific community, data science will build its systematic scientific foundations, disciplinary structure, theoretical systems, technological families, and engineering tool sets as an independent science. It will consist of a systematic family of scientific methodologies and methods and self-contained disciplinary systems and curricula that are not merely a relabeled 'salad' created by mixing existing disciplinary components. Consequently, data science and technologies will intrinsically and fundamentally disrupt science, reorder world-empowering mechanisms, displace professions, and reform socio-economic growth, and drive the formation of the fourth scientific, technological and industrial revolution.

12.2 The Fourth Revolution: Data+Intelligence Science, Technology and Economy

12.2.1 Data Science, Technology and Economy: An Emerging Area

In the G7 Academies's Joint Statements 2017 [268], data science is recognized by the G7 Academies of Science as one of two emerging technologies with great potential to impact virtually all economic activities. In addition, artificial intelligence is regarded as one of five technological drivers with accelerating impact on changing "our work-life balance" and many fields. As data science essentially forms the foundational enabler of this round of machine intelligence growth, data science demonstrates and will further demonstrate unprecedented impact and potential on all scientific, technological, economic and social fields and activities.

In the OECD report [313] on "Data-Driven Innovation: Big Data for Growth and Well-Being", OECD suggests that "countries could be getting much more out of data analytics in terms of economic and social gains" and "data analytics is increasingly driving innovation". OECD urges countries to train more and better data scientists, enable cross-border data flows, and invest in business processes to apply data analytics. A more recent OECD report on "Key Issues in Digital Transformation in the G20" further emphasizes the importance "to best maximise the benefits of an increasingly digitalised global economy" and raises the warning of new technologies to "displace workers doing specific tasks" and "increase existing gaps in access and use, resulting in digital divides and greater inequality" when no appropriate and instant policies, attention and investment are made to address the opportunities and issues.

Specifically, many governmental science councils, professional bodies, and large vendors have made attempt to discuss and predict future directions and opportunities in specific data-related areas, especially in big data analytics, artificial intelligence, and digital transformation. For example, the US National Science Foundation released its solicitation of "Critical Techniques and Technologies for

Advancing Foundations and Applications of Big Data Science and Engineering (BIGDATA)."[401] in 2015, which highlighted two major areas:

- *foundations*: the development of novel and widely applicable techniques or theoretical analysis, or experimental evaluation of techniques to address the fundamental theoretical and methodological issues related to Big Data, and
- *innovative applications*: the development of innovative methodologies and technologies for specific application areas or innovative adaptations of existing techniques, methodologies, and technologies to new application areas for the analysis of Big Data.

The research areas and challenges listed include

- reproducibility, replicability, and uncertainty quantification;
- data confidentiality, privacy, and security issues related to Big Data;
- generating hypotheses, explanations, and models from data;
- prioritizing, testing, scoring, and validating hypotheses;
- interactive data visualization techniques;
- scalable machine learning, statistical inference, and data mining;
- eliciting causal relations from observations and experiments; and
- addressing foundational mathematical and statistical principles at the core of the new big data technologies.

In 2017, the US NSF further announced a cross-cutting or disciplinary program solicitation of "Computational and Data-Enabled Science and Engineering (CDS&E)" to "identify and capitalize on opportunities for major scientific and engineering breakthroughs through new computational and data analysis approaches" [402] for

- the creation, development, and application of the next generation of mathematical, computational and statistical theories and tools essential for addressing the challenges in computational modeling and simulation and the explosion and production of digital experimental and observational data;
- creation, development and application of novel computational, mathematical and statistical methods, algorithms, software, data curation, analysis, visualization and mining tools to address major, heretofore intractable questions in core science and engineering disciplines, including large-scale simulations and analysis of large and heterogeneous collections of data;
- generation of new paradigms and novel techniques, and generation and utilization of digital data in innovative ways to complement or dramatically enhance traditional computational, experimental, observational, and theoretical tools for scientific discovery and application; and
- the interface between scientific frameworks, computing capability, measurements and physical systems that enable advances well beyond the expected natural progression of individual activities, including development of science-driven algorithms to address pivotal problems in science and engineering and efficient methods to access, mine, and utilize large data sets.

12.2.2   The Fourth Scientific, Technological and Economic Revolution

While data science, technology, and economy show their pillar role and potential in transforming science, technology and economy, their values and applications for new technology and innovation are still underused by businesses [314]. Policy-makers, decision-makers and business owners need to think outside the existing box and the way to do business, at least thinking broadly about basic yet critical question like "there is a lot of and more data in my organization, what can you make from it?", and come up with new visions and ways to use and transform data.

Although it is highly difficult at this early stage of data science to predict specific future data science innovation and research in the next 50 years, the thinking and exploration of the various scientific, economic, social, and other unknowns in the unknown world of data science (see Fig. 3.7) and X-complexities and X-intelligence might indicate instant prospects in the coming years in data research, innovation, economy and society. These may include, but are not limited to: disrupting science, disordering world-powering mechanisms, displacing profession, and deforming economic and social growth.

12.2.2.1   Disrupting Science

Big data science is challenging the existing ways of doing science by theoretical, computational and experimental exploration. It enables to explore the unknowns in the universe by data-driven discovery, and to discover new unknowns for science breakthroughs within and across disciplines. New opportunities discussed in previous chapters, relevant government initiatives, and media predictions include solving global challenges such as enabling sustainable development and inclusive world growth; disclosing unknown universal spaces and working mechanisms; creating new materials, energy, and synthetic biology; managing climate change, resource, food and water shortage; discovering new medicines for curing cancers and managing life and health quality; and developing political and social sciences for managing the existing societal and social problems and emerging problems in data-translated societies and politics such as ensuring cyber openness and security, and safeguarding intelligent weaponry, super-intelligent hard and soft robots, and new additive materials and synthetic biology. Another possible impact is to disrupt the bad scientific behaviors such as plagiarism and irreproducibility.

12.2.2.2   Reordering World-Powering Mechanisms

The existing driving power and working mechanisms in economic, social, cultural and technological worlds are facing increasingly powerful challenge from new science and technology emerging in recent years. This disordering and reordering of world power and mechanisms form the fourth scientific, technological and industrial revolution. This fourth science and technology revolution started from 2010 or so, marked by landscape revolution driven by data-driven science, technology and economy. Typical science and technologies include big data analytics, data-driven machine learning, artificial intelligence and robotics (e.g., deep learning, self-driving cars, IBM Watson, AlphaGo), Internet of Things, nano, bio and quantum technologies, intelligent manufacturing and materials, and open and share science, innovation and economy. New power and mechanisms are under development, which, on one hand, may eventually enable people's long-lasting dream of building just, fair, open, and efficient worlds to some degree; while, on the other hand, may also trigger new inequality, unfairness, dictation, and inefficiency.

This fourth science and technology revolution is incomparable by the three previous industrial revolutions in terms of depth, width, and length of impact and potential. The previous revolutions were on industrial power that experienced

- the First Industrial Revolution, characterized by the replacement of human production methods and intensive laborious jobs by large-scale adoption and manufacture of stream-powered transport and factories in 1760–1870 or so in Great Britain [444];
- the Second Industrial Revolution (also known as the Technological Revolution): characterized by the mass production and applications of electrical and electronic technologies and systems (i.e, electrification), railroads, telegraph, large-scale iron, steel and petroleum production in 1870–1940 or so in Europe, United States and Japan [446]; and
- the Third Industrial Revolution: represented by the technological, industrial and social power transformation to computerized, digitized and networked/Internet technologies, renewable and green energy and electricity, and space and atomic energy technologies [341].

Figure 12.1 summarizes the main enablers and impact of the four industrial revolutions.

12.2.2.3 Displacing Profession

Another fundamental revolution driven by data+intelligence science, technology and economy is the displacement of many professional roles by automated and intelligent machines, robots and software. On one hand, many existing professional roles and types of jobs will disappear or decline substantially, for example, law, health, accounting, sales and marketing consultants, staff working at department stores and retail businesses, workers for designing, making and repairing private cars and bicycles, institutions for training students and professionals, and builders, farmers and manufacturing workers. On the other hand, new professional roles will emerge, requiring foundations, capabilities and experiences in data+intelligence science, technology and business.

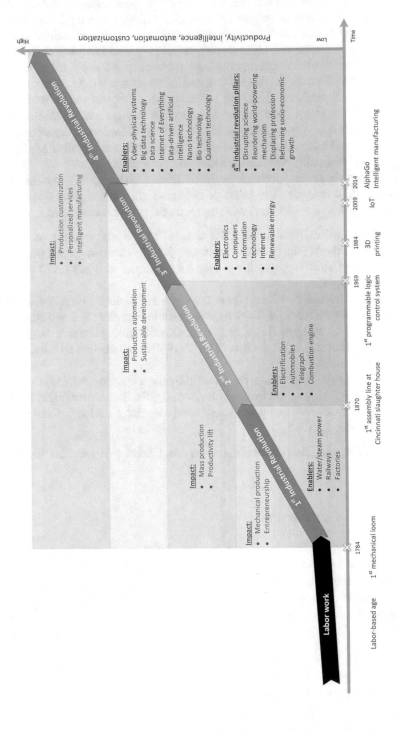

Fig. 12.1 Four industrial revolutions

12.2.2.4 Reforming Economic Growth and Social Goodness

While we are in this wave of deep recession of global economy and social unrest, the global economy has been increasingly transformed to the digital and data era. In a not too distant future (10–15 years or so), many existing industrial sectors, business and service activities will experience significant evolution from their current ways and states of operations to being data+intelligence-driven, and new data+intelligence-driven businesses emerge to drive the new economic growth and social goodness.

A significant data-driven business mode is that increasing data-driven service providers are in charge of traditional businesses and services and globalize their activities through Internet/wireless and social networks, such as Google retrieving and managing global data, Alibaba Taobao connecting and managing retail shops and payments, Airbnb connecting and managing hotels and leased residential accommodations, Uber connecting and managing taxis and private cars. More and more traditional businesses will be connected and taken over by data-based service providers who do not own the businesses but partially or fully take in charge of the business operations, sales, marketing, customer relationship management, payment and reformation. This new data-driven business mode is fundamentally reshaping traditional manufacturing, service industries, logistics and payment systems, and further changing people's consuming behaviors, living, working, traveling, and entertaining activities.

Another fundamental data-driven business type replaces traditionally human-centered economic and social activities by data and intelligent products. Typical examples are the replacement of cash by Bitcoin, banks by e-payment providers, human-driving taxis and cars by self-driving taxis and cars, postman-led door-to-door and postage delivery by unmanned aerial vehicle-based delivery, on-campus and in-class training and education by online courses, general and tailored product design and manufacturing by 3D-based design and printing, and general practitioners and consultants-based healthcare services by web-based intelligent analytics-driven expert and decision-support systems. This will reform businesses, displace many classic businesses, and create numerous new businesses. Traditional taxis, cars, bikes, training organizations, factories, etc. will significantly decline or disappear, replaced by intelligent and shared ones.

At the same time, as discussed in Chap. 8 on data economy and industrialization and in Sect. 12.6 on advanced data science and technologies, many new data-driven businesses will emerge and quickly create and dominate the market.

12.3 Data Science of Sciences

Science owns, manages and generates the largest proportion of the universal data. Scientific data consists of every aspect of natural, artificial and social systems, which form the sources of scientific data, and the transformed data through undertaking scientific activities on these sources of data. This scientific data enables unlimited opportunities and prospects for data science, forming the area of data science of science.

Opportunities and prospects of data science on scientific data are enormous, for example,

- data-driven discovery of unknown scientific fields with large volumes of data and possible solutions or directions, such as the working mechanisms of the outer spaces and unknown universe, and the semiconductor technology after carbon nanotube transistors reaching their nano limit (1 nm) and whether any new law to replace the Moore Law;
- data-driven discovery of unknown scientific problems in existing science and possible solutions or directions, such as precision medicine on health data, medical data, and gnomic data, and reasons for causing cancers and climate changes;
- discovery of high potential and high impact scientific areas, problems, and directions and their corresponding possible solutions or research directions;
- data-driven discovery of unscientific activities and problems in science, for example, to discover irreproducible, misinformed, false, and misleading results reported or submitted in the literature; and
- data-driven evaluation of science quality and impact.

12.4 Data Brain

For the research and applications of data science, it would be valuable to build data brain for a country, organization, enterprise, research center, or as an open data platform. Building on data center and a data-driven network, *data brain* serves the fundamental roles and responsibilities like human brain and domain-specific expert team for complex intelligent data-driven discovery, decision-support, and action-taking.

Such data brain may be built with the following facilities and functionalities:

- High-performance computing infrastructure: to enable high-performance computing, analytics, cognitive activities, and decision-support;
- High-performance networks: to enable on-demand networking within the brain and with external machines, e.g., to Internet, dedicated cloud platforms, mobile networks, social networks, proprietary networks, and various intelligent agents (here intelligent agents broadly refer to intelligent systems, e.g., a learning system or a human expert in a system);
- Intelligent communication: to ensure on-demand single or multiple media-based communications within the brain and with external entities, entity groups, centralized and distributed (to individuals) communications, group and individual communications, mobile communications, wireless personal instant messaging, etc.

- Large-scale heterogeneous sources of data management: to connect, collect, store, manage, and access large data from different sources within the brain and from external resources;
- Strong data collection and fusion: to retrieve and crawl data within an enterprise and from public domain, Internet including wireless networks-based resources such as instant messaging networks and social media on demand;
- Strong knowledge encyclopedia: in addition to collect common knowledge, to collect, organize, and manage relevant knowledge to form encyclopedia about the problem domain, especially from existing digital libraries and public encyclopedia systems such as Wikipedia;
- Intelligent data processing: to prepare (clean and transform) data and make data ready for various decision-support objectives, tasks and jobs;
- Intelligent data analysis and learning systems: to create, maintain, call and manage (e.g., delete, activate, update) learning algorithms and modules, and to create, maintain and manage libraries of algorithms and tool kits;
- Seamless human-machine interaction: to support interactions between humans and between humans and machines, including facilities of visual interactions, virtual reality, and augmented reality;
- Decision-support facilities: to generate decision-making recommendations and to support various decision-making modes, such as automated decisions, human-machine-cooperative decisions, and human-centered decisions;
- System assurance facilities: to enable the governance of data, computing, analytics and decision quality, ethics (including privacy), security, performance and assessment.

Accordingly, data brain is not just a data center or data-driven intelligent decision-support system. It is a metasynthetic platform that meta-synthesizes

- various sources of data, information and knowledge;
- various X-complexities and X-intelligence;
- various X-analytics, X-informatics and learning facilities and capabilities;
- various computing and engineering facilities and capabilities;
- various networking and communication facilities and capabilities; and
- various decision-support and human-machine interaction facilities and capabilities.

On one hand, a data brain should own and fulfill the basic functions and responsibilities, including data management, data engineering, computing and programming, data exploration, knowledge discovery, machine learning, information retrieval, recommendation, and communications and decision-support facilities. On the other hand, the brain should cater for and/or develop high-level cognitive activities and intelligence for enabling various machine intelligence and machine thinking, which will be discussed in Sect. 12.5.

Data brain is driven by data and intelligence science, technologies, infrastructure, services and applications. Data brain converts data to values by sensing and analyzing the input and requirements, understanding problem complexities, learning data intelligence, and delivering insight for decision-making.

IBM Watson [221] may be an early-stage[1] example of a general data brain. IBM Watson machine is artificial intelligence-driven. It is embedded with capabilities for data understanding, reasoning about a user's personality, tone and emotion, machine learning of data, and interactions with people by creating chat bots for dialog with users. The Watson machine extends machine computing to cognitive computing for purposes such as analytics, engagement adviser, explorer, discovery adviser, knowledge studio, ontology, and clinical trial matching.

12.5   Machine Intelligence and Thinking

Machine intelligence will experience a significant evolution from shallow modeling and simulation to deep representation, simulation, construction and emergence of human sensing, recognition, reasoning, inference, neural processing, learning, behaving, and decision-making. The evolution will see increasingly powerful *hard intelligence* and *soft intelligence*, as well as the fusion of other X-intelligence with machine and data intelligence.

First, *hard intelligence* is embedded or emerges from (1) hardware, physical/computerized intelligent systems, devices, and infrastructure; and (2) hard-coded intelligence through pre-defined sensing, reacting, planning, reasoning, inference, analytics, learning, optimization, and aggregation of various intelligence at the design time.

Second, *soft intelligence* is embedded or emerges from (1) software and software-based computerized intelligent engines, systems, services and applications; (2) autonomous analysis, learning and reasoning-based intelligence obtained from existing systems and environment, and (3) autonomous evolutionary intelligence obtained via evolution and mutation. Soft intelligence may emerge from hard-coded intelligent agents at design time and soft-coded intelligent agents at running time. Soft intelligence is built on real-time and online data acquisition, integration and synthesis, analysis and learning, reasoning and inference, conflict resolution and risk management, and decision-making and action-taking.

In addition, the formation and emergence of hard and soft machine intelligence rely on hierarchical intelligent building blocks consisting of low-level and high-level intelligent units and their connected intelligence. Data-driven intelligent systems go beyond the individual and local intelligence built in system constituents themselves, but the system and network intelligence that emerges through self- and co-connections, interactions, cooperation, learning, and evolution within and between intelligent units. The networking-based connections of relevant individual and local

---

[1] Here early-stage indicates far more opportunities in terms of developing human-like intelligence and thinking.

system/machine intelligence play irreplaceable roles and even unpredictable and uncontrollable scenarios in super-intelligent systems.

To revolutionize machine intelligence, machine thinking must be built. *Machine thinking* refers to the thought processes, approaches, and tools to enable machine to understand problem complexities and generate corresponding intelligent solutions. Traditionally, machine thinking research has been focused on aspects of mathematical and statistical thinking, computational thinking, and logical thinking. With the emergence of data intelligence and the meta-synthesis of X-intelligence, machine thinking is expected to be imaginative and human-machine-cooperative.

These various types and stages of machine thinking are briefly explained below.

- Mathematical and statistical thinking: a typical statistical thinking approach is statistical inference, which learns (estimates) the characteristics (properties, or parameters) of a population and the reliability of statistical relationships typically by analysis of randomly drawn samples;
- Computational thinking: the thought and problem-solving processes and approaches for formulating a problem and expressing the solution(s) by engaging data brain's computational power;
- Logical thinking: the thought and problem-solving processes and approaches for data brain to derive a conclusion or consequence based on a given precondition or premise, usually by following certain logics, rules, or logical reasoning;
- Imaginative thinking: the thought processes and approaches for data brain to make intuitive, innovative, creative, and challenging assumptions or solutions, by learning from, cooperated with or inspired by domain experts and domain expert experience, or making strategic choices through balancing global and local, and short term and long term objectives;
- Human-machine-cooperative cognitive computing and thinking: combining computerized learners for computational and logical thinking with human learners for human imaginative thinking through human-machine cooperation and interaction by integratively making both machine-impossible and human-impossible possible.

12.6 Advancing Data Science and Technology and Economy

Many areas and directions of data science and technology will be substantially developed or upgraded in the coming years. Considering the intrinsic challenges and nature of data science [64, 65], below several such directions are listed.

- Develop *enabling theories and technologies* for substantially enhancing machine thinking and intelligence, especially human-like cognitive thinking and imaginative thinking;
- Design and develop *data brain*: that can autonomously mimic human brain working mechanisms while recognize, understand, analyze and learn data and

environment, infer and reason about knowledge and insight, and correspondingly decide actions;

- Deepen our *understanding of data invisibility*: about invisible data characteristics, complexities, intelligence and value, as shown in Spaces B and D in Fig. 3.6, in particular, to understand their X-complexities and X-intelligence [64]. The exploration of what we do not know about what we do not know will strengthen our understanding of the unknown capabilities, limitations of existing work, and future directions of data science;
- Broaden *conceptual, theoretical and technological foundation for data science*: by enabling cross-disciplinary and trans-disciplinary research, innovation and education. This will address existing issues such as the variations in statistics hypotheses and will discover and propose problems that are currently invisible to broad science or specific fields;
- Invent *new data representation capabilities*: including designs, structures, schemas and algorithms to make invisible data complexities and unknown characteristics in complex data visible, and more easily understood or explored;
- Design *new storage, access and management mechanisms*: including memory, disk and cloud-based mechanisms, to enable the acquisition, storage, access, sampling, and management of richer characteristics and properties in the physical world that have been simplified and filtered by existing systems, and to support scalable, transparent, flexible, interpretable and personalized data manipulation and analytics in real time;
- Create *new analytical and learning capabilities*: including original mathematical, statistical and analytical theories, algorithms and models, to disclose the unknown knowledge in unknown space;
- Develop *data authentication technologies*: including techniques for recognizing and certifying misinformation, ownership, sourcing, and tracking digital information and its transportation in networks especially the Internet.

The relevant technical innovation opportunities in data technologies and economy are enormous in the coming decades, for example,

- *human-like robots and automated systems*: which can sense, collect, process, analyze, reason about, collaborate and make decisions based on context, scenarios, and needs. Such systems will replace most of routine work currently taken by human workers, and substantially reform the human-robot-cooperative workplace. Human brain-computer interaction will enable robots to be as smart as humans and/or to be able to undertake intelligent tasks that would previously only be taken by human.
- *mobile+cloud-based service systems*: mobile devices-based services sense, respond, and act on health, living, entertaining, studying and working conditions, activities and needs, and cloud-based computing and services stay at the back end to make recommendations and decisions for action-taking.
- *new intelligent systems and services*: including corporate and Internet-based collaborative platforms and services, to support the automated, or human-data-

machine cooperative, collaborative and collective exploration of invisible and unknown challenges in unknown space.

- *precision medicine and personalized health services*: by analyzing gnomic, health, medical and personal data, medicines and services will be customized for a specific patient that will best fit the patient's situation and need, make best possible positive effect and avoid mistreatments and negative effect that may be caused due to mismatches of medicine and patient's scenario.
- *smart world*: with the widespread access to next-generation broadband services, and the applications of Internet of Things with city facilities, residential service facilities, mobile services, sensor networks, utility services including water, electricity, gas, transport, entertainment, healthcare, and living services, all relevant data and business are connected and analyzed, everything will be connected to a smart world, smart cities, smart homes, smart healthcare, working and other aspects are connected, recommendations are made for smarter services.
- *open and shared systems and services*: open model, open innovation, open economic mode, and open systems and services overtake the traditional services and economy; new design, manufacturing, marketing, sales, distribution, delivery, and customer service modes emerge to cater for open and shared economy; research and innovation rely more on 3D printing-based design and manufacturing, and data analytics of customer preferences, needs, feedback, and changes; economy and business expand from local, regional to global scale and become increasingly distributed in terms of markets but dominated with respect to provision and services.
- *digitized world*: with the applications of social networks, smart cities and home, Internet of Things, mobile, wireless and sensor network services, embedded systems, and human brain-computer communication systems, physical worlds are increasingly digitized, digital worlds become equally or even more important than physical worlds. Opportunities and challenges are with how and what to digitize, making them interconnected and intelligent, and competing and developing in a highly open, connected, transparent, efficient, and evolving digital-physical worlds. In addition to economic opportunities and challenges, existing social, political, cultural, religious, and military systems are significantly challenged, with substantial new opportunities and problems emerging.

12.7 Advancing Data Education and Profession

From the disciplinary, professional and practical perspectives, the coming decade will see significant development in data education and profession, such as

- Forming the discipline of data science: by building its own foundations, technological fields, subject areas, research map and agenda, and practical methodologies and processes;

- Training the next-generation data scientists and data professionals: who are qualified for complex and independent data science problem-solving, with data literacy, thinking, competency, consciousness, curiosity, communication and cognitive intelligence, to work on the above data science agenda;
- Fostering capabilities and professionals to assure cross-domain and trans-disciplinary cooperation, collaborations and alliance in complex data science problem-solving. This requires the education of competent data scientists who are multi-disciplinary experts, as well as collaboration between data scientists and domain-specific experts; and
- Creating talents who can discover and invent *data power* as yet unknown to current understanding and imagination, such as new data economy, mobile applications, social applications, and data-driven business.

12.8 Summary

Our world is experiencing intrinsic and fundamental landscape evolution from one paradigm of science and generation of technologies and economy to another. Data science, technology, economy and society are driving the fourth scientific and technological revolution and industrial, economic and globalization transformation. This remarkable new revolution is characterized by data+intelligence-driven

- breakthroughs of unknown scientific, technological, social and economic problems, challenges and opportunities;
- disruption of existing world-empowering mechanisms and drivers for science, technologies and economy;
- transformation of old-generation of research, innovation, productivity, services and operations; and
- improvement of research and innovation quality and impact.

As a result, data science, technology, economy and society will lead to high socio-economic impact through

- creating a wide range of data-driven businesses for producing innovative and practical data products and services;
- transforming existing and classic business and social activities toward faster, cheaper, globalized and personalized operations and services; and
- upgrading and improving the quality and performance of business activities and public sector polices and services.

The emerging data science evolution means opportunities for breakthrough research, technological innovation, and a new data economy. If parallels are drawn between the evolution of the Internet and the evolution of data science, the future and the socio-economic and cultural impact of data science will be unprecedented indeed, though as yet unquantifiable.

References

1. ACEMS: The Australian research council (arc) centre of excellence for mathematical and statistical frontiers (2014). URL www.acems.org.au/
2. Agarwal, R., Dhar, V.: Editorial-Big data, data science, and analytics: The opportunity and challenge for IS research. Information Systems Research **25**(3), 443–448 (2014)
3. Agency, X.N.: The 13th five-year plan for the national economic and social development of the peoples' republic of China (2016). URL http://news.xinhuanet.com/politics/2016lh/2016-03/17/c_1118366322.htm
4. AGIMO: AGIMO big data strategy - issues paper (2013). URL www.finance.gov.au/files/2013/03/Big-Data-Strategy-Issues-Paper1.pdf
5. Anderson, P.E., Bowring, J.F., McCauley, R., Pothering, G., Starr, C.W.: An undergraduate degree in data science: Curriculum and a decade of implementation experience. In: Computer Science Education: Proceedings of the 45th ACM Technical Symposium (SIGCSE'14), pp. 145–150 (2014)
6. Anderson, P.E., Turner, C., Dierksheide, J., McCauley, R.: An extensible online environment for teaching data science concepts through gamification. In: 2014 IEEE Frontiers in Education Conference (FIE), pp. 1–8 (2014)
7. Anya, O., Moore, B., Kieliszewski, C., Maglio, P., Anderson, L.: Understanding the practice of discovery in enterprise big data science: An agent-based approach. In: 6th International Conference on Applied Human Factors and Ergonomics (AHFE 2015) and the Affiliated Conferences, vol. 3, pp. 882–889 (2015)
8. Apache: Apache opennlp (2016). URL https://opennlp.apache.org/
9. Apache: Apache spark mllib (2016). URL http://spark.apache.org/mllib/
10. Apache: The apache software foundation (2018). URL https://www.apache.org/
11. ARC: Codes and guidelines | Australian research council (2017). URL www.arc.gov.au/codes-and-guidelines
12. ASA: ASA views on data science (2015). URL http://magazine.amstat.org/?s=data+science&x=0&y=0
13. ASA: Ethical guidelines for statistical practice, American statistical association (2016). URL https://www.certifiedanalytics.org/ethics.php
14. AU: Data-matching program (1990). URL http://www.comlaw.gov.au/Series/C2004A04095
15. AU: Declaration of open government (2010). URL http://agimo.gov.au/2010/07/16/declaration-of-open-government/
16. AU: Attorney-General's department (2013). http://www.attorneygeneral.gov.au/Mediareleases/Pages/2013/Second%20quarter/22May2013-AustraliajoinsOpenGovernmentPartnership.aspx

17. AU: Australia big data (2016). URL http://www.finance.gov.au/big-data/
18. Auschitzky, E., Hammer, M., Rajagopaul, A.: How big data can improve manufacturing? (2016). URL http://www.mckinsey.com/business-functions/operations/our-insights/how-big-data-can-improve-manufacturing
19. Ayankoya, K., Calitz, A., Greyling, J.: Intrinsic relations between data science, big data, business analytics and datafication. ACM International Conference Proceeding Series **28**, 192–198 (2014)
20. Bailer, J., Hoer, R., Madigan, D., Montaquila, J., Wright, T.: Report of the asa workgroup on master's degrees (2012). URL http://magazine.amstat.org/wp-content/uploads/2013an/masterworkgroup.pdf
21. Barber, D.: Bayesian Reasoning and Machine Learning. Cambridge University Press (2012)
22. Batini, C., Scannapieco, M.: Data and Information Quality: Dimensions, Principles and Techniques. Springer (2016)
23. Baumer, B.: A data science course for undergraduates: Thinking with data. The American Statistician **69**(4), 334–342 (2015)
24. BBC: Facebook-Cambridge analytica data scandal (2018). URL http://www.bbc.com/news/topics/c81zyn0888lt/facebook-cambridge-analytica-data-scandal
25. BDL: Big data landscape (2016). URL www.bigdatalandscape.com
26. BDL: Big data landscape 2016 (version 3.0) (2016). URL http://mattturck.com/2016/02/01/big-data-landscape/
27. BDSS: Big data social science. URL http://bdss.psu.edu/
28. BESC: Behavioral, economic, socio-cultural computing.   URL   http://www.behaviorinformatics.org/
29. Beyer, M.A., Laney, D.: The importance of 'big data': A definition (2012). URL https://www.gartner.com/doc/2057415. Gartner
30. Bhardwaj, A., Bhattacherjee, S., Chavan, A., Deshp, A., Elmore, A.J., Madden, S., Parameswaran, A.: Datahub: Collaborative data science & dataset version management at scale. In: In CIDR (2015)
31. BI: Behavioral insights (2014). URL http://www.behaviouralinsights.co.uk/
32. BigML: Bigml (2016). URL https://bigml.com/
33. BIID: Beyond iid in information theory (biid) workshop (2013). URL https://sites.google.com/site/beyondiid4/biid-conference-series
34. Boccara, N.: Modeling Complex Systems. Springer (2003)
35. Bono, E.D.: Lateral Thinking: Creativity Step by Step. Harper & Row (1970)
36. Borne, K.D., Jacoby, S., Carney, K., Connolly, A., Eastman, T., Raddick, M.J., Tyson, J.A., Wallin, J.: The revolution in astronomy education: Data science for the masses (2010). URL http://arxiv.org/pdf/0909.3895v1.pdf
37. Bothun, D., Vollmer, C.A.H.: 2016 entertainment & media industry trends (2016). URL https://www.strategyand.pwc.com/media/file/2016-Entertainment-and-Media-Trends.pdf
38. Boulding, K.: Notes on the information concept. Exploration (Toronto) **6**(103-112, CP IV), 21–32 (1955)
39. Boyer, S., Gelman, B.U., Schreck, B., Veeramachaneni, K.: Data science foundry for MOOCs. In: IEEE International Conference on Data Science and Advanced Analytics (DSAA), pp. 1–10 (2015)
40. BPMM: Business process maturity model^{TM} (bpmm^{TM}) (2008). URL http://www.omg.org/spec/BPMM/
41. Brain, G.: Tensorflow (2016). URL https://www.tensorflow.org/
42. Breiman, L.: Statistical modeling: The two cultures. Statist. Sci. **16**(3), 199–231 (2001)
43. Brockman, J.: What to Think About Machines That Think: Today's Leading Thinkers on the Age of Machine Intelligence. Harper Perennial (2015)
44. Broman, K.: Data science is statistics (2013). URL https://kbroman.wordpress.com/2013/04/05/data-science-is-statistics/

45. Brown, G.: Review of education in mathematics, data science and quantitative disciplines: Report to the group of eight universities (2009). URL https://go8.edu.au/publication/go8-review-education-mathematics-data-science-and-quantitative-disciplines

46. Burnes, J.: 10 steps to improve your communication skills. URL https://www.aim.com.au/blog/10-steps-improve-your-communication-skills

47. Burtch, L.: The burtch works study: Salaries of data scientists (2014). URL http://www.burtchworks.com/files/2014/07/Burtch-Works-Study_DS_final.pdf

48. Bussaban, K., Waraporn, P.: Preparing undergraduate students majoring in computer science and mathematics with data science perspectives and awareness in the age of big data. In: 7th World Conference on Educational Sciences, vol. 197, pp. 1443–1446 (2015)

49. Bynum, T.: Computer and information ethics. In: The Stanford encyclopedia of philosophy (ed. Zalta EN) (2015). URL See http://plato.stanford.edu/archives/win2015/entries/ethics-computer/

50. CA: Canada capitalizing on big data (2016). URL http://www.sshrc-crsh.gc.ca/news_room-salle_de_presse/latest_news-nouvelles_recentes/big_data_consultation-donnees_massives_consultation-eng.aspx

51. Campbell, J.D., Molines, K., Swarth, C.: Data mining for ecological field research: Lessons learned from amphibian and reptile activity analysis. In: NSF Symposium on Next Generation of Data Mining and Cyber-Enabled Discovery for Innovation (2007)

52. Campbell, J.: *Environmental Informatics.* http://www.environmentalinformatics.com/Bibliography.php

53. Cao, L.: Data Mining and Multi-agent Integration (edited). Springer (2009)

54. Cao, L.: Domain driven data mining: Challenges and prospects. IEEE Trans. on Knowledge and Data Engineering **22**(6), 755–769 (2010)

55. Cao, L.: In-depth behavior understanding and use: The behavior informatics approach. Information Science **180**(17), 3067–3085 (2010)

56. Cao, L.: Strategic recommendations on advanced data industry and services for the yanhuang science and technology park (2011)

57. Cao, L.: Actionable knowledge discovery and delivery. Wiley Interdisc. Rew.: Data Mining and Knowledge Discovery **2**(2), 149–163 (2012)

58. Cao, L.: Social security and social welfare data mining: An overview. IEEE Trans. Systems, Man, and Cybernetics, Part C **42**(6), 837–853 (2012)

59. Cao, L.: Combined mining: Analyzing object and pattern relations for discovering and constructing complex but actionable patterns. WIREs Data Mining and Knowledge Discovery **3**(2), 140–155 (2013)

60. Cao, L.: Non-iidness learning in behavioral and social data. The Computer Journal **57**(9), 1358–1370 (2014)

61. Cao, L.: Coupling learning of complex interactions. J. Information Processing and Management **51**(2), 167–186 (2015)

62. Cao, L.: Metasynthetic Computing and Engineering of Complex Systems. Springer (2015)

63. Cao, L.: Data science: A comprehensive overview. Submitted to ACM Computing Survey pp. 1–37 (2016)

64. Cao, L.: Data science: Challenges and directions (2016). Technical Report, UTS Advanced Analytics Institute

65. Cao, L.: Data science: Nature and pitfalls (2016). Technical Report, UTS Advanced Analytics Institute

66. Cao, L.: Data science: Profession and education (2016). Technical Report, UTS Advanced Analytics Institute

67. Cao, L.: Data Science: Techniques and Applications (2018)

68. Cao, L.: Data Science Thinking: The Next Scientific, Technological and Economic Revolution. Springer (2018)

69. Cao, L., Dai, R.: Open Complex Intelligent Systems. Post & Telecom Press (2008)

70. Cao, L., Dai, R., Zhou, M.: Metasynthesis: M-Space, M-Interaction and M-Computing for open complex giant systems. IEEE Trans. On Systems, Man, and Cybernetics–Part A **39**(5), 1007–1021 (2009)

71. Cao, L., Dong, X., Zheng, Z.: e-NSP: Efficient negative sequential pattern mining. Artificial Intelligence **235**, 156–182 (2016)

72. Cao, L., (Eds), P.S.Y.: Behavior Computing: Modeling, Analysis, Mining and Decision. Springer (2012)

73. Cao, L., Gorodetsky, V., Mitkas, P.A.: Agent mining: The synergy of agents and data mining. IEEE Intelligent Systems **24**(3), 64–72 (2009)

74. Cao, L., Ou, Y., Yu, P.S.: Coupled behavior analysis with applications. IEEE Trans. on Knowledge and Data Engineering **24**(8), 1378–1392 (2012)

75. Cao, L., Weiss, G., Yu, P.S.: A brief introduction to agent mining. Autonomous Agents and Multi-Agent Systems **25**(3), 419–424 (2012)

76. Cao, L., Yu, P.S., Kumar, V.: Nonoccurring behavior analytics: A new area. IEEE Intelligent Systems **30**(6), 4–11 (2015)

77. Cao, L., Yu, P.S., Zhang, C., Zhao, Y.: Domain Driven Data Mining. Springer (2010)

78. Cao, L., Zhang, C.: Knowledge actionability: Satisfying technical and business interestingness. International Journal of Business Intelligence and Data Mining **2**(4), 496–514 (2007)

79. Cao, L., Zhao, Y., Zhang, C.: Mining impact-targeted activity patterns in imbalanced data. IEEE Trans. on Knowledge and Data Engineering **20**(8), 1053–1066 (2008)

80. Cao, W., Cao, L.: Financial crisis forecasting via coupled market state analysis. IEEE Intelligent Systems **30**(2), 18–25 (2015)

81. Capterra: Top project management tools (2016). URL http://www.capterra.com/project-management-software/

82. Capterra: Top reporting software products (2016). URL http://www.capterra.com/reporting-software/

83. Casey, E.: The growing importance of data science in digital investigations. Digital Investigation **14**, A1–A2 (2015)

84. CBDIO: China big data industrial observation (2016). URL www.cbdio.com

85. CCF-BDTF: China computer federation task force on big data (2013). URL http://www.bigdataforum.org.cn/

86. Ceglar, A., Roddick, J.: Association mining. ACM Computing Surveys **38**(2), 5 (2006)

87. Chambers, J.M.: Greater or lesser statistics: A choice for future research. Statistics and Computing **3**(4), 182–184 (1993)

88. Chandrasekaran, S.: Becoming a data scientist (2013). URL http://nirvacana.com/thoughts/becoming-a-data-scientist/

89. Chawla, S., Hartline, J., Nekipelov, D.: Mechanism design for data science. In: Economics and computation: Proceedings of the Fifteenth ACM Conference, pp. 711–712 (2014)

90. Chemuturi, M.: Mastering Software Quality Assurance: Best Practices, Tools and Techniques for Software Developers. J. Ross Publishing (2010)

91. Chen, H., Chiang, R.H.L., Storey, V.C.: Business intelligence and analytics: From big data to big impact. MIS Quarterly **36**(4), 1165–1188 (2012)

92. Chen, Z., Liu, B.: Lifelong Machine Learning. Synthesis Lectures on Artificial Intelligence and Machine Learning. Morgan & Claypool (2016)

93. Clancy, T.R., Bowles, K.H., Gelinas, L., Androwich, I., Delaney, C., Matney, S., Sensmeier, J., Warren, J., Welton, J., Westra, B.: A call to action: Engage in big data science. Nursing Outlook **62**(1), 64–65 (2014)

94. Classcentral: Data science and big data | free online courses (2016). URL https://www.class-central.com/subject/data-science

95. Clauset, A., Larremore, D.B., Sinatra, R.: Data-driven predictions in the science of science. Science **355**, 477–480 (2017)

96. Clay, K.: Ces 2013: The year of the quantified self? (2013). URL http://www.forbes.com/sites/kellyclay/2013/01/06/ces-2013-the-year-of-the-quantified-self/#4cf4d2b55e74

97. Cleveland, W.S.: Data science: An action plan for expanding the technical areas of the field of statistics. International Statistical Review **69**(1), 21–26 (2001). doi:10.1111/j.1751-5823.2001.tb00477.x. URL http://dx.doi.org/10.1111/j.1751-5823.2001.tb00477.x

98. CMIST: China will establish a series of national labs (2016). URL http://news.sciencenet.cn/htmlnews/2016/4/344404.shtm

99. CNSF: National natural science foundation of China (2015). URL http://www.nsfc.gov.cn/

100. Cohen, R., Havlin, S.: Complex Networks: Structure, Robustness and Function, 1st Edition. Cambridge University Press (2010)

101. Commission, E.: Commission urges governments to embrace potential of big data (2014). URL www.europa.eu/rapid/press-release_IP-14-769_en.htm

102. Commission, E.: Towards a thriving data-driven economy (2014). URL https://ec.europa.eu/digital-single-market/en/towards-thriving-data-driven-economy

103. Coursera: Coursera (2016). URL www.coursera.org/data-science

104. CRISP-DM: CRISP-DM (2016). URL www.sv-europe.com/crisp-dm-methodology

105. Crowston, K., Qin, J.: A capability maturity model for scientific data management: Evidence from the literature. In: Proceedings of the American Society for Information Science and Technology, vol. 48, pp. 1–9 (2011)

106. CSC: Big data universe beginning to explode (2012). URL http://www.csc.com/insights/flxwd/78931-big_data_growth_just_beginning_to_explode

107. CSNSTC: Harnessing the power of digital data for science and society (2009). URL https://www.nitrd.gov/About/Harnessing_Power_Web.pdf. Report of the Interagency Working Group on Digital Data to the Committee on Science of the National Science and Technology Council

108. CSS: Computational social science. URL http://www.gesis.org/en/institute/gesis-scientific-departments/computational-social-science/

109. Csurka, G.: Domain adaptation for visual applications: A comprehensive survey. CoRR **abs/1702.05374** (2017). URL http://arxiv.org/abs/1702.05374

110. CTM: Defining critical thinking (1987). URL https://www.criticalthinking.org/pages/defining-critical-thinking/766

111. Cummings, T., Jones, Y.: Creating actionable knowledge. URL http://meetings.aomonline.org/2004/theme.htm

112. Cuzzocrea, A., Gaber, M.M.: Data science and distributed intelligence: Recent developments and future insights. Studies in Computational Intelligence **446**, 139–147 (2013)

113. DABS: Data analytics book series (2016). URL http://www.springer.com/series/15063

114. Dai, R., Wang, J., Tian, J.: Metasynthesis of Intelligent Systems. Zhejiang Sci. Technol. Press, Hangzhou, China (1995)

115. DARPA: DARPA xdata program (2016). URL www.darpa.mil/program/xdata

116. Data61: Data61 (2016). URL https://www.data61.csiro.au/

117. DataRobot: Datarobot (2016). URL https://www.datarobot.com/

118. Datasciences.org: Datasciences.org (2005). URL www.datasciences.org

119. Dataversity: How data scientists can improve communications skills. URL http://www.dataversity.net/how-data-scientists-can-improve-communications-skills/

120. Daumé III, H., Marcu, D.: Domain adaptation for statistical classifiers. J. Artif. Int. Res. **26**(1), 101–126 (2006)

121. Davenport, T.H., Patil, D.: Data scientist: The sexiest job of the 21st century. Harvard Business Review pp. 70–76 (2012)

122. Davis, J.: 10 programming languages and tools data scientists used (2016). URL http://www.informationweek.com/devops/programming-languages/10-programming-languages-and-tools-data-scientists-use-now/d/d-id/1326034

123. Deeplearning: Deeplearning (2016). URL www.deeplearning.net/

124. Deo, N.: Graph Theory with Applications to Engineering and Computer Science (Prentice Hall Series in Automatic Computation). Prentice-Hall, Upper Saddle River, NJ, USA (1974)

125. Desale, D.: Top 30 social network analysis and visualization tools (2015). URL http://www. kdnuggets.com/2015/06/top-30-social-network-analysis-visualization-tools.html
126. Dhar, V.: Data science and prediction. Communications of the ACM **56**(12), 64–73 (2013)
127. Dierick, H.A., Gabbiani, F.: Drosophila neurobiology: No escape from 'big data' science. Current Biology **25**(14), 606–608 (2015)
128. Diggle, P.J.: Statistics: A data science for the 21st century. Journal of the Royal Statistical Society: Series A (Statistics in Society) **178**(4), 793–813 (2015)
129. Donoho, D.: 50 years of data science (2015). URL http://courses.csail.mit.edu/18.337/2015/docs/50YearsDataScience.pdf
130. Dorr, B.J., Greenberg, C.S., Fontana, P., Przybocki, M.A., Bras, M.L., Ploehn, C.A., Aulov, O., Michel, M., Golden, E.J., Chang, W.: The NIST data science initiative. In: 2015 IEEE International Conference on Data Science and Advanced Analytics (DSAA), pp. 1–10 (2015)
131. Dowden, B.H.: Logical Reasoning. Philosophy Department, California State University Sacramento (2017). URL http://www.csus.edu/indiv/d/dowdenb/4/logical-reasoning.pdf
132. Drew, C.: Data science ethics in government. Phil. Trans. R. Soc. A **374** (2016)
133. DSA: Data science association (2016). URL http://www.datascienceassn.org/
134. DSA: Data science code of professional conduct, data science association (2016). URL http://www.datascienceassn.org/code-of-conduct.html
135. DSAA: IEEE/ACM/ASA international conference on data science and advanced analytics (2014). URL www.dsaa.co
136. DSC: College & university data science degrees (2016). URL http://datascience.community/colleges (accessed on 16 April 2016.)
137. DSC: The data science community (2016). URL http://datasciencebe.com/
138. DSCentral: Data science central (2016). URL http://www.datasciencecentral.com/
139. DSE: Data science and engineering (2015). URL http://link.springer.com/journal/41019
140. DSJ: Data science journal (2014). URL www.datascience.codata.org
141. DSKD: Data science and knowledge discovery lab, university of technology Sydney (2007). URL http://www.uts.edu.au/research-and-teaching/our-research/quantum-computation-and-intelligent-systems/data-sciences-and
142. DSSG: Data science for social good fellowship. URL https://dssg.uchicago.edu/
143. Duncan, D.E.: Experimental Man: What One Man's Body Reveals about His Future, Your Health, and Our Toxic World. New York: Wiley & Sons (2009)
144. Dunleavy, P.: The social science of human-dominated & human-influenced systems: Annual lecture report (2014). URL http://www.acss.org.uk/news/annual-lecture-report/
145. van Dyk, D., Fuentes, M., Jordan, M.I., Newton, M., Ray, B.K., Lang, D.T., Wickham, H.: ASA statement on the role of statistics in data science (2015). URL http://magazine.amstat.org/blog/2015/10/01/asa-statement-on-the-role-of-statistics-in-data-science/
146. (Ed.), M.P.: Similarity-based pattern analysis and recognition. Springer (2013)
147. Edx: EDX courses (2016). URL https://www.edx.org/course?search_query=data+science
148. Ehling, M., Korner, T.: Handbook on Data Quality Assessment Methods and Tools (eds.). EUROSTAT, Wiesbaden (2007)
149. Elder, L., Paul, R.: The Thinker's Guide to Analytic Thinking: How to Take Thinking Apart and What to Look for When You Do (2nd ed.). Foundation for Critical Thinking (2007)
150. EMC: Data science revealed: A Data-Driven glimpse into the burgeoning new field (2011). URL www.emc.com/collateral/about/news/emc-data-science-study-wp.pdf
151. EPJDS: EPJ data science (2012). URL http://epjdatascience.springeropen.com/
152. ESF: Research integrity : European science foundation (2016). URL www.archives.esf.org/coordinating-research/mo-fora/research-integrity.html
153. ESRC: Big data and the social sciences: A perspective from the esrc. URL https://www2.warwick.ac.uk/fac/soc/economics/research/centres/cage/events/conferences/peuk/peter_elias_big_data_and_the_social_sciences_pe_final.pdf
154. EU-DSA: The European data science academy (2016). URL www.edsa-project.eu
155. EU-OD: The European union open data portal (2016). URL https://open-data.europa.eu/
156. Facebook: Facebook data (2016). URL https://www.facebook.com/careers/teams/data/

157. Faghmous, J.H., Kumar, V.: A big data guide to understanding climate change: The case for theory-guided data science. Big Data **2**(3), 155–163 (2014)
158. Fairfielda, J., Shteina, H.: Big data, big problems: Emerging issues in the ethics of data science and journalism. Journal of Mass Media Ethics **29**(1), 38–51 (2014)
159. Faris, J., Kolker, E., Szalay, A., Bradlow, L., Deelman, E., Feng, W., Qiu, J., Russell, D., Stewart, E., Kolker, E.: Communication and data-intensive science in the beginning of the 21st century. A Journal of Integrative Biology **15**(4), 213–215 (2011)
160. Fawcett, T.: Mining the quantified self: Personal knowledge discovery as a challenge for data science. Big Data **3**(4), 249–266 (2016)
161. Fayyad, U., Piatetsky-Shapiro, G., Smyth, P.: From data mining to knowledge discovery in databases. AI Magazine **17**(3), 37–54 (1996)
162. Feng, J.X., Kusiak, A.: Data mining applications in engineering design, manufacturing and logistics. International Journal of Production Research **44**(14), 2689–2694 (2007)
163. Fichman, P., Rosenbaum, H.: Social Informatics: Past, Present and Future. Cambridge Scholars Publishing (2014)
164. Finzer, W.: The data science education dilemma. Technology Innovations in Statistics Education **7**(2) (2013). URL http://escholarship.org/uc/item/7gv0q9dc#page-1
165. Floreano, D., Mattiussi, C.: Bio-Inspired Artificial Intelligence: Theories, Methods, and Technologies. The MIT Press (2008)
166. Floridi, L.: The ethics of information. Oxford University Press (2013)
167. Floridi, L., Taddeo, M.: What is data ethics. Phil. Trans. R. Soc. A **374**(2083) (2016)
168. Forbes: The world's biggest public companies (2016). URL https://www.forbes.com/global2000/list
169. Fox, G., Maini, S., Rosenbaum, H., Wild, D.J.: Data science and online education. In: 2015 IEEE 7th International Conference on Cloud Computing Technology and Science (CloudCom), pp. 582–587 (2015)
170. Freedman, D., Pisani, R., Purves, R.: Statistics (4th edn.). W. W. Norton (2007)
171. G. Szkely, e.a.: Measuring and testing independence by correlation of distances. Annals of Statistics **35**(6), 2769–2794 (2007)
172. Galetto, M.: Top 50 data science resources (2016). URL http://www.ngdata.com/top-data-science-resources/?
173. Galin, D.: Software Quality Assurance: From Theory to Implementation. Pearson (2003)
174. Ganis, M., Kohirkar, A.: Social Media Analytics: Techniques and Insights for Extracting Business Value Out of Social Media. IBM Press (2015)
175. Ganiz, M., George, C., Pottenger, W.: Higher order naive bayes: A novel non-iid approach to text classification. IEEE Transactions on Knowledge and Data Engineering **23**(7), 1022–1034 (2011)
176. Gavrilovski, A., Jimenez, H., Mavris, D.N., Rao, A.H., Shin, S., Hwang, I., Marais, K.: Challenges and opportunities in flight data mining: A review of the state of the art. In: AIAA Infotech Aerospace, AIAA SciTech Forum, (AIAA 2016-0923), pp. 1–66 (2016)
177. Geczy, P.: Big data characteristics. The Macrotheme Review **3**(6), 94–104 (2014)
178. Gentle, J.E.: Computational Statistics. Springer Publishing Company (2009)
179. GEO: Gene expression omnibus (2016). URL http://www.ncbi.nlm.nih.gov/geo/
180. George, G., Osinga, E., Lavie, D., Scott, B.: Big data and data science methods for management research. Academy of Management Journal **59**(5), 1493–1507 (2016). URL https://aom.org/uploadedFiles/Publications/AMJ/Oct_2016_FTE.pdf
181. Ghodke, D.: Bye bye 2015: What lies ahead for bi? (2015). URL www.ciol.com/bye-bye-2015-what-lies-ahead-for-bi/
182. Github: Data science colleges (2016). URL https://github.com/ryanswanstrom/awesome-datascience-colleges. (retrieved on 4 April 2016)
183. Github: List of recommender systems (2016). URL https://github.com/grahamjenson/list_of_recommender_systems
184. Gitub, D.: Data science for social good. URL https://github.com/dssg

185. Globalsecurity: Worldwide military command and control system (1996). URL http://www.globalsecurity.org/wmd/systems/wwmccs.htm

186. Gold, M., McClarren, R., Gaughan, C.: The lessons oscar taught us: Data science and media & entertainment. Big Data **1**(2), 105–109 (2013)

187. Golge, E.: Brief history of machine learning, a blog from a human-engineer-being. retrieved 21 march 2017 (2017). URL http://www.erogol.com/brief-history-machine-learning/

188. Goodfellow, I., Bengio, Y., Courville, A.: Deep Learning. MIT Press (2016). URL http://www.deeplearningbook.org

189. Google: Deepmind (2016). URL https://deepmind.com/

190. Google: Google bigquery and cloud platform (2016). URL https://cloud.google.com/bigquery/

191. Google: Google cloud prediction api (2016). URL https://cloud.google.com/prediction/docs/

192. Google: Google online open education (2016). URL https://www.google.com/edu/openonline/

193. Google: Google trends (2016). URL https://www.google.com.au/trends/explore#q=data%20\science%2C%20data%20analytics%2C%20big%20data%2C%20data%20analysis%2C%20\advanced%20analytics&cmpt=q&tz=Etc%2FGMT-11. Retrieved on 14 November 2016

194. Google: Open mobile data (2016). URL https://console.developers.google.com/storage/\browser/openmobiledata_public/

195. Government, B.M.: Beijing big data and cloud computing development action plan (2016). URL http://zhengwu.beijing.gov.cn/gh/dt/t1445533.htm

196. Government, C.: China big data (2015). URL http://www.gov.cn/zhengce/content/2015-09/05/content_10137.htm

197. Graham, M.J.: The art of data science. In: Astrostatistics and Data Mining, Volume 2 of the series Springer Series in Astrostatistics, pp. 47–59 (2012)

198. Gray, J.: escience – a transformed scientific method (2007). URL http://research.microsoft.com/en-us/um/people/gray/talks/NRC-CSTB_eScience.ppt

199. GTD: Global terrorism database (2016). URL https://www.start.umd.edu/gtd/

200. Gupta, A., Cecen, A., Goyal, S., Singh, A.K., Kalidindi, S.R.: Structure-property linkages using a data science approach: Application to a non-metallic inclusion/steel composite system. Acta Mater **91**, 239–254 (2015)

201. H. Lu, e.a.: Beyond intratransaction association analysis. ACM Transactions on Information Systems **18**(4), 423–454 (2000)

202. Han, J., Kamber, M., Pei, J.: Data Mining: Concepts and Techniques, 3rd edn. Morgan Kaufmann Publishers Inc., San Francisco, CA, USA (2011)

203. Hand, D.J.: Statistics and computing: The genesis of data science. Statistics and Computing **25**(4), 705–711 (2015)

204. Hardin: Github (2016). URL www.hardin47.github.io/DataSciStatsMaterials/

205. Hardin, J., Hoerl, R., Horton, N.J., Nolan, D.: Data science in statistics curricula: Preparing students to "think with data". The American Statistician **69**(4), 343–353 (2015)

206. Harris, H., Murphy, S., Vaisman, M.: Analyzing the Analyzers: An Introspective Survey of Data Scientists and Their Work. O'Reilly Media (2013)

207. Hastie, T., Tibshirani, R., Friedman, J.H.: The elements of statistical learning: data mining, inference, and prediction, 2nd Edition. Springer (2009)

208. Hazena, B.T., Booneb, C.A., Ezellc, J.D., Jones-Farmer, L.A.: Data quality for data science, predictive analytics, and big data in supply chain management: An introduction to the problem and suggestions for research and applications. International Journal of Production Economics **154**, 72–80 (2014)

209. Hey, T., Tansley, S., (Eds.), K.T.: The Fourth Paradigm: Data-Intensive Scientific Discovery. Microsoft Research (2009). URL http://research.microsoft.com/en-us/collaboration/fourthparadigm/

210. Hey, T., Trefethen, A.: The Data Deluge: An e-Science Perspective, pp. 809–824. John Wiley & Sons (2003)

211. HLSG: Final report of the high level expert group on scientific data. In: Riding the wave: How Europe can gain from the rising tide of scientific data (2010). URL http://ec.europa.eu/information_society/newsroom/cf/document.cfm?action=display&doc_id=707

212. HLSG: An rda europe report. In: The Data Harvest: How sharing research data can yield knowledge, jobs and growth (2014). URL http://www.e-nformation.ro/wp-content/uploads/2014/12/TheDataHarvestReport_-Final.pdf

213. Hofmann, T., Schölkopf, B., Smola, A.J.: Kernel methods in machine learning (2008)

214. Horizon: European commission horizon 2020 big data private public partnership (2014). URL http://ec.europa.eu/programmes/horizon2020/en/h2020-section/information-and-communication-technologies

215. Horton, N.J., Baumer, B.S., Wickham, H.: Setting the stage for data science: Integration of data management skills in introductory and second courses in statistics. arXiv preprint arXiv:1502.00318 (2015)

216. Huber, P.J.: Data Analysis: What Can Be Learned From the Past 50 Years. John Wiley & Sons (2011)

217. IASC: International association for statistical computing (1977). URL http://www.iasc-isi.org/

218. IBM: The value of analytics in healthcare. URL https://www.ibm.com/services/us/gbs/thoughtleadership/ibv-healthcare-analytics.html

219. IBM: Capitalizing on complexity (2010). URL http://www-935.ibm.com/services/us/ceo/ceostudy2010/multimedia.html

220. IBM: Ibm analytics and big data (2016). URL http://www.ibm.com/analytics/us/en/ or http://www-01.ibm.com/software/data/bigdata/

221. IBM: Ibm watson (2016). URL https://www.ibm.com/watson/

222. IBM: What is a data scientist? (2016). URL http://www-01.ibm.com/software/data/infosphere/data-scientist/

223. IEEEBD: IEEE big data initiative (2014). URL http://bigdata.ieee.org/

224. IEMSS: The international environmental modelling & software society. URL http://www.iemss.org/society/

225. IFSC-96: Data science, classification, and related methods. In: IFSC-96 (1996). URL http://d-nb.info/955715512/04

226. IJDS: International journal of data science (2016). URL http://www.inderscience.com/jhome.php?jcode=ijds

227. IJRDS: International journal of research on data science (2017). URL http://www.sciencepublishinggroup.com/journal/index?journalid=310

228. INFORMS: Informs code of ethics for certified analytics professionals. URL https://www.certifiedanalytics.org/ethics.php

229. INFORMS: Candidate handbook (2014). URL https://www.informs.org/Certification-Continuing-Ed/Analytics-Certification/Candidate-Handbook

230. INFORMS: Institute for operations research and the management sciences (2016). URL https://www.informs.org/

231. Iwata, S.: Scientific "agenda" of data science. Data Science Journal 7(5), 54–56 (2008)

232. J. Hair, e.a.: Multivariate data analysis (7th Edition). Prentice Hall (2009)

233. Jagadish, H., Gehrke, J., Labrinidis, A., Papakonstantinou, Y., Patel, J.M., Ramakrishnan, R., Shahabi, C.: Big data and its technical challenges. Communications of the ACM 57(7), 86–94 (2014)

234. Jagadish, H.V.: Big data and science: Myths and reality. Big Data Research 2(2), 49–52 (2015)

235. JDS: Journal of data science (2002). URL http://www.jds-online.com/

236. JDSA: International journal of data science and analytics (JDSA) (2015). URL http://www.springer.com/41060

237. JFDS: The journal of finance and data science (2016). URL http://www.keaipublishing.com/en/journals/the-journal-of-finance-and-data-science/

238. Johnstone, I., Roberts, F.: Data science at nsf (2014). URL http://www.nsf.gov/attachments/130849/public/Stodden-StatsNSF.pdf

239. Jones, R.P.: Foundations of Critical Thinking. Cengage Learning (2000)

240. Josephson, R., J. & G. Josephson, S.: Abductive Inference: Computation, Philosophy, Technology. Cambridge University Press, New York & Cambridge (1994)

241. Kaggle: Kaggle competition data (2016). URL https://www.kaggle.com/competitions

242. Kalidindi, S.R.: Data science and cyberinfrastructure: critical enablers for accelerated development of hierarchical materials. International Materials Reviews **60**(3), 150–168 (2015)

243. Kan, S.H.: Metrics and Models in Software Quality Engineering, 2nd Edition. Addison-Wesley Professional (2002)

244. Kanter, J.M., Veeramachaneni, K.: Deep feature synthesis: Towards automating data science endeavors. In: 2015 IEEE International Conference on Data Science and Advanced Analytics (DSAA), pp. 1–10 (2015)

245. KDD89: IJCAI-89 workshop on knowledge discovery in databases (1989). URL http://www.kdnuggets.com/meetings/kdd89/index.html

246. KDnuggets: Visualization software (2015).   URL http://www.kdnuggets.com/software/visualization.html

247. Kdnuggets: Kdnuggets (2016). URL http://www.kdnuggets.com/

248. KDnuggets: Software suites/platforms for analytics, data mining, & data science (2017). URL http://www.kdnuggets.com/software/suites.html

249. Keller, J.M., Liu, D., Fogel, D.B.: Fundamentals of Computational Intelligence: Neural Networks, Fuzzy Systems, and Evolutionary Computation. Wiley-IEEE Press (2016)

250. Kelly, K.: The quantified century.   In: Quantified Self Conference (2012).   URL http://quantifiedself.com/conference/Palo-Alto-2012

251. Kenett, R.S., Shmueli, G.: Information Quality: The Potential of Data and Analytics to Generate Knowledge. Wiley (2016)

252. Khan, N., Yaqoob, I., Hashem, I.A.T., et al: Big data: Survey, technologies, opportunities, and challenges. The Scientific World Journal **2014**, 18 (2014)

253. King, J., Magoulas, R.: 2015 data science salary survey (2015). URL http://duu86o6n09pv.cloudfront.net/reports/2015-data-science-salary-survey.pdf

254. Kirk, A.: Data Visualisation: A Handbook for Data Driven Design. SAGE Publications (2016)

255. Kirk, R.E.: Experimental Design: Procedures for the Behavioral Sciences (4th Edition. SAGE Publications (2012)

256. Kirkpatrick, K.: Putting the data science into journalism. Communications of the ACM **58**(5), 15–17 (2015)

257. Kohavi, R.: Mining e-commerce data: the good, the bad, and the ugly. In: SIGKDD, pp. 8–13 (2001)

258. Kohavi, R., Mason, L., Parekh, R., Zheng, Z.: Lessons and challenges from mining retail e-commerce data. Mach. Learn. **57**(1-2), 83–113 (2004)

259. Kohavi, R., Rothleder, N.J., Simoudis, E.: Emerging trends in business analytics. Communications of the ACM **45**(8), 45–48 (2002)

260. Koller, D., Friedman, N.: Probabilistic Graphical Models: Principles and Techniques. The MIT Press (2009)

261. Kramer, A., Guillory, J., Hancock, J.: Experimental evidence of massive-scale emotional contagion through social networks. Proc. Natl. Acad. Sci. **111**(24), 8788–8790 (2014)

262. Kung, S.Y.: Kernel Methods and Machine Learning. Cambridge University Press (2014)

263. Kurzweil, R.: How to Create a Mind: The Secret of Human Thought Revealed. Penguin Books (2013)

264. Lab, A.: Mlbase (2016). URL http://mlbase.org/

265. Labrinidis, A., Jagadish, H.V.: Challenges and opportunities with big data. Proceedings of the VLDB Endowment **5**(12), 2032–2033 (2012)

266. Laney, D.: 3D data management: Controlling data volume, velocity and variety (2001). Technical Report, META Group

267. Larder, B., Summerhayes, N.: Application of smiths aerospace data mining algorithms to british airways 777 and 747 fdm data (2004). URL https://flightsafety.org/files/FDM_data_mining_report.pdf

268. Lassonde, M., Candel, S., Hacker, J., Quadrio-Curzio, A., Onishi, T., Ramakrishnan, V., McNutt, M.: G7 academies' joint statements 2017, new economic growth: The role of science, technology, innovation and infrastructure (2017)

269. Lazer, D., Kennedy, R., King, G., Vespignani, A.: The parable of google flu: Traps in big data analysis. Science **343**, 1203–1205 (2014)

270. LDC: Linguistic data consortium (2016). URL https://www.ldc.upenn.edu/about

271. Lehmann, E.L., Lehmann, J.P.: Testing Statistical Hypotheses. Springer (2010)

272. Lencioni, P.: The Five Dysfunctions of a Team: A Leadership Fable. Jossey-Bass (2002)

273. Leonelli, S.: Locating ethics in data science: responsibility and accountability in global and distributed knowledge production systems. Phil. Trans. R. Soc. A **374** (2016)

274. Leuphana: Master's programme in management & data science (2017). URL http://www.leuphana.de/en/graduate-school/master/course-offerings/management-data-science.html

275. Li, R., Shengjie Wang, e.a.: Towards social user profiling: unified and discriminative influence model for inferring home locations. In: Proceedings of KDD2012, pp. 1023–1031 (2012)

276. LinkedIn: Linkedin jobs (2016). URL https://www.linkedin.com/jobs/data-scientist-jobs

277. Loshin, D.: Enterprise Knowledge Management. Morgan Kaufmann (2001)

278. Loukides, M.: The Evolution of Data Products. O'Reilly, Cambridge (2011)

279. Loukides, M.: What is data science? O'Reilly Media, Sebastopol, CA (2012). URL http://radar.oreilly.com/2010/06/what-is-data-science.html#data-scientists

280. MacKay, D.J.C.: Information Theory, Inference & Learning Algorithms. Cambridge University Press, New York (2002)

281. Manieri, A., Brewer, S., Riestra, R., Demchenko, Y., Hemmje, M., Wiktorski, T., Ferrari, T., Frey, J.: Data science professional uncovered: How the EDISON project will contribute to a widely accepted profile for data scientists. In: 2015 IEEE 7th International Conference on Cloud Computing Technology and Science (CloudCom), pp. 588–593 (2015)

282. Manieri, A., Nucci, F.S., Femminella, M., Reali, G.: Teaching Domain-Driven data science: Public-Private co-creation of Market-Driven certificate. In: 2015 IEEE 7th International Conference on Cloud Computing Technology and Science (CloudCom), pp. 569–574 (2015)

283. Matsudaira, K.: The science of managing data science. Communications of the ACM **58**(6), 44–47 (2015)

284. Mattison, R.: Data warehousing and data mining for telecommunications. Artech House (1997)

285. Mattmann, C.A.: Computing: A vision for data science. Nature **493**(7433), 473–475 (2013)

286. Mattmann, C.A.: Cultivating a research agenda for data science. Journal of Big Data **1**(1), 1–8 (2014)

287. McCartney, P.R.: Big data science. The American Journal of Maternal/Child Nursing **40**(2), 130–130 (2015)

288. McKinsey: Big data: The next frontier for innovation, competition, and productivity (2011). McKinsey Global Institute

289. Microsoft: Azure. URL www.azure.microsoft.com/

290. Miller, C.C.: Data science: The numbers of our lives. New York Times (2013). URL http://www.nytimes.com/2013/04/14/education/edlife/universities-offer-courses-in-a-hot-new-field-data-science.html?pagewanted=all&_r=0

291. Miller, K., Taddeo, M.: The ethics of information technologies. In: Library of Essays on the Ethics of Emerging Technologies (ed.). NY: Routledge (2017)

292. MIT: Data analytics in urban transportation. URL http://dusp.mit.edu/transportation/project/data-analytics-urban-transportation

293. MIT: Checklist for software quality (2011). URL http://web.mit.edu/~6.170/www/quality.html

294. Mitchell, M.: Complexity: A Guided Tour. Oxford University Press (2011)

295. Mitchell, T.: Machine Learning. McGraw Hill (1997)

296. Mittelstadt, B., Floridi, L.: The ethics of big data: current and foreseeable issues in biomedical contexts. Sci. Eng. Ethics **22**, 303–341 (2015)
297. de Moraes, R.M., Martinez, L.: Computational intelligence applications for data science. Knowledge-Based Systems **87**, 1–2 (2015)
298. Morrell, A.J.H.: Information processing 68 (ed.). In: Proceedings of IFIP Congress 1968. Edinburgh, UK (1968)
299. Murray-Rust, P.: Data-Driven science: A scientist's view. In: NSF/JISC 2007 Digital Repositories Workshop (2007). URL http://www.sis.pitt.edu/repwkshop/papers/murray.pdf
300. Naur, P.: 'datalogy', the science of data and data processes. In: Proceedings of the IFIP Congress 68, pp. 1383–1387 (1968)
301. Naur, P.: Concise Survey of Computer Methods. Studentlitteratur, Lund, Sweden (1974)
302. NCSU: Institute for advanced analytics, north carolina state university (2007). URL http://analytics.ncsu.edu/
303. NCSU: Master of science in analytics, institute for advanced analytics, north carolina state university (2007). URL http://analytics.ncsu.edu/
304. Neapolitan, R.E.: Learning Bayesian Networks. Prentice-Hall, Upper Saddle River, NJ, USA (2003)
305. Nelson, M.L.: Data-driven science: A new paradigm? EDUCAUSE Review **44**(4), 6–7 (2009)
306. von Neumann, J., Kurzweil, R.: The Computer and the Brain, 3rd Edition. Yale University Press (2012)
307. Neville, J., Jensen, D.: Relational dependency networks. The Journal of Machine Learning Research **8**, 653–692 (2007)
308. NICTA: National ict Australia (2016). URL https://www.nicta.com.au/
309. NIST: NIST text retrieval conference data (2015). URL http://trec.nist.gov/data.html
310. NSB: Long-lived digital data collections: Enabling research and education in the 21st century. In: US National Science Board (2005). URL http://www.nsf.gov/pubs/2005/nsb0540/
311. NSF: US NSF07-28. In: Cyberinfrastructure Vision for 21st Century Discovery (2007). URL http://www.nsf.gov/pubs/2007/nsf0728/nsf0728.pdf
312. OECD: OECD principles and guidances for access to research data from public funding (2007). URL https://www.oecd.org/sti/sci-tech/38500813.pdf
313. OECD: Data-driven innovation: Big data for growth and well-being (2015). doi:http://dx.doi.org/10.1787/9789264229358-en
314. OECD: The next production revolution: Implications for governments and business (2017). doi:http://dx.doi.org/10.1787/9789264271036-en
315. O'Leary, D.E.: Ethics for big data and analytics. IEEE Intelligent Systems **31**(4), 81–84 (2016)
316. O'Neil, C., Schutt, R.: Doing data science: Straight talk from the frontline. O'Reilly Media, Sebastopol, CA (2013)
317. OpenCV: Open source computer vision library (2016). URL www.opencv.org/
318. OPENedX: OPENedX online education platform (2016). URL https://open.edx.org/
319. O'Reilly, T.: What is web 2.0 (2005). URL http://oreilly.com/pub/a/web2/archive/what-is-web-20.html?page=3
320. Pal, S.K., Meher, S.K., Skowron, A.: Data science, big data and granular mining. Pattern Recognition Letters **67**(2), 109–112 (2015)
321. Pan, S.J., Yang, Q.: A survey on transfer learning. IEEE Trans. on Knowl. and Data Eng. **22**(10), 1345–1359 (2010)
322. Patil, D.: Building Data Science Teams. O'Reilly Media (2011)
323. Patterson, K., Grenny, J.: Crucial Conversations Tools for Talking When Stakes Are High (Second Edition). McGraw-Hill Education (2011)
324. Paul, R., Elder, L.: The Thinker's Guide to Scientific Thinking Based on Critical Thinking Concepts & Principles. Foundation for Critical Thinking (2008)
325. Paulk, M.C., Curtis, B., Chrissis, M.B., Weber, C.V.: Capability maturity model version 1.1. IEEE Software **10**(4), 18–27 (1993)

326. Pearson, K.: Report on certain enteric fever inoculation statistics. Br Med J. 2(2288), 1243–1246 (1904)
327. Peter, F., James, H.: The science of data science. Big Data 2(2), 68–70 (2014)
328. Philip, J.C.: Computer Generated Artificial Life: A Biblical And Logical Analysis (Integrated Apologetics), 10th edition. Philip Communications (2015)
329. Pike, J.: Global command and control system (2003). URL https://fas.org/nuke/guide/usa/c3i/gccs.htm
330. Press, G.: A very short history of data science (2013). URL http://www.forbes.com/sites/gilpress/2013/05/28/a-very-short-history-of-data-science/#61ae3ebb69fd
331. Priebe, T., Markus, S.: Business information modeling: A methodology for data-intensive projects, data science and big data governance. In: 2015 IEEE International Conference on Big Data (Big Data), pp. 2056–2065 (2015)
332. Provost, F., Fawcett, T.: Data science and its relationship to big data and Data-Driven decision making. Big Data 1(1), 51–59 (2013)
333. Qian, X.: Revisiting issues on open complex giant systems. Pattern Recognit. Artif. Intell. 4(1), 5–8 (1991)
334. Qian, X.: Building Systematism. ShanXi Sci. Technol Press, Taiyuan, China (2001)
335. Qian, X., Yu, J., Dai, R.: A new discipline of science-the study of open complex giant system and its methodology. Chin. J. Syst. Eng. Electron. 4(2), 2–12 (1993)
336. Raghavan, S.N.: Data mining in e-commerce: A survey. Sadhana 30(2 & 3), 275–289 (2005)
337. RapidMiner: Rapidminer (2016). URL https://rapidminer.com/
338. Redman, T.: Data Quality: The Field Guide. Digital Press (2001)
339. Renae, S.: Data analytics: Crunching the future. Bloomberg Businessweek (2011). September 8
340. Review, S.: Data integration and application integration solutions directory (2016). URL http://solutionsreview.com/data-integration/data-integration-solutions-directory/
341. Rifkin, J.: The Third Industrial Revolution: How Lateral Power is Transforming Energy, the Economy, and the World. Palgrave MacMillan (2011)
342. Rowley, J.: The wisdom hierarchy: representations of the DIKW hierarchy. Journal of Information and Communication Science 33(2), 163–180 (2007)
343. Rudin, C., Dunson, D., Irizarry, R., Ji, H., Laber, E., Leek, J., McCormick, T., Rose, S., Schafer, C., van der Laan, M., Wasserman, L., Xue, L.: Discovery with data: Leveraging statistics with computer science to transform science and society (2014). URL http://www.amstat.org/policy/pdfs/BigDataStatisticsJune2014.pdf. A Working Group of the American Statistical Association
344. Russell, S.J., Norvig, P.: Artificial Intelligence: A Modern Approach, 2 edn. Pearson Education (2003)
345. SAS: Big data analytics: An assessment of demand for labour and skills, 2012-2017 (2013). URL https://www.thetechpartnership.com/globalassets/pdfs/research-2014/bigdata_report_nov14.pdf. Report. SAS/The Tech Partnership
346. SAS: Sas enterprise miner (2016). URL http://www.sas.com
347. SAS: SAS insights (2016). URL http://www.sas.com/en_us/insights.html
348. Sayama, H.: Introduction to the Modeling and Analysis of Complex Systems. Open SUNY Textbooks (2015)
349. Schadt, E., Chilukuri, S.: The role of big data in medicine (2015). URL http://www.mckinsey.com/industries/pharmaceuticals-and-medical-products/our-insights/the-role-of-big-data-in-medicine
350. Schoenherr, T., Speier-Pero, C.: Data science, predictive analytics, and big data in supply chain management: Current state and future potential. Journal of Business Logistics 36(1), 120–132 (2015)
351. Schulmeyer, G.G., Mcmanus, J.I.: Handbook of Software Quality Assurance, 3rd Edition. Prentice Hall PTR (1998)
352. SCJ: Science council of Japan - code of conduct for scientists (2017). URL www.scj.go.jp/en/report/code.html

353. Scott, J.: Social Network Analysis (4th Edition). SAGE Publications (2017)

354. SDS: Social data science lab. URL http://socialdatalab.net/

355. Sebastian-Coleman, L.: Measuring Data Quality for Ongoing Improvement: A Data Quality Assessment Framework. Morgan Kaufmann (2013)

356. Security, C.I.: Big data strategies and actions in major countries (2015). URL http://www. cac.gov.cn/2015-07/03/c_1115812491.htm

357. Shi, C., Yu, P.S.: Heterogeneous Information Network Analysis and Applications. Springer (2017)

358. SIAM: Siam career center (2016). URL http://jobs.siam.org/home/

359. Siart, C., Kopp, S., Apel, J.: The interface between data science, research assessment and science support - highlights from the German perspective and examples from Heidelberg university. In: 2015 IIAI 4th International Congress on Advanced Applied Informatics (IIAI-AAI), pp. 472–476 (2015)

360. Silk: Data science university programs (2016). URL http://data-science-university-programs. silk.co/

361. Simovici, D.A., Djeraba, C.: Mathematical Tools for Data Mining: Set Theory, Partial Orders, Combinatorics. Springer Publishing Company (2008)

362. Siroker, D., Koomen, P.: A / B Testing: The Most Powerful Way to Turn Clicks Into Customers. Wiley (2015)

363. Smarr, L.: Quantifying your body: A how-to guide from a systems biology perspective. Biotechnology Journal **7**(8), 980–991 (2012). doi:10.1002/biot.201100495. URL http://dx. doi.org/10.1002/biot.201100495

364. Smith, F.J.: Data science as an academic discipline. Data Science Journal **5**, 163–164 (2006)

365. SMU: Living analytics research centre (2017). URL https://larc.smu.edu.sg/

366. Sobel, C., Li, P.: The Cognitive Sciences: An Interdisciplinary Approach (2nd Edition). SAGE Publications (2013)

367. Society, B.R.A.: Astronomical databases and archives. URL https://www.ras.org.uk/ education-and-careers/for-everyone/126-astronomical-databases-and-archives

368. Sonnenburg, S., Raetsch, G.: Shogun (2016). URL http://www.shogun-toolbox.org/

369. SSDS: Springer series in the data sciences (2015). URL http://www.springer.com/series/ 13852

370. Stanford: Stanford data science initiatives, Stanford university (2014). URL https://sdsi. stanford.edu/

371. Stanton, J.: An introduction to data science (2012). URL http://surface.syr.edu/istpub/165/

372. Stevens, M.L.: An ethically ambitious higher education data science. Research & Practice in Assessment **9**, 96–97 (2014)

373. Stewart, T.R., McMillan, J.C.: Descriptive and prescriptive models for judgment and decision making: Implications for knowledge engineering. In: J.L. Mumpower, O. Renn, L.D. Phillips, V.R.R.U. (Eds.) (eds.) Expert Judgment and Expert Systems, pp. 305–320. Springer-Verlag, London (1987)

374. Stonebraker, M., Madden, S., Dubey, P.: Intel 'big data' science and technology center vision and execution plan. SIGMOD Record **42**(1), 44–49 (2013)

375. Suchma, L.: Human-Machine Reconfigurations: Plans and Situated Actions. Cambridge University Press (2006)

376. Swan, A., Brown, S.: The skills, role & career structure of data scientists & curators: Assessment of current practice & future needs. In: UK Joint Information Systems Committee (2008). Technical Report. University of Southampton

377. Swan, M.: The quantified self: Fundamental disruption in big data science and biological discovery. Big Data **1**(2), 85–99 (2013)

378. Taddeo, M., (eds.), L.F.: The ethical impact of data science. Phil. Trans. R. Soc. A **374** (2016). URL http://rsta.royalsocietypublishing.org/content/374/2083

379. Taleb, N.N.: The Black Swan: The Impact of the Highly Improbable. Random House, New York (2007)

380. Tang, L., Liu, H.: Community Detection and Mining in Social Media. Morgan and Claypool (2010)

381. Technavio: Top 10 healthcare data analytics companies (2016). URL http://www.technavio.com/blog/top-10-healthcare-data-analytics-companies

382. TFDSAA: IEEE task force on data science and advanced analytics (2013). URL http://dsaatf.dsaa.co/

383. Thrun, S., Pratt, L.e.: Learning to learn. Boston, Mass.: Kluwer Academic (1998)

384. Tilburg: Msc specialization data science: Business and governance (2017). URL https://www.tilburguniversity.edu/education/masters-programmes/data-science-business-and-governance/

385. TOBD: IEEE transactions on big data (2015). URL https://www.computer.org/web/tbd

386. Today, P.A.: 29 data preparation tools and platforms (2016). URL http://www.predictiveanalyticstoday.com/data-preparation-tools-and-platforms/

387. Tukey, J.W.: The future of data analysis. Ann. Math. Statist. **33**(1), 1–67 (1962)

388. Tukey, J.W.: Exploratory Data Analysis. Pearson (1977)

389. Tutiempo: Global climate data (2016). URL http://en.tutiempo.net/climate

390. U-Waikato: Weka (2016). URL http://www.cs.ubc.ca/labs/beta/Projects/autoweka/

391. UCI: UCI machine learning repository (2016). URL www.archive.ics.uci.edu/ml/

392. UCL: Msin105p: Critical analytical thinking (2015). URL https://www.mgmt.ucl.ac.uk/module/msin105p-critical-analytical-thinking

393. Udacity: Udacity courses (2016). URL https://www.udacity.com/courses/data-science

394. Udemy: Udemy courses (2016). URL https://www.udemy.com/courses/search/?ref=home&\src=ukw&q=data+science&lang=en

395. UK: Uk big data (2016). URL https://www.ukri.org

396. UK-HM: Uk hm government. In: Open Data White Paper: Unleashing the Potential (2012). URL http://data.gov.uk/sites/default/files/Open_data_\White_Paper.pdf

397. UK-OD: UK open data (2016). URL http://data.gov.uk/

398. UMichi: Michigan institute for data science, university of Michigan (2015). URL http://midas.umich.edu/

399. UN: United nation global pulse projects (2010). URL http://www.unglobalpulse.org/

400. Uprichard, E.: Big data, little questions? (2013). URL http://discoversociety.org/2013/10/01/focus-big-data-little-questions/

401. US National Science Foundation: Critical techniques and technologies for advancing foundations and applications of big data science & engineering (bigdata) (2015). URL https://www.nsf.gov/funding/pgm_summ.jsp?pims_id=504767

402. US National Science Foundation: Computational and data-enabled science and engineering (cds&e) (2017). URL https://www.nsf.gov/funding/pgm_summ.jsp?pims_id=504813&org=CISE&sel_org=CISE&from=fund

403. US-OD: US government open data (2016). URL https://www.data.gov/

404. USAID: Usaid recommended data quality assessment (dqa) checklist (2016). URL https://usaidlearninglab.org/sites/default/files/resource/files/201sae.pdf

405. USD2D: US national consortium for data science (2016). URL www.data2discovery.org

406. USDSC: US degree programs in analytics and data science (2016). URL http://analytics.ncsu.edu/?page_id=4184

407. USNSF: US big data research initiative (2012). URL http://www.nsf.gov/cise/news/bigdata.jsp

408. UTS: Master of analytics (research) and doctor of philosophy thesis: Analytics, Advanced Analytics Institute, University of Technology Sydney (2011). URL http://www.uts.edu.au/research-and-teaching/our-research/advanced-analytics-institute/education-and-research-opportuniti-1

409. UTSAAI: Advanced analytics institute, university of technology Sydney (2011). URL https://analytics.uts.edu.au/

410. Vapnik, V.N.: The Nature of Statistical Learning Theory. Springer-Verlag New York, New York, USA (2000)

411. Vast: Visual analytics community (2016). URL http://vacommunity.org/HomePage
412. Veaux, R.D.D., Agarwal, M., Averett, M., Baumer, B.S., Bray, A., Bressoud, T.C., Bryant, L., Cheng, L.Z., Francis, A., Gould, R., Kim, A.Y., Kretchmar, M., Lu, Q., Moskol, A., Nolan, D., Pelayo, R., Raleigh, S., Sethi, R.J., Sondjaja, M., Tiruviluamala, N., Uhlig, P.X., Washington, T.M., Wesley, C.L., White, D., Ye, P.: Curriculum guidelines for undergraduate programs in data science. Annu. Rev. Stat. Appl. 4(2), 1–16 (2017). URL https://www.amstat.org/asa/files/pdfs/EDU-DataScienceGuidelines.pdf
413. Vesset, D., Woo, B., Morris, H.D., Villars, R.L., Little, G., Bozman, J.S., Borovick, L., Olofson, C.W., Feldman, S., Conway, S., Eastwood, M., Yezhkova, N.: Worldwide big data technology and services 2012-2015 forecast (2012). IDC
414. Viseu, A., Suchman, L.: Wearable Augmentations: Imaginaries of the Informed Body, pp. 161–184. Berghahn Books, New York (2010)
415. Walker, M.A.: The professionalisation of data science. Int. J. of Data Science 1(1), 7–16 (2015)
416. Wang, C., Cao, L., Chi, C.: Formalization and verification of group behavior interactions. IEEE Trans. Systems, Man, and Cybernetics: Systems 45(8), 1109–1124 (2015)
417. WEF: The global competitiveness report 2011-2012: An initiative of the world economic forum (2011)
418. Wei, W.: Copula-based high dimensional dependence modelling. Ph.D. thesis, University of Technology Sydney (2014)
419. Wei Wei Junfu Yin, J.L., Cao, L.: Modeling asymmetry and tail dependence among multiple variables by using partial regular vine. In: SDM2014 (2014)
420. Weiss, K., Khoshgoftaar, T.M., Wang, D.: A survey of transfer learning. Journal of Big Data 3(1) (2016)
421. Whitehouse: The white house names dr. DJ patil as the first U.S. chief data scientist (2015). URL https://www.whitehouse.gov/blog/2015/02/18/white-house-names-dr-dj-patil-first-us-chief-data-scientist
422. Wikipedia: Bioinformatics. URL https://en.wikipedia.org/wiki/Bioinformatics
423. Wikipedia: Computational trust. URL https://en.wikipedia.org/wiki/Computational_trust
424. Wikipedia: Computing. URL https://en.wikipedia.org/wiki/Computing
425. Wikipedia: Dikw pyramid. URL https://en.wikipedia.org/wiki/DIKW_Pyramid
426. Wikipedia: Genetic linkage. URL https://en.wikipedia.org/wiki/Genetic_linkage
427. Wikipedia: Health care & analytics. URL http://analytics-magazine.org/health-care-a-analytics/
428. Wikipedia: Intelligent transportation system. URL https://en.wikipedia.org/wiki/Intelligent_transportation_system
429. Wikipedia: Social influence. URL https://en.wikipedia.org/wiki/Social_influence
430. Wikipedia: Social network analysis. URL https://en.wikipedia.org/wiki/Social_network_analysis
431. Wikipedia: Statistical relational learning. URL https://en.wikipedia.org/wiki/Statistical_relational_learning
432. Wikipedia: Sustainability. URL https://en.wikipedia.org/wiki/Sustainability
433. Wikipedia: Targeted advertising. URL https://en.wikipedia.org/wiki/Targeted_advertising
434. Wikipedia: Comparison of cluster software (2016). URL https://en.wikipedia.org/wiki/Comparison_of_cluster_software
435. Wikipedia: General data protection regulation (2016). URL https://en.wikipedia.org/wiki/General_Data_Protection_Regulation
436. Wikipedia: Informatics (2016). URL https://en.wikipedia.org/wiki/Informatics
437. Wikipedia: List of reporting software (2016). URL https://en.wikipedia.org/wiki/List_of_reporting_software
438. Wikipedia: National data protection authority (2016). URL https://en.wikipedia.org/wiki/National_data_protection_authority
439. Wikipedia: Sports analytics (2016). URL https://en.wikipedia.org/wiki/Sports_analytics
440. Wikipedia: Accuracy, precision, recall and specificity (2017). URL https://en.wikipedia.org/wiki/Precision_and_recall

441. Wikipedia: Capability maturity model (cmm) (2017). URL https://en.wikipedia.org/wiki/Capability_Maturity_Model
442. Wikipedia: Complexity (2017). URL https://en.wikipedia.org/wiki/Complexity
443. Wikipedia: Data quality (2017). URL https://en.wikipedia.org/wiki/Data_quality
444. Wikipedia: Industrial revolution (2017). URL https://en.wikipedia.org/wiki/Industrial_Revolution
445. Wikipedia: List of statistical packages (2017). URL https://en.wikipedia.org/wiki/List_of_statistical_packages
446. Wikipedia: Second industrial revolution (2017). URL https://en.wikipedia.org/wiki/Second_Industrial_Revolution
447. Wikipedia: Timeline of machine learning. retrieved 21 march 2017 (2017). URL https://en.wikipedia.org/wiki/Timeline_of_machine_learning
448. Wikipedia: Agile software development (2018). URL https://en.wikipedia.org/wiki/Agile_software_development
449. Wikipedia: Industry 4.0 (2018). URL https://en.wikipedia.org/wiki/Industry_4.0
450. Wikipedia: Internet of things (2018). URL https://en.wikipedia.org/wiki/Internet_of_things
451. Wikipedia: Open access (2018). URL https://en.wikipedia.org/wiki/Open_access
452. Wikipedia: Open data (2018). URL https://en.wikipedia.org/wiki/Open_data
453. Wikipedia: Open education (2018). URL https://en.wikipedia.org/wiki/Open_education
454. Wikipedia: Open peer review (2018). URL https://en.wikipedia.org/wiki/Open_peer_review
455. Wikipedia: Open science (2018). URL https://en.wikipedia.org/wiki/Open_science
456. Wikipedia: Open source (2018). URL https://en.wikipedia.org/wiki/Open-source_software
457. Wikipedia: Smart manufacturing (2018). URL https://en.wikipedia.org/wiki/Smart_manufacturing
458. Wikipedia: Waterfall model (2018). URL https://en.wikipedia.org/wiki/Waterfall_model
459. Williamson, J.: Big data analytics is transforming manufacturing (2016). URL http://www.themanufacturer.com/articles/big-data-analytics-is-transforming-manufacturing/
460. WIRED: How europe can seize the starring role in big data (2014). URL www.wired.com/insights/2014/09/europe-big-data/
461. Wladawsky-Berger, I.: Why do we need data science when we've had statistics for centuries? The Wall Street Journal (2014). URL http://blogs.wsj.com/cio/2014/05/02/why-do-we-need-data-science-when-weve-had-statistics-for-centuries/
462. Wolf, G.: The data-driven life. New York Times (2012). URL www.nytimes.com/2010/05/02/magazine/02self-measurement-t.html
463. Woodall P., B.A., Parlikad, A.: Data quality assessment: The hybrid approach. Information & Management 50(7), 369–382 (2013)
464. Woodall P., O.M., A., B.: A classification of data quality assessment and improvement methods. International Journal of Information Quality 3(4), 298–321 (2014)
465. Works, B.: Burtch works flash survey (2014). URL http://www.burtchworks.com/category/flash-survey/
466. WTTC: Big data - the impact on travel & tourism (2014). URL https://www.wttc.org/research/other-research/big-data-the-impact-on-travel-tourism/
467. Wu, J.: Statistics = data science? (1997). URL http://www2.isye.gatech.edu/~jeffwu/presentations/datascience.pdf
468. Xie, T., Thummalapenta, S., Lo, D., Liu, C.: Data mining for software engineering. Computer 42(8) (2009)
469. Yahoo: Yahoo finance (2016). URL www.finance.yahoo.com
470. Yau, N.: Rise of the data scientist (2009). URL http://flowingdata.com/2009/06/04/rise-of-the-data-scientist/
471. Yin, J., Zheng, Z., Cao, L.: Uspan: An efficient algorithm for mining high utility sequential patterns. In: KDD 2012, pp. 660–668 (2012)
472. Yiu, C.: The big data opportunity (2012). URL http://www.policyexchange.org.uk/images/publications/the%20big%20data%20opportunity.pdf
473. Yu, B.: IMS presidential address: Let us own data science. IMS Bulletin Online (2014). 1 Oct 2014

Index